Lecture Notes in Mathematics

Edited by A. Dold and B. Eckmann

1162

S.J.L. van Eijndhoven
J. de Graaf

Trajectory Spaces, Generalized Functions and Unbounded Operators

Springer-Verlag
Berlin Heidelberg New York Tokyo

Authors

S. J. L. van Eijndhoven
J. de Graaf
Eindhoven University of Technology
Den Dolech 2, P.O. Box 513
5700 MB Eindhoven, The Netherlands

Mathematics Subject Classification (1980): 46A12, 46F05, 46F10, 47D30, 81B05

ISBN 3-540-16065-5 Springer-Verlag Berlin Heidelberg New York Tokyo
ISBN 0-387-16065-5 Springer-Verlag New York Heidelberg Berlin Tokyo

Library of Congress Cataloging-in-Publication Data. Eijndhoven, Stephanus van, 1956- Trajectory Spaces, generalized functions, and unbounded operators. (Lecture notes in mathematics; 1162) Bibliography: p. Includes index. 1. Linear topological spaces. 2. Mappings (Mathematics) 3. Quantum theory. I. Graaf, Johannes de, 1942-. II. Title. III. Series: Lecture notes in mathematics (Springer-Verlag); 1162.
QA3.L28 no. 1162 [QA322] 510 s [515.7'3] 85-27810
ISBN 0-387-16065-5 (U.S.)

Printing and binding: Beltz Offsetdruck, Hemsbach/Bergstr.
2146/3140-543210

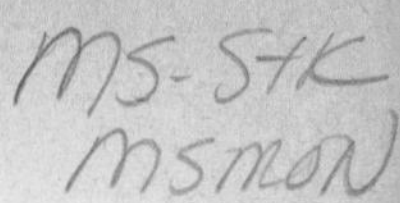

CONTENTS

PROLOGUE

The introduction of generalized functions has considerably advanced mathematical analysis, in particular harmonic analysis and the theory of partial differential equations. In a non rigorous way, electrical engineers and physicists have been using generalized functions for almost a century. Still they employ generalized functions such as the Heaviside step function and the Dirac delta function as if they were genuine C^{∞}-functions. But it took some time before mathematical justification has been taken up.

The first mathematical concepts which started up a theory of generalized functions were the finite parts of divergent integrals used by Hadamard and the Riemann-Liouville integrals due to Riesz. Later Sobolev defined generalized derivatives by means of integration by parts, and Bochner developed the theory of the Fourier transform for functions increasing as some power of their argument. Many of these results were unified by Schwartz in his monograph: Théorie des Distributions. Here the unifying concept is the notion of dual system of locally convex topological vector spaces. Generalized functions (distributions) can be regarded as continuous linear functionals on such spaces of well behaved functions.
Later on, also Gelfand and Shilov defined many classes of generalized functions. They introduced the notion of countable Hilbert space (a functional analytic generalization of the theory of tempered distributions) and the notion of Gelfand triple. But, as importantly, they showed how to use generalized functions in mathematical analysis. It turned out that generalized functions connect many aspects of classical analysis, of functional analysis, of the theory of partial differential equations and of the representation theory of locally compact Lie groups.
Thus, generalized functions have gained wide popularity among mathematicians.

The theories of Schwartz and of Gelfand-Shilov can roughly be described as follows. One starts with a vector space $\mathcal{S}$ of 'good' functions. Here we can take the set $\mathcal{D}$ of infinitely differentiable functions with compact support

or the set S of infinitely differentiable functions of rapid decrease. The vector space S is called the test space. It carries a suitable Hausdorff topology which makes S a locally convex topological vector space. The choice of the topology is not arbitrary; an extra condition will be imposed. The topological dual of S is denoted by S'. In order to consider S' or a representation of S' as a space of generalized functions we construct a space S^* that can be identified with S' and that contains S. Therefore, let X be a Hilbert space (e.g. $L_2(\mathbb{R})$ or a Sobolev space) such that S is a dense subspace of X and such that the embedding of S in X is continuous. We observe that the existence of X is equivalent to the existence in S of a positive non-degenerate continuous sesquilinear form. By means of the inner product, the dense subspace S of X induces the weak Hausdorff topology $\sigma(X,S)$ on X. Next, one considers the sequential completion S^* of X with respect to this topology. The mentioned condition one has to impose on the topology for S is the following: each member of S' can be represented by an element of S^* by means of the canonical pairing of S and S^*. Thus S' and S^* can be identified. Since $S \subset X \subset S^*$ and since the members of S are functions, S^*, and hence S' can be regarded as a space of improper functions. In this way, for instance $\mathcal{D}'$ can be interpreted as a space of improper functions which are derivatives of some order of continuous functions on the real line.
Also, the Lighthill-Temple approach of generalized functions can be described in the above functional analytic setup. They consider so-called regular sequences in S which converge in a weak sense. It turns out that a sequence is regular if it is $\sigma(X,S)$-convergent. Two regular sequences are equivalent if the difference of these sequences is a null sequence. A generalized function in the sense of Lighthill is just an equivalence class of regular sequences. So the theory based on the triplet $S \subset X \subset S^*$ and the theory based on regular sequences are equivalent.

In this book we present a new theory of generalized functions in a completely functional analytic setting. So our test space and our distribution space are not described by means of classical analysis but in the language of functional analysis. In fact, we introduce a model for a particular type of distribution theories.
The first inspiring ideas which have led to the construction of this model can be found in a paper by De Bruijn, [Br]. In the paper [Br], De Bruijn proposed a new distribution theory based on complex analysis. In this theory for the test space the space $S_{\frac{1}{2}}^{\frac{1}{2}}$ of Gelfand-Shilov is taken. Three

kinds of functions occur: smooth functions, smoothed functions and generalized functions. A function is said to be smooth if it belongs to the space $S^{\frac{1}{2}}_{\frac{1}{2}}$ which consists of a specific growth class of entire functions. A smoothed function f is derived from a smooth function g through application to g of an operator from a set of smoothing operators. This set of smoothing operators is a one-parameter semigroup denoted by $(N_\alpha)_{\alpha>0}$. De Bruijn proved that each smooth function is smoothed and that each smoothed function is smooth, i.e.,

$$\bigcup_{\alpha>0} N_\alpha(S^{\frac{1}{2}}_{\frac{1}{2}}) = S^{\frac{1}{2}}_{\frac{1}{2}} .$$

Now, a generalized function is a mapping F from $(0,\infty)$ into the set of smooth functions which satisfies $N_\alpha F(\beta) = F(\alpha+\beta)$ for all positive α and β. De Bruijn established a pairing between the spaces of smoothed functions and of generalized functions. However, no topologies are introduced for these spaces and questions about duality and continuity of linear mappings can be linked to sequential convergence only.

In [G], one of the authors of the present monograph has generalized De Bruijn's theory considerably by replacing the semigroup $(N_\alpha)_{\alpha>0}$ by an arbitrary one-parameter semigroup $(e^{-tA})_{t>0}$ where A denotes a nonnegative self-adjoint operator in a separable Hilbert space X. The series of papers [G] contain the fundamental concepts on which the contents of our book are based. These papers describe the topological spaces $S_{X,A}$ and $T_{X,A}$ which establish a dual system. The space $S_{X,A}$ is an inductive limit of Hilbert spaces. This inductive limit is not strict! So the theory on inductive limits, in which always strictness is assumed, cannot be applied. The main feature in [G] is the introduction of the set of Borel functions $\mathbf{B}(\mathbb{R})$. Because of the Spectral Theorem each element of $\mathbf{B}(\mathbb{R})$ gives rise to a normal operator in X. As a major result it has been proved that $S_{X,A}$ is the maximal common dense domain of the operators $\varphi(A)$, $\varphi \in \mathbf{B}(\mathbb{R})$ and that the inductive limit topology for $S_{X,A}$ is the coarsest topology for which all operators $\varphi(A)\colon S_{X,A} \to X$, $\varphi \in \mathbf{B}(\mathbb{R})$, are continuous. These observations have led to complete characterizations of null sequences, of bounded subsets and of compact subsets of $S_{X,A}$, just as for strict inductive limits. Furthermore, the full strength of Hilbert space theory became available. In [G] this has led already to a detailed exposition of continuous linear mappings, of topological tensor products and of so called Kernel theorems. Considerations of this type are not current in classical distribution theory.

In the thesis [E1] we continued the description of the theory of the spaces $S_{X,A}$ and $T_{X,A}$. A major source of inspiration for this thesis was the functional analytic approach in [G] to continuous linear mappings. It became clear that our theory resembles Hilbert space theory in many aspects. This reflects e.g. in the treatment of algebras of continuous linear mappings as presented in this work.
Also in [E1] we have started our mathematical interpretation of Dirac's formalism, searching both for a justification of the bold way in which Dirac treats the (generalized) eigenfunctions corresponding to the continuous spectrum of a self-adjoint operator and for a mathematical apparatus which does full justice to Dirac's bracket formalism. Our interpretation goes further than any other interpretation of Dirac's formalism that is known to us.
The present treatise is a coupling and further elaboration of the contents both of [G] and of [E1]. Further, we devote a great deal of attention to the description or discovery of spaces of analytic functions which can be typified as a space of type $S_{X,A}$ for a well chosen A and X.

The second part of this prologue is devoted to a short survey of each chapter of this book.

For a nonnegative unbounded self-adjoint operator A in a separable Hilbert space X the analyticity space $S_{X,A}$ is the dense subspace of X defined by

$$S_{X,A} = \bigcup_{t>0} e^{-tA}(X) .$$

On $S_{X,A}$ a (non-strict) inductive limit topology is imposed. The space $S_{X,A}$ contains all analytic vectors of the operator A, i.e.,

$$w \in S_{X,A} \Leftrightarrow \exists_{a>0} \exists_{b>0} \forall_{n\in\mathbb{N}}: w \in \mathcal{D}(A^n) \wedge \|A^n w\| \leq n!\, a^n b .$$

We call $S_{X,A}$ the analyticity space. The elements of the space $T_{X,A}$ are mappings $F: (0,\infty) \to X$ which satisfy

$$\forall_{t>0} \forall_{\tau>0}: F(t+\tau) = e^{-\tau A} F(t) .$$

They are called trajectories and the space $T_{X,A}$ is called the trajectory space. Examples of such trajectories are $t \mapsto A^m e^{-tA} x$ with $x \in X$ and $m \geq 0$. By a suitable choice of seminorms the space $T_{X,A}$ becomes a Frechet space.

The Hilbert space X can be embedded in $T_{X,A}$ by means of the embedding emb: $X \to T_{X,A}$ defined by

$$\mathrm{emb}(x): t \mapsto e^{-tA} x \,, \qquad t > 0,\ x \in X \,.$$

Thus we obtain the triplet $S_{X,A} \subset X \subset T_{X,A}$.
For each $w \in S_{X,A}$, there exists $\tau > 0$ such that $w \in \mathcal{D}(e^{\tau A})$. So it makes sense to define a pairing between $S_{X,A}$ and $T_{X,A}$ in the following way:

$$\langle w,G\rangle = (e^{\tau A}w, G(\tau)) \,, \quad w \in S_{X,A} \,, \quad G \in T_{X,A} \,.$$

Here $(\cdot,\cdot)$ denotes the inner product of X. Due to the trajectory property of the elements of $T_{X,A}$, the definition of $\langle\cdot,\cdot\rangle$ does not depend on the choice of $\tau > 0$. With this pairing the spaces $S_{X,A}$ and $T_{X,A}$ can be regarded as each other's strong duals.

Chapter I contains detailed discussions of several topological features of analyticity spaces and trajectory spaces, and of their duality. Furthermore, this chapter contains precise characterizations of five types of continuous linear mappings. One of these types is established by the so called extendible mappings, i.e., all continuous mappings from $S_{X,A}$ into $S_{X,A}$ which have an extension as a continuous mapping from $T_{X,A}$ into $T_{X,A}$.

The theory of analyticity spaces yields a functional analytic description for a new type of distribution theories. If $X = \mathbf{L}_2(M,\mu)$ with M some measure space, then loosely speaking we could say that $T_{X,A}$ consists of improper functions on M.
In Chapter II we discuss a number of classical distribution theories which can be obtained by a suitable choice of A and X. This chapter is based on our papers [EG 1-4] in which we have characterized certain analyticity spaces in terms of classical analysis. The following types are discussed:

- Analyticity spaces based on the diffusion equation $\frac{\partial u}{\partial t} = -\Delta^{\nu}u$, $\nu \geq \frac{1}{2}$. We will restrict ourselves to $\mathbb{R}$, $(0,\infty)$, $[-\pi,\pi]$ and $[0,\pi]$, and for $\nu = \frac{1}{2}$ to the unit sphere S^q in $\mathbb{R}^{q+1}$. So we consider $X = \mathbf{L}_2(\mathbb{R})$, $\mathbf{L}_2((0,\infty))$, $\mathbf{L}_2([-\pi,\pi])$, $\mathbf{L}_2([0,\pi])$ and $\mathbf{L}_2(S^q)$, and $A = \Delta^{\nu}$ where Δ denotes the Laplacian. We give precise characterizations of the analytic functions corresponding to the elements of the respective analyticity spaces $S_{X,A}$. Moreover, a number of examples of continuous linear mappings on these spaces is discussed. These examples are based on simple geometric and analytical considerations.

- Analyticity spaces connected to the Gelfand-Shilov spaces S_α^β.
  For a number of spaces S_α^β we give an operator $A_{\alpha,\beta}$ in the Hilbert space $L_2(\mathbb{R})$ such that $S_\alpha^\beta = S_{L_2(\mathbb{R}),A_{\alpha,\beta}}$.
- Analyticity spaces based on classical polynomials.
  The Hermite, Laguerre and Jacobi polynomials give rise to orthonormal bases in the Hilbert spaces $L_2(\mathbb{R})$, $L_2(\mathbb{R},x^{2\alpha+1}dx)$, $\alpha > -1$, and $L_2([-1,1],(1+x)^\alpha(1-x)^\beta dx)$, $\alpha,\beta > -1$, respectively. Each orthonormal basis consists of the normalized eigenfunctions of an appropriate self-adjoint operator. The elements of the associated analyticity spaces can be described by series expansions with respect to the corresponding basis. It leads to conditions on the order of growth of expansion coefficients. We characterize spaces of analytic functions with precisely these kinds of expansion coefficients.

Further, in this chapter we discuss certain relations between the representation theory of Lie groups and our theory.

In Chapter III we continue the description of the functional analytic theory An important device in distribution theory is the existence of Kernel theorems. In classical distribution theory there are considered two Kernel theorems: One for the continuous linear mappings from S into S' and one for the continuous linear mappings from S' into S. Here we present a Kernel theorem for each of the considered five types of continuous linear mappings. In our view, a Kernel theorem gives conditions such that all continuous linear mappings of a specific type arise from the elements (kernels) out of a suitable topological tensor product.
The natural description of a complete topological product of two analyticity spaces or two trajectory spaces is again an analyticity space or a trajectory space. However, in order to describe a complete topological tensor product of an analyticity space and a trajectory space, we have to introduce two new types of spaces. We have called them analyticity-trajectory spaces and trajectory-analyticity spaces. These spaces can be regarded as generalizations of analyticity and trajectory spaces. For these compound spaces we study the usual topological features. We introduce a pairing between an analyticity-trajectory and a trajectory-analyticity space, and we show their

duality. Moreover, we describe certain intersections and algebraic sums. In a separate section we present an inclusion diagram which contains all relevant spaces. The last section of Chapter III contains the five announced Kernel theorems and their proofs.

In Chapter IV we study the following operator algebras: the algebra $\mathbf{L}(S_{X,A})$ of continuous linear mappings from $S_{X,A}$ into itself, the algebra $\mathbf{L}(T_{X,A})$ of continuous linear mappings from $T_{X,A}$ into itself and the algebra $\mathbf{E}(S_{X,A})$ of extendible linear mappings. The algebra $\mathbf{E}(S_{X,A})$ can be regarded as a $*$-algebra of unbounded operators with common dense domain. We use this algebra in a mathematical model for a description of quantum statistical mechanics. We proceed our research with respect to the introduced algebras along the lines of operator theory for Hilbert spaces.
In this chapter we devote also some attention to infinite matrices. If $S_{X,A}$ is a nuclear space, then to every element of $\mathbf{L}(S_{X,A})$, $\mathbf{L}(T_{X,A})$ and $\mathbf{E}(S_{X,A})$, respectively, there can be associated an infinite matrix. We derive simple characterizations of these matrices. As a class of examples we discuss the continuous linear mappings which have a matrix with only one nonzero co-diagonal. These mappings are usually called weighted shifts. In fact, weighted shifts and their finite combinations frequently appear in applied mathematics and in special function theory.
Last but not least we apply the mathematical calculus in the construction of nuclear analyticity spaces on which the operators out of a given set act continuously.

Chapter V is the self-contained part of this monograph in which we present a mathematical interpretation of Dirac's formalism. It consists of two parts. In the first part we present a theory of generalized eigenfunctions based on our Gelfand triple $S_{X,A} \subset X \subset T_{X,A}$. To this end we introduce the notion of Dirac basis. It is the continuous analogue of the usual notion of orthonormal basis. In a well-specified way any element of $T_{X,A}$ can be expanded towards a Dirac basis. We prove that to any self-adjoint operator P in X there can be associated a Dirac basis. If P can be extended to a closed mapping in $T_{X,A}$, then this Dirac basis consists of (generalized) eigenvectors. A main role in the proof of this result is played by a so called measure theoretical Sobolev lemma. Also we insert commutative multiplicity theory for self-adjoint operators.

In the second part of this chapter we present a mathematical apparatus which can cope with the formal computation with bras and kets in Dirac's formalism. A reinterpretation of Dirac's bracket notion leads to a mathematical theory which involves a pairing between any bra and any ket, Fourier expansion of kets, orthogonality of complete sets of eigenkets and matrices of unbounded linear mappings, all in the spirit of Dirac.

In the epilogue we indicate related results which have been achieved in the past five years.

We conclude this introduction with a summary of advantages of the functional analytic approach employed in this monograph and of possibilities for applications.

- Generalized functions are represented by trajectories, a concept which is very close to the physical intuition of what a generalized function should be.
- Test function spaces can be constructed that are invariant under a set of given operators. We can always do this in an abstract way. However, the characterization of thus obtained test spaces in terms of classical analysis may be a hard job. For results in this direction see Chapter II.
- Many of the test spaces of Gelfand-Shilov are special examples of analyticity spaces. So our general theory applies to them. See Chapter II.
- Spaces of real analytic functions on compact analytic manifolds as considered by Hashizume et al. are special examples of analyticity spaces. See [HMO].
- Hyperfunctions of fixed bounded support can be represented by trajectories. So our general theory also applies here. See Chapter II.
- Generalized functions (and δ-functions in particular) can be introduced on geometrical measure spaces. See Chapter II.
- Our functional analytic approach makes it possible to transfer large pieces of Hilbert space theory to distribution theory. This has led to a detailed exposition of continuous linear mappings, of topological tensor products and of five kernel theorems. Systematic considerations on continuous linear mappings and operator algebras are not current in distribution theory! See Chapter III, IV.
- For solving the generalized eigenvalue problem for any arbitrary self adjoint operator an adapted theory of generalized "functions" is needed. Our approach has all flexibility required in such cases. The corresponding

generalized eigenvectors constitute a "Dirac basis". See Chapter V.

- A matrix calculus for continuous operators on nuclear analyticity/trajectory spaces has been developed. See Chapter IV.5.
- With the aid of trajectory spaces an Ansatz to a mathematical rigorization of Dirac's formalism has been given which goes much beyond the traditional (attempts to) rigorizations. See Chapter V.
- A functional analytic model for quantum statistical mechanics with unbounded observables has been constructed. See Chapter IV.

CHAPTER I

Analyticity spaces, trajectory spaces and linear mappings between them

Introduction

In this chapter we describe the Gelfand triple $S_{X,A} \subset X \subset T_{X,A}$ which is fixed by a nonnegative self-adjoint operator A in a Hilbert space X. The space $S_{X,A}$ consists of all vectors $v \in \mathcal{D}^{\infty}(A) = \bigcap_{n=1}^{\infty} \mathcal{D}(A^n)$ with the property that there are $a,b > 0$ such that for all $n \in \mathbb{N}$

$$\|A^n v\| \leq n!\, a^n b \ .$$

These vectors are called analytic vectors of A and therefore we have named the space $S_{X,A}$ the analyticity space. Let $(e^{-tA})_{t>0}$ denote the semigroup generated by $-A$. The vector v is an analytic vector of A if and only if there exists $t > 0$ such that $v \in e^{-tA}(X)$. So it is natural to introduce $S_{X,A}$ as the union

$$S_{X,A} = \bigcup_{t>0} e^{-tA}(X)$$

and to impose on $S_{X,A}$ the locally convex inductive limit topology originating from the spaces $e^{-tA}(X)$. In Section I.1 we give explicit seminorms on $S_{X,A}$ which generate a locally convex topology for $S_{X,A}$ which is equivalent to the inductive limit topology. Furthermore, these seminorms lead to a description of $S_{X,A}$ as a projective limit. Although our inductive limit $S_{X,A}$ is not strict, it has many properties of a strict inductive limit.
The space $T_{X,A}$ consists of mappings F from the open interval $(0,\infty)$ into X with the property that $F(t+\tau) = e^{-\tau A} F(t)$ for all $t,\tau > 0$. We have called these mappings trajectories, and hence $T_{X,A}$ the trajectory space. A trajectory F is a solution of the evolution equation

$$\frac{du}{dt} = -\,Au\ , \qquad t > 0\ ,$$

where the 'initial condition' $\lim_{t\downarrow 0} u(t)$ does not necessarily exist in X. In Section I.2 we describe the set of 'initial conditions' which give rise to a trajectory. The set $T_{X,A}$ is in one to one correspondence with this set of permitted initial conditions. Thus we are able to describe $T_{X,A}$ as a locally convex topological vector space in two equivalent ways. One way leads to a description as a projective limit, the other leads to a description of $T_{X,A}$ as an inductive limit. We remark that there exists a great symmetry between the structures of $S_{X,A}$ and $T_{X,A}$.

The sections I.1 and I.2 are devoted to a detailed description of the topological vector spaces $S_{X,A}$ and $T_{X,A}$. In Section I.3 we introduce a pairing on $S_{X,A} \times T_{X,A}$, and we show that $S_{X,A}$ and $T_{X,A}$ are in duality. Finally, in Section I.4 we characterize all types of continuous linear mappings between these spaces. One of these types concerns the extendible mappings, i.e., the continuous linear mappings on $S_{X,A}$ which can be extended to a continuous linear mapping on $T_{X,A}$.

I.1. The analyticity space $S_{X,A}$

Let A be a nonnegative self-adjoint operator in a separable Hilbert space X, and let $(E_\lambda)_{\lambda\in\mathbb{R}}$ denote the spectral resolution of the identity with respect to A. In virtue of the Spectral Theorem (Cf. [Y]) for each everywhere finite Borel function f, the operator

$$f(A) = \int_{\mathbb{R}} f(\lambda)\,dE_\lambda$$

is well defined on its maximal domain $\mathcal{D}(f(A))$

$$\mathcal{D}(f(A)) = \left\{ x \in X \;\middle|\; \int_{\mathbb{R}} |f(\lambda)|^2 \, d(E_\lambda x, x) < \infty \right\} .$$

For convenience we introduce the following convention: By

$$\int_a^b f(\lambda)dE_\lambda \;, \qquad -\infty \leqq a < b \leqq \infty$$

we mean

$$\int_{\mathbb{R}} f(\lambda)\chi_{a,b}(\lambda)dE_\lambda$$

with

$$\chi_{a,b}(\lambda) = \begin{cases} 1 & \text{on } [a,b) \\ 0 & \text{elsewhere.} \end{cases}$$

For each $t > 0$, the operator e^{-tA} is positive and bounded. Its inverse e^{tA} is a self-adjoint operator in X. The dense subspace $e^{-tA}(X)$ gets its own Hilbert space structure by the sesquilinear form

$$(v,w)_t := (e^{tA}v, e^{tA}w) \ , \quad v,w \in e^{-tA}(X) \ .$$

Here $(\cdot,\cdot)$ denotes the inner product of X.

(I.1.1) Definition

The analyticity space $S_{X,A}$ is introduced as

$$S_{X,A} := \bigcup_{t>0} e^{-tA}(X) = \bigcup_{n\in\mathbb{N}} e^{-\frac{1}{n}A}(X) \ .$$

We topologize $S_{X,A}$ with the inductive limit topology induced by the Hilbert spaces $e^{-tA}(X)$. It is the finest locally convex topology for which all injections $i_t: e^{-tA}(X) \hookrightarrow S_{X,A}$ are continuous. This inductive limit topology is denoted by σ_{ind}.

We note that a set O is open in $S_{X,A}$ if and only if $O \cap e^{-tA}(X)$ is open in X_t for each $t > 0$. Because the bounded operators e^{-tA}, $t > 0$, establish a semigroup, the inductive limit topology is already brought about by the countable family $\{e^{-(1/n)A}(X) \mid n \in \mathbb{N}\}$. It is not hard to check that the inductive limit topology is not strict if A is unbounded. So the general literature on inductive limits of Hilbert spaces which assumes strictness, is not applicable in our situation. However, for our special space $S_{X,A}$ we are able to give explicit seminorms which generate a topology for $S_{X,A}$ which is equivalent to the inductive limit topology.
To this end we introduce the function algebra $\mathbf{B}(\mathbb{R})$.

(I.1.2) Definition

- $\mathbf{B}(\mathbb{R})$ consists of all everywhere finite Borel functions ψ on $\mathbb{R}$ with the property that for all $t > 0$

$$\sup_{\lambda>0} |\psi(\lambda)e^{-t\lambda}| < \infty .$$

- $\mathbf{B}_+(\mathbb{R})$ denotes the strict positive part of $\mathbf{B}(\mathbb{R})$, i.e. $\psi \in \mathbf{B}_+(\mathbb{R})$ iff $\psi \in \mathbf{B}(\mathbb{R})$ and $\psi(x) \geq \varepsilon$ for some $\varepsilon > 0$.

Let $\psi \in \mathbf{B}(\mathbb{R})$. Then $\psi(A)$ is an unbounded normal operator. For all $t > 0$ the operator $\psi(A)e^{-tA}$ is bounded. So each $w \in S_{X,A}$ is a member of $\mathcal{D}(\psi(A))$. Therefore it makes sense to introduce the following seminorms.

(I.1.3) Definition

Let $\psi \in \mathbf{B}_+(\mathbb{R})$. Then the norm s_ψ on $S_{X,A}$ is defined by

$$s_\psi(w) = \|\psi(A)w\| , \qquad w \in S_{X,A} .$$

The following lemma is a very fundamental result which gives a better understanding of the topological structure of $S_{X,A}$.

(I.1.4) Lemma

(i) The seminorms s_ψ, $\psi \in \mathbf{B}_+(\mathbb{R})$, are continuous with respect to the topology σ_{ind} for $S_{X,A}$.

(ii) Let Ω be a convex set in $S_{X,A}$ with the property that for all $t > 0$, $\Omega \cap e^{-tA}(X)$ contains an open neighbourhood of zero in $e^{-tA}(X)$. Then there exists $\psi \in \mathbf{B}_+(\mathbb{R})$ and $\varepsilon > 0$ such that

$$\{w \in S_{X,A} \mid s_\psi(w) < \varepsilon\} \subset \Omega .$$

Proof

(i) Because of the inequality $\|\psi(A)u\| \leq \|\psi(A)e^{-tA}\| \|e^{tA}u\|$ the norm s_ψ is continuous on $e^{-tA}(X)$ for each $t > 0$. Hence s_ψ is continuous on $(S_{X,A}, \sigma_{ind})$.

(ii) Introduce the operator

$$P_n = \int_{n-1}^{n} dE_\lambda , \qquad n \in \mathbb{N} .$$

Let r_n denote the radius of the largest open ball in $P_n(X)$ which fits in $\Omega \cap P_n(X)$, i.e.

$$r_n := \sup \{\rho \mid [u \in P_n(X) \wedge \|P_n u\| < \rho] \Rightarrow u \in \Omega\} .$$

Define the function χ as follows

$$\chi(\lambda) = \begin{cases} 0 & \text{if } \lambda < 0 \\ 2 \max \left(\frac{n^2}{r_n} ,1\right) & \text{if } \lambda \in [n-1,n) , \quad n \in \mathbb{N} . \end{cases}$$

We prove that $\chi \in \mathbf{B}_+(\mathbb{R})$. Let $t > 0$. Then there is $\varepsilon > 0$ such that

$$\left\{u \mid \int_0^\infty e^{\lambda t} \, d(E_\lambda u,u) < \varepsilon^2\right\} \subset \Omega \cap e^{-\frac{1}{2}tA}(X)$$

because (by assumption) $\Omega \cap e^{-\frac{1}{2}tA}(X)$ contains an open neighbourhood of 0 in $e^{-\frac{1}{2}tA}(X)$. Thus we find $r_n > \varepsilon e^{-\frac{1}{2}nt}$ for all $n \in \mathbb{N}$. So for all $\lambda \in [n-1,n)$

$$\chi(\lambda)e^{-\lambda t} \leq 2e^{-(n-1)t} \max \left(\frac{n^2}{r_n} ,1\right) \leq 2\left(\frac{n^2}{\varepsilon} e^{-\frac{1}{2}nt} + 1\right)e^t .$$

It follows that $\chi \in \mathbf{B}_+(\mathbb{R})$.
We now show

(*) $$\|\chi(A)u\| < 1 \Rightarrow u \in \Omega .$$

Suppose $u \in e^{-tA}(X)$ for some $t > 0$. Then $\sum_{n=1}^\infty \|P_n u\|_t^2 < \infty$, and, for some τ with $0 < \tau < t$

(**) $$\|P_n u\|_\tau^2 \leq e^{-2(n-1)(t-\tau)} \|u\|_t^2 .$$

Further, because of our assumption (*)

$$\|P_n u\| \leq \tfrac{1}{2} \min(n^{-2} r_n ,1) .$$

So we have $2n^2 P_n u \in \Omega \cap e^{-\tau A}(X)$ for every $n \in \mathbb{N}$. Now we can present u by

$$u = \sum_{n=1}^{N} \frac{1}{2n^2} (2n^2 P_n u) + \left(\sum_{n=N+1}^{\infty} \frac{1}{2n^2} \right) u_N$$

with

$$u_N = \left(\sum_{j=N+1}^{\infty} \frac{1}{2j^2} \right)^{-1} \sum_{n=N+1}^{\infty} P_n u .$$

Then with (**) we obtain the estimate

$$\|u_N\|_\tau^2 \leq 4N^4 \sum_{n=N+1}^{\infty} \|P_n u\|_\tau^2 \leq 4N^4 e^{-2(N+1)(t-\tau)} \|u\|_t^2 .$$

It follows that $u_N \to 0$ in $e^{-\tau A}(X)$ if $N \to \infty$, and hence $u_N \in \Omega \cap e^{-\tau A}(X)$ for N sufficiently large.

Finally, we observe that u is a subconvex combination of elements in Ω, so that $u \in \Omega$. □

We have observed already that $S_{X,A}$ is contained in the domain $\mathcal{D}(\psi(A))$ of any $\psi \in \mathbf{B}(\mathbb{R})$. We want to prove that $S_{X,A} = \bigcap_{\psi \in \mathbf{B}_+(\mathbb{R})} \mathcal{D}(\psi(A))$. Therefore, the following lemma is essential.

(I.1.5) Lemma

Let χ be a bounded Borel function on $\mathbb{R}^+$ with the property that

$$\sup_{\lambda \geq 0} |\chi(\lambda)\psi(\lambda)| < \infty$$

for all $\psi \in \mathbf{B}_+(\mathbb{R})$. Then there exists $t > 0$ and $C > 0$ such that

$$\chi(\lambda) \leq Ce^{-t\lambda} , \quad \lambda \geq 0 .$$

Proof. Suppose the contrary, i.e.

$$\forall_{t>0} \forall_{C>0} \exists_{\lambda \geq 0} : |\chi(\lambda)| > Ce^{-t\lambda} .$$

Then for each $k \in \mathbb{N}$ there exists $\lambda_k \geq 0$ such that

$$|\chi(\lambda_k)| \exp\left(\frac{1}{k+1} \lambda_k\right) > k$$

We may as well assume that $\lambda_k < \lambda_{k+1}$ for all $k \in \mathbb{N}$. Put $\lambda_0 = 0$, and define

$$f(\lambda) = \exp\left(\frac{1}{k+1} \lambda\right) , \quad \lambda \in [\lambda_k, \lambda_{k+1}) , \quad k \in \mathbb{N} \cup \{0\} .$$

Then for each t > 0

$$\sup_{\lambda\geq 0} (e^{-t\lambda} f(\lambda)) = \sup_{k\in\mathbb{N}\cup\{0\}} \sup_{\lambda\in[\lambda_k,\lambda_{k+1})} (\exp(-(t-\frac{1}{k+1})\lambda) < \infty$$

and hence $f \in \mathbf{B}_+(\mathbb{R})$. However, for each $k \in \mathbb{N}$

$$\sup_{\lambda\geq 0} (\chi(\lambda)f(\lambda)) \geq \chi(\lambda_k)f(\lambda_k) > k$$

which is a contradiction. □

(I.1.6) Theorem

Let $w \in \mathcal{D}(\psi(A))$ for each $\psi \in \mathbf{B}_+(\mathbb{R})$. Then $w \in S_{X,A}$, i.e., there exists $t > 0$ such that $w \in e^{-tA}(X)$.

Proof. Put

$$P_n = \int_{n-1}^{n} dE_\lambda \quad \text{and} \quad r_n = \|P_n w\| , \quad n \in \mathbb{N} .$$

We define the function χ by

$$\chi(\lambda) = nr_n , \quad \lambda \in [n-1,n) , \quad n \in \mathbb{N} .$$

For each $\psi \in \mathbf{B}_+(\mathbb{R})$ the function $\hat{\psi}$ defined by

$$\hat{\psi}(\lambda) = \sup_{\mu\in[n-1,n)} \psi(\mu) , \quad \lambda \in [n-1,n) , \quad n \in \mathbb{N}$$

is again in $\mathbf{B}_+(\mathbb{R})$. So we obtain for all $\lambda \in [n-1,n)$

$$\chi(\lambda)\psi(\lambda) \leq \chi(\lambda)\hat{\psi}(\lambda) \leq \|A\hat{\psi}(A)P_n w\| \leq \|A\hat{\psi}(A)w\| .$$

It follows that $\chi\cdot\psi$ is a bounded Borel function for all $\psi \in \mathbf{B}_+(\mathbb{R})$, and hence that there exists $C > 0$ and $t > 0$ such that $\chi(\lambda) \leq Ce^{-t\lambda}$, $\lambda \geq 0$, following the previous lemma.
We obtain for all $N \in \mathbb{N}$

$$\sum_{n=1}^{N} \|e^{tA} P_n w\|^2 \leq e^{2t} \sum_{n=1}^{N} e^{2(n-1)t}\|P_n w\|^2 \leq$$

$$\leq C^2 e^{2t} \sum_{\substack{n=1\\ r_n\neq 0}}^{\infty} \frac{1}{n^2} \left[\frac{\|P_n w\|^2}{r_n^2}\right] \leq C^2 e^{2t} \frac{\pi^2}{6} . \quad □$$

(I.1.7) Definition

The topology σ_{proj} on $S_{X,A}$ is the locally convex topology induced by the seminorms $w \mapsto \|\psi(A)w\|$, $\psi \in B_+(\mathbb{R})$.

Let $\psi \in B_+(\mathbb{R})$. Consider the Hilbert space $\mathcal{D}(\psi(A))$ in which the inner product is given by $(\psi(A)x,\psi(A)y)$, $x,y \in \mathcal{D}(\psi(A))$. Then the injection $i_\psi : S_{X,A} \hookrightarrow \mathcal{D}(\psi(A))$ is continuous with respect to σ_{proj}. It can easily be seen that σ_{proj} is the coarsest topology for which all injections i_ψ, $\psi \in B_+(\mathbb{R})$, are continuous. So σ_{proj} is the projective limit topology on $S_{X,A} = \bigcap_{\psi \in B_+(\mathbb{R})} \mathcal{D}(\psi(A))$. From Lemma (I.1.4) we obtain

(I.1.8) Theorem

The projective limit topology σ_{proj} for $S_{X,A}$ is equivalent to the inductive limit topology σ_{ind} for $S_{X,A}$.

Remark: Apparently the space $S_{X,A}$ has two equivalent descriptions as a locally convex topological vector space. On one hand, it is an inductive limit of Hilbert spaces, $S_{X,A} = \bigcup_{t>0} e^{-tA}(X)$, with locally convex topology σ_{ind}, on the other hand, it is a projective limit of Hilbert spaces, $S_{X,A} = \bigcap_{\psi \in B_+(\mathbb{R})} \mathcal{D}(\psi(A))$ with topology σ_{proj}.

These observations enable us to study the topological structure of $S_{X,A}$ rather exhaustingly. The thus obtained results have a greater subtlety than the general properties of strict inductive limits of Hilbert spaces have.

(I.1.9) Lemma

Let B be a subset of $S_{X,A}$. Then B is bounded iff there exists $t > 0$ such that B is a bounded subset of $e^{-tA}(X)$.

Proof.

$\Leftarrow$) By assumption there exist $t > 0$ and $K > 0$ such that for all $w \in B$

$$\|e^{tA}w\| \leq K .$$

Let $\psi \in B_+(\mathbb{R})$. Then for all $w \in B$

$$\|\psi(A)w\| \leq K\|\psi(A)e^{-tA}\| .$$

Hence B is a bounded subset of $S_{X,A}$.

$\Rightarrow$) The proof is similar to the proof of Theorem (I.1.7).

Define $r_n := \sup_{w\in B} \|P_n w\|$, where $P_n = {}_{n-1}\int^n dE_\lambda$, $n \in \mathbb{N}$, and then the function χ on $[0,\infty)$ by

$$\chi(\lambda) = nr_n , \qquad \lambda \in [n-1,n) , \quad n \in \mathbb{N} .$$

Then for each $\psi \in \mathbf{B}_+(\mathbb{R})$, and all $\lambda \in [n-1,n)$

$$\chi(\lambda)\psi(\lambda) \leqq \chi(\lambda)\hat{\psi}(\lambda) \leqq \sup_{w\in B} \|A\hat{\psi}(A)P_n w\| \leqq \sup_{w\in B} \|A\hat{\psi}(A)w\| .$$

The latter expression is finite, because B is assumed to be a bounded subset of $S_{X,A}$. From Lemma (I.1.6) we obtain constants $C > 0$ and $t > 0$ such that

$$\chi(\lambda) \leqq Ce^{-t\lambda} , \qquad \lambda \geqq 0 .$$

Hence for all $w \in B$

$$\sum_{n=1}^{\infty} \|e^{tA} P_n w\|^2 \leqq e^{2t} \sum_{n=1}^{\infty} e^{2nt} \|P_n w\|^2 \leqq$$

$$\leqq C^2 e^{2t} \sum_{\substack{n=1 \\ r_n \neq 0}}^{\infty} \frac{1}{n^2} \frac{\|P_n w\|^2}{r_n^2} \leqq C^2 e^{2t} \frac{\pi^2}{6} .$$

Since the latter expression does not depend on w, the proof is complete.□

(I.1.10) <u>Theorem</u>

Let B be a bounded subset of $S_{X,A}$. On B put the relative topology $\sigma_{\text{ind}}|_B$. Then there exists a bounded subset B_0 of X and $t > 0$ such that $B = e^{-tA}(B_0)$, and the mapping $e^{-tA}: B_0 \to B$ is a homeomorphism.
In other words, any bounded subset of $S_{X,A}$ is homeomorphic to a bounded subset of the Hilbert space X.

<u>Proof</u>. The set B - B is also bounded in $S_{X,A}$. So there exists $t' > 0$ such that B - B is a bounded subset of $e^{-t'A}(X)$. Now let $t > 0$ be any number smaller than t'. Then $B_0 = e^{tA}(B)$ is bounded in X.
Let $(w_\alpha) \subset B$ be a net convergent to an element w of B. Then the net $(w_\alpha - w)$ converges to zero in B - B. It means that

$$\|\psi(A)(w_\alpha - w)\| \to 0 \quad \text{as } \alpha \uparrow .$$

for all $\psi \in \mathbf{B}_+(\mathbb{R})$.

Now let $\varepsilon > 0$. Take $K > 0$ so large that $\|e^{t'A}u\| \leqq K$ for all $u \in B - B$. Next, let $L > 0$ so large that

$$K \exp(-(t'-t)L) < \varepsilon/2 .$$

Since $\lambda \mapsto \chi_{[0,L)}(\lambda)e^{t\lambda}$ belongs to $\mathcal{B}(\mathbb{R})$ there exists α_0 such that

$$\|\chi_{[0,L)}(A)e^{tA}(w_\alpha - w)\| < \varepsilon/2$$

for all $\alpha \geqq \alpha_0$. So we have

$$\|e^{tA}(w_\alpha - w)\|^2 = \|\chi_{[0,L)}(A)e^{tA}(w_\alpha - w)\|^2 + \|\chi_{[L,\infty)}(A)e^{tA}(w_\alpha - w)\|^2 \leqq$$

$$\leqq (\varepsilon/2)^2 + e^{-2(t'-t)L}\|e^{t'A}(w_\alpha - w)\|^2 \leqq$$

$$\leqq (\varepsilon/2)^2 + (\varepsilon/2)^2 = \tfrac{1}{2}\varepsilon^2 .$$

So the net $(e^{tA}w_\alpha)$ tends to $e^{tA}w$ in $B_0 \subset X$. Since an arbitrary convergent net in B has been taken, the mapping $e^{tA}: B \to B_0$ is continuous. It is trivial that $e^{-tA}: B_0 \to B$ is continuous. □

The preceding theorem has a number of interesting consequences.

(I.1.11) Corollary

(i) $K \subset S_{X,A}$ is compact iff there exists $t > 0$ such that K is a compact subset of the Hilbert space $e^{-tA}(X)$.

(ii) $K \subset S_{X,A}$ is compact iff K is sequentially compact.

(I.1.12) Corollary

(i) Let $(u_n)_{n\in\mathbb{N}}$ be a sequence in $S_{X,A}$. Then $(u_n)_{n\in\mathbb{N}}$ is a Cauchy sequence in $S_{X,A}$ iff there exists $t > 0$ such that $(u_n)_{n\in\mathbb{N}}$ is a Cauchy sequence in the Hilbert space $e^{-tA}(X)$.

(ii) $S_{X,A}$ is sequentially complete.

Finally, using standard terminology (cf. [SCH]) of topological vector spaces we list the following properties of $S_{X,A}$.

(I.1.13) Theorem

(i) $S_{X,A}$ is complete.

(ii) $S_{X,A}$ is bornological.

(iii) $S_{X,A}$ is barreled.

(iv) $S_{X,A}$ is a Montel iff for every $t > 0$ the operator e^{-tA} is compact on X.

(v) $S_{X,A}$ is nuclear iff for every $t > 0$ the operator e^{-tA} is Hilbert-Schmidt on X.

Proof

(i) Let (w_α) be a Cauchy net in $S_{X,A}$. Then $(\psi(A)w_\alpha)$ is a Cauchy net in X for each $\psi \in B_+(\mathbb{R})$. Since X is complete there exists $w(\psi)$ in X such that $\psi(A)w_\alpha$ tends to $w(\psi)$ in the Hilbert space X. Put $\mathbb{1}(\lambda) = 1$, $\lambda \in \mathbb{R}$. Then $\mathbb{1}(A)$ is the identity operator on X and $w_\alpha \to w(\mathbb{1})$ in X. Since the operators $\psi(A)$ are closed one has $w(\mathbb{1}) \in \mathcal{D}(\psi(A))$ and $w(\psi) = \psi(A)w(\mathbb{1})$. The result follows by applying Theorem (I.1.6).

(ii) Every circled convex subset $\Omega \subset S_{X,A}$ that absorbs every bounded subset $W \subset S_{X,A}$ has to be a neighbourhood of 0. Let B_t denote the unit ball in $e^{-tA}(X)$, $t > 0$. Then clearly one has $\varepsilon B_t \subset \Omega \cap e^{-tA}(X)$ for some $\varepsilon > 0$. Applying Theorem (I.1.4) leads to the wanted result.

(iii) Let V be a barrel, i.e., an absorbing convex circled and closed subset of $S_{X,A}$. Then $V \cap e^{-tA}(X)$ is a barrel in the Hilbert space $e^{-tA}(X)$ for each $t > 0$. So $V \cap e^{-tA}(X)$ contains an open neigbourhood of zero, since $e^{-tA}(X)$ is barreled being a Hilbert space. Following Theorem (I.1.4), V contains an open set $\{w \mid s_\psi(w) < \varepsilon\}$ for some $s > 0$ and some $\psi \in B_+(\mathbb{R})$. Hence $S_{X,A}$ is barreled.

(iv) $\Leftarrow$) Consider the following diagram

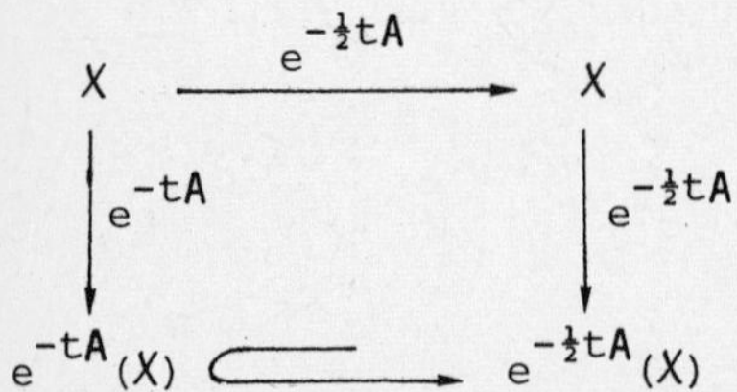

From this diagram the following observations become clear.

Let B a closed and bounded set in $S_{X,A}$. Then B is closed and bounded in $e^{-tA}(X)$ for some $t > 0$. Cf. Lemma (I.1.9). Since $e^{tA}(B)$ is closed and bounded in X and since $e^{-\frac{1}{2}tA}$ is a compact operator on X it follows that B is a compact subset of $e^{-\frac{1}{2}tA}(X)$.

⇒) Let $(u_n)_{n\in\mathbb{N}}$ be a bounded sequence in X. Then $(e^{-tA}u_n)_{n\in\mathbb{N}}$ is a bounded sequence in $e^{-tA}(X)$. So the closure of the set

$$\{e^{-tA}u_n \mid n \in \mathbb{N}\}$$

is compact and hence sequentially compact in $S_{X,A}$. It follows that the sequence $(e^{-tA}u_n)_{n\in\mathbb{N}}$ contains an $S_{X,A}$- and hence X-convergent subsequence.

(v) ⇐) Let e^{-tA} be a Hilbert-Schmidt operator for all $t > 0$. By the Spectral Theorem the operator A has a discrete spectrum $0 \leqq \lambda_1 \leqq \lambda_2 \leqq \leqq \ldots$ with $\lambda_n \uparrow \infty$. We denote the corresponding eigenvectors by v_n, $n \in \mathbb{N}$. So $(v_n)_{n\in\mathbb{N}}$ is an orthonormal basis for X.
First we construct a function $\chi \in B_+(\mathbb{R})$ such that $\chi(A)^{-1}$ is a nuclear operator. To this end we choose the sequence of natural numbers $(N_n)_{n\in\mathbb{N}}$ as follows:

- $\lambda_{N_n} + 1 < \lambda_{N_{n+1}}$

- $\displaystyle\sum_{k=N_n}^{\infty} e^{-\frac{1}{n}\lambda_k} < \frac{1}{n^2}$.

Then the function χ is defined by

$$\chi(\lambda) = \exp(\tfrac{1}{n}\lambda)\ , \quad \lambda \in [\lambda_{N_n}, \lambda_{N_{n+1}})$$

$$\chi(\lambda) = 1 \quad , \quad \lambda \in [0, \lambda_{N_1})\ .$$

So we get

$$\sum_{k=1}^{\infty} \chi(\lambda_k)^{-1} = \sum_{n=1}^{\infty} \left(\sum_{k=N_n}^{N_{n+1}-1} e^{-\frac{1}{n}\lambda_k} \right) + N_1 \leqq N_1 + \sum_{n=1}^{\infty} \frac{1}{n^2}\ .$$

Hence the operator $\chi(A)^{-1}$ is trace class.
Now let $\psi \in B_+(\mathbb{R})$, and let $\hat{S}_\psi$ denote the completion of $S_{X,A}$ with respect to the norm s_ψ. Put $\nu = \psi\cdot\chi$. Then $\nu \geqq \psi$. The injection $J_{\nu,\psi}: \hat{S}_\nu \hookrightarrow \hat{S}_\psi$ is nuclear, because it can be written as

$$J_{\nu,\psi}\, w = \sum_{k=1}^{\infty} \chi(\lambda_k)^{-1} (w, \nu(\lambda_k)^{-1} v_k)_\nu \, \psi(\lambda_k)^{-1} v_k \, .$$

Note that $(\nu(\lambda_k)^{-1} v_k)_{k\in\mathbb{N}}$ is an orthonormal basis in $\hat{S}_\nu$ and that $(\psi(\lambda_k)^{-1} v_k)_{k\in\mathbb{N}}$ is an orthonormal basis in $\hat{S}_\psi$.

$\Rightarrow$) Let $S_{X,A}$ be a nuclear space. Take $\psi = 1$. Then there exists $\chi \in B_+(\mathbb{R})$ such that the injection

$$J : \hat{S}_\chi \hookrightarrow X$$

is nuclear. Hence $\chi(A)^{-1}$ is a nuclear operator from X into X. But then $e^{-tA} = (e^{-tA}\chi(A))\chi(A)^{-1}$ is a nuclear operator for all $t > 0$. □

I.2. The trajectory space $T_{X,A}$

In this section we describe the third component in the Gelfand triple $S_{X,A} \subset X \subset T_{X,A}$. In the definition of the space $T_{X,A}$ the semigroup $(e^{-tA})_{t>0}$ plays an important role.

(I.2.1) Definition

The space $T_{X,A}$ consists of mappings F from the open interval $(0,\infty)$ into X with the additional property that

$$\forall_{t>0}\ \forall_{\tau>0} : F(t+\tau) = e^{-\tau A} F(t) \, .$$

The elements of $T_{X,A}$ are called trajectories and the space $T_{X,A}$ is called the trajectory space. We note that, for all $t > 0$, $F(t) \in S_{X,A}$.

The space $T_{X,A}$ consists of all solutions of the evolution equation

$$\frac{dF}{dt} = -AF$$

$$F(t) \in X \, , \qquad t > 0 \, .$$

The initial condition 'F(0)' need not be an element of X. Consider e.g. the trajectory $F : t \mapsto A^k e^{-tA} x$ with $x \in X \setminus \mathcal{D}(A)$, $k \in \mathbb{N}$. Loosely speaking, the notion of trajectory gives a mathematical meaning to certain initial conditions which are beyond Hilbert space.

(I.2.2) Definition

The topology τ_{proj} on $T_{X,A}$ is the locally convex topology induced by the seminorms $q_t(F) = \|F(t)\|$, $t > 0$.

The seminorms q_t are in fact norms on $T_{X,A}$. To see this, let $\|F(t)\| = 0$ for some $t > 0$. Then $F(t) = 0$, and $F(\tau) = 0$ for all $\tau > 0$ because $e^{-\tau A}$ is injective.

The Hilbert space X can be embedded into $T_{X,A}$ in the following way. Define the mapping emb: $X \to T_{X,A}$, $\text{emb}(x): t \to e^{-tA}x$, $t > 0$, $x \in X$.

(I.2.3) Proposition

emb(X) is a dense subspace of $T_{X,A}$.

Proof. Let $F \in T_{X,A}$. Then $F(t) \in X$ for each $t > 0$. For all $\tau > 0$ we have

$$\lim_{t\downarrow 0} \|\text{emb}(F(t))(\tau) - F(\tau)\| = \lim_{t\downarrow 0} \|F(t+\tau) - F(\tau)\| = 0 .$$

Hence the net $(F(t))_{t>0}$ converges to F in the topology τ_{proj}. □

Since

$$\|F(t)\| \leq \|\exp(-(t - \tfrac{1}{n})A)\| \|F(\tfrac{1}{n})\|$$

for all $n \in \mathbb{N}$ with $0 < \frac{1}{n} \leq t$, the topology τ_{proj} is already generated by the countable number of seminorms $q_{1/n}$, $n \in \mathbb{N}$. So the space $T_{X,A}$ with topology τ_{proj} is metrizable. It is also complete.

(I.2.4) Theorem

The space $T_{X,A}$ with the topology τ_{proj} is a Frechet space.

Proof. By standard arguments it can be proved that the metric of

$$d(F) = \sum_{n=1}^{\infty} 2^{-n} \frac{\|F(\frac{1}{n})\|}{1 + \|F(\frac{1}{n})\|}$$

gives rise to a locally convex topology which is equivalent to the topology τ_{proj}.

We prove the completeness of $T_{X,A}$.

Let $(F_n)_{n\in\mathbb{N}}$ be a Cauchy sequence in $T_{X,A}$. Then for all $t > 0$ the sequence $(F_n(t))_{n\in\mathbb{N}}$ is Cauchy in X. So there exists $F(t)$ in X with

$$\lim_{n\to\infty} \|F_n(t) - F(t)\| = 0 .$$

Define $F: t \mapsto F(t)$. Then

$$F(t+\tau) = \lim_{n\to\infty} F_n(t+\tau) = e^{-\tau A} \lim_{n\to\infty} F_n(t) = e^{-\tau A} F(t) .$$

Hence $F \in T_{X,A}$, and the sequence $(F_n)_{n\in\mathbb{N}}$ converges to F. □

In the beginning of this section we indicated that the space $T_{X,A}$ is connected to a certain set of initial conditions which extend Hilbert space in some sense. Now we want to give a precise meaning to this heuristic assertion.

Let $\psi \in B_+(\mathbb{R})$ and let $F \in T_{X,A}$. Then $\psi(A)F$ denotes the trajectory $t \mapsto \psi(A)F(t)$, $t > 0$. By abuse of notation we write $\psi(A)x$ instead of $\psi(A)\,\mathrm{emb}(x)$ for each $x \in X$. Here we prove that all elements of $T_{X,A}$ can be written as $\psi(A)x$ for certain $x \in X$ and $\psi \in B_+(\mathbb{R})$.

(I.2.5) <u>Proposition</u>

Let $F \in T_{X,A}$. Then there exist $\psi \in B_+(\mathbb{R})$ and $x \in X$ such that $F = \psi(A)x$.

<u>Proof</u>. Let $t > 0$ be fixed. Put $P_n = \int_{n-1}^{n} dE_\lambda$, $n \in \mathbb{N}$. We observe that the operator $e^{tA} P_n$ is bounded for each $n \in \mathbb{N}$. So the numbers $r_n = \|e^{tA} P_n F(t)\|$ are well defined; they do not depend on t. Now put

$$\psi(\lambda) = \max\,(nr_n, 1) , \qquad \lambda \in [n-1,n) , \quad n \in \mathbb{N} .$$

Then for each $\tau > 0$ and all $\lambda \in [n-1,n)$

$$\psi(\lambda)e^{-\lambda t} \leqq \max\,(nr_n, 1)e^{-(n-1)\tau} \leqq \max\,(1, e^{\tau}\|F(\tau)\|) .$$

Hence $\psi \in B_+(\mathbb{R})$. Put

$$x = \sum_{n=1}^{\infty} \left[\frac{1}{\max\,(nr_n, 1)}\, e^{tA} P_n F(t)\right] .$$

Then $x \in X$ and $F = \psi(A)x$. □

The correspondence between F and its initial condition $\psi(A)x$ gives rise to an alternative description of $T_{X,A}$. Therefore, for each $\psi \in B_+(\mathbb{R})$ we introduce the space $\psi(A)X$ which consists of all trajectories $\psi(A)x$, $x \in X$. With the inner product

$$(\psi(A)x,\psi(A)y)_\psi = (x,y) , \qquad x,y \in X$$

$\psi(A)X$ becomes a Hilbert space. We note that $\psi(A)X$ can be seeen as a completion of X with respect to the norm $\|\psi(A)^{-1}\cdot\|$. Following the previous proposition $T_{X,A}$ equals $\bigcup_{\psi\in B_+(\mathbb{R})} \psi(A)X$ as a set.

(I.2.6) <u>Definition</u>

On the set $\bigcup_{\psi\in B_+(\mathbb{R})} \psi(A)X$ $(=T_{X,A})$ we impose the inductive limit topology denoted by τ_{ind}, originating from the Hilbert spaces $\psi(A)X$.

We shall prove that the topology τ_{proj} for $T_{X,A}$ is equivalent to the topology τ_{ind}. The following lemma is similar to Lemma (I.1.4).

(I.2.7) <u>Lemma</u>

(i) The seminorms q_t, $t > 0$, are continuous with respect to the topology τ_{ind} for $T_{X,A}$.

(ii) Let a convex set $\Omega \subset T_{X,A}$ be such that for each $\psi \in B_+(\mathbb{R})$ the set $\Omega \cap \psi(A)X$ contains an open neighbourhood of zero in the Hilbert space $\psi(A)X$. Then Ω contains a set

$$V_{t,\varepsilon} := \{F \in T_{X,A} \mid \|F(t)\| < \varepsilon\}$$

for certain $\varepsilon > 0$ and $t > 0$.

<u>Proof</u>:

(i) Let $t > 0$. Then for all $F \in \psi(A)X$, $\psi \in B_+(\mathbb{R})$,

$$\|F(t)\| \leq \|\psi(A)e^{-tA}\| \, \|F\|_\psi .$$

From this observation the wanted continuity of q_t follows.

(ii) The proof of this part is similar to the proof of part (ii) of Lemma (I.1.4).

For each $n \in \mathbb{N}$ we put $P_n = {}_{n-1}\int^n dE_\lambda$. For each $F \in \mathcal{T}_{X,A}$ we interprete $P_n F$ as an element of X through the definition

$$P_n F = e^A P_n F(1) .$$

We define the positive numbers r_n by

$$r_n := \sup \{\rho > 0 \mid [F \in P_n(X) \wedge \|P_n F\| < \rho] \Rightarrow F \in \Omega\}$$

and the function χ on $\mathbb{R}^+$ by

$$\chi(\lambda) = \frac{2n^2}{r_n} , \qquad \lambda \in [n-1,n) , \quad n \in \mathbb{N}$$

Let $\psi \in B_+(\mathbb{R})$. Then also the function $\widetilde{\psi}$ defined by

$$\widetilde{\psi}(\lambda) = n^2 \sup_{\mu \in [n-1,n)} \psi(\mu) , \qquad \lambda \in [n-1,n) , \quad n \in \mathbb{N}$$

belongs to $B_+(\mathbb{R})$. So there exists $\delta > 0$ such that

$$\{F \in \widetilde{\psi}(A)X \mid \|F\|_{\widetilde{\psi}} < \delta\} \subset \Omega \cap \psi(A)X$$

because $\Omega \cap \widetilde{\psi}(A)X$ contains an open neighbourhood of 0. We obtain

$$r_n > \delta n^2 \sup_{\lambda \in [n-1,n)} \psi(\lambda) , \qquad n \in \mathbb{N} .$$

So for each $n \in \mathbb{N}$ and all $\lambda \in [n-1,n)$

$$\chi(\lambda)\psi(\lambda) \leqq \frac{2n^2}{r_n} \sup_{\lambda \in [n-1,n)} \psi(\lambda) \leqq \frac{2}{\delta} .$$

Since $\psi \in B_+(\mathbb{R})$ has been taken arbitrarily, it follows from Lemma (I.1.6) that there exists $C > 0$ and $t > 0$ such that $\chi(\lambda) \leqq Ce^{-t\lambda}$, $\lambda \geqq 0$. We next prove that

$$(*) \qquad F(t) \leqq \frac{1}{C} \Rightarrow F \in \Omega .$$

So let $F \in \psi(A)X$ for some $\psi \in B_+(\mathbb{R})$. Then

$$\sum_{n=1}^{\infty} \|P_n F\|_\psi^2 < \infty$$

and

$$(**) \qquad \|(I + A^2)^{-3} P_n F\|_\psi \leqq (1 + (n-1)^2)^{-3} \|F\|_\psi , \qquad n \in \mathbb{N} .$$

Because of (*) we have

$$\|e^{tA} P_n F(t)\| \leqq C\|\chi(A)^{-1} P_n F(t)\| \leqq \frac{r_n}{2n^2}$$

and hence $2n^2 P_n F \in \Omega$ for all $n \in \mathbb{N} \cup \{0\}$. Now we express F as

$$F = \sum_{n=1}^{N} \frac{1}{2n^2} (2n^2 P_n F) + \left(\sum_{n=N+1}^{\infty} \frac{1}{2n^2} \right) F_N$$

where

$$F_N = \left(\sum_{j=N+1}^{\infty} \frac{1}{2j^2} \right)^{-1} \left(\sum_{n=N+1}^{\infty} P_n F \right) .$$

Put $\varphi(\lambda) = (1 + \lambda^2)^3 \psi(\lambda)$, $\lambda \geqq 0$. Then $\varphi \in B_+(\mathbb{R})$, and

$$\|F_N\|_\varphi \leqq 4N^4 (1 + N^2)^{-3} \|F\|_\psi .$$

So $\|F_N\|_\varphi$ tends to zero as $N \uparrow \infty$. It follows that $F_N \in \Omega \cap \varphi(A)X$ for sufficiently large N. Since F is a subconvex combination of elements in the convex set Ω, F belongs to Ω. □

(I.2.8) Theorem

The topology τ_{proj} for $T_{X,A}$ is equal to the topology τ_{ind} for $T_{X,A}$.

In fact we now proceed in the same way as in the previous section. The space $S_{X,A}$ and the space $T_{X,A}$ both have the structure of both an inductive limit and of a projective limit of Hilbert spaces.

(I.2.9) Theorem

(i) A set $B \subset T_{X,A}$ is bounded iff the set $\{F(t) \mid t > 0\}$ is bounded in X for all $t > 0$.

(ii) A set $B \subset T_{X,A}$ is bounded iff there exists $\psi \in B_+(\mathbb{R})$ and a bounded set $B_0 \subset X$ such that $B = \psi(A)B_0 = \{\psi(A)x \mid x \in B_0\}$.

Proof.

(i) The proof of (i) follows trivially from the definition of τ_{proj}.

(ii) ⇐) Again take $P_n = {}_{n-1}\int^n dE_\lambda$, $n \in \mathbb{N}$. We have

$$\|P_n F\| \leq \|e^{A} P_n\| \|F(1)\| \ .$$

So it makes sense to define $r_n = \sup_{F \in B} \|P_n F\|$, and then

$$\psi(\lambda) = \max\,(nr_n, 1) \ , \qquad \lambda \in [n-1,n) \ , \quad n \in \mathbb{N} \ .$$

Then for all $t > 0$ and all $\lambda \in [n-1,n)$

$$nr_n e^{-\lambda t} \leq nr_n e^{-(n-1)t} \leq 2e^{t} \|Ae^{-\frac{1}{2}tA}\| \sup_{F \in B} \|F(\tfrac{1}{2}t)\| \ .$$

It follows that $\psi \in \mathbf{B}_+(\mathbb{R})$.
Let $F \in B$. The series $\sum_{n=1}^{\infty} [(\max(nr_n,1))^{-2} \|P_n F\|^2]$ converges, hence $x_F := \sum_{n=0}^{\infty} (\max(nr_n,1))^{-1} P_n F$ belongs to X. Note that $F = \psi(A) x_F$. Now the set $B_0 = \{x_F \mid F \in B\}$ is bounded in X, because

$$\|x_F\|^2 = \sum_{n=1}^{\infty} [(\max(nr_n,1))^{-2} \|P_n F\|^2] \leq \sum_{n=1}^{\infty} \frac{1}{n^2} \ .$$

⇒) Trivial. □

(I.2.10) Lemma

Let B be a bounded subset of $\mathcal{T}_{X,A}$. Then there exists $\psi \in \mathbf{B}_+(\mathbb{R})$ with the following properties:

- There exists a bounded subset $B_0 \subset X$ such that $B = \psi(A) B_0$.
- Let (F_α) be a net in B. Then (F_α) converges to $F \in B$ in the topology τ_{proj} iff $F_\alpha = \psi(A) x_\alpha$, $F = \psi(A) x$ and $\|x_\alpha - x\| \to 0$.

Proof. By Theorem (I.2.8) there exists $\varphi \in \mathbf{B}_+(\mathbb{R})$ and a bounded subset $B_1 \subset X$ such that $B = \varphi(A) B_1$. Put $\psi(\lambda) = (1 + \lambda^2)\varphi(\lambda)$, $\lambda \geq 0$, and put $B_0 = (I + A^2)^{-1}(B_1)$. Then clearly we have $B = \psi(A) B_0$.

⇒) For each α there exist $(x_\alpha - x), (y_\alpha - y) \in X$ such that $F_\alpha - F = \varphi(A)(y_\alpha - y) = \psi(A)(x_\alpha - x)$. Let $\varepsilon > 0$, and take $L > 0$ so large that $(1 + \lambda^2)^{-1} < \varepsilon$ for $\lambda > L$. Then we obtain

$$(*) \qquad \int_{\lambda > L} d(E_\lambda (x_\alpha - x), x_\alpha - x) \leq \varepsilon^2 \int_{\lambda > L} d(E_\lambda (y_\alpha - y), y_\alpha - y) \leq$$

$$\leq \varepsilon^2 \|y_\alpha - y\|^2 \leq 2\varepsilon^2 \sup_{z \in B_1} \|z\| \ .$$

Moreover, let $t > 0$. Then

$$\int_{\lambda\leq L} d(E_\lambda(x_\alpha - x), x_\alpha - x) \leq \sup_{\lambda\in[0,L]} (\psi(\lambda)^{-1} e^{\lambda t})^2 \|F_\alpha(t) - F(t)\|^2 .$$

So there exists α_0 such that for all $\alpha > \alpha_0$

$$(**) \qquad \int_{\lambda\leq L} d(E_\lambda(x_\alpha - x), x_\alpha - x) < \varepsilon^2 .$$

Now (*) and (**) yield the wanted result.

$\Leftarrow$) Trivial. □

(I.2.11) Theorem

Each bounded subset B of $\mathcal{T}_{X,A}$ is homeomorphic to a bounded subset of X. The homeomorphism is established by an operator $\psi(A)$ for suitable $\psi \in B_+(\mathbb{R})$.

(I.2.12) Theorem

Let K be a subset of $\mathcal{T}_{X,A}$. Then the following statements are equivalent:

(i) K is compact.

(ii) There exists $\psi \in B_+(\mathbb{R})$ and a compact subset $K_0 \subset X$ such that $K = \psi(A)K_0$.

(iii) For all $t > 0$ the set $\{F(t) \mid F \in K\}$ is compact.

Proof. The equivalence (i) $\Leftrightarrow$ (ii) is an immediate consequence of the previous theorem. We only prove (i) $\Leftrightarrow$ (iii).

(i) $\Rightarrow$ (iii): If K is compact then each sequence $(F_n)_{n\in\mathbb{N}} \subset K$ has a convergent subsequence. This means that in the set $\{F(t) \mid F \in K\}$ each sequence has a convergent subsequence.

(iii) $\Rightarrow$ (i): Let $(F_n)_{n\in\mathbb{N}}$ be a sequence in K. Consider the sequence $(F_n(1))_{n\in\mathbb{N}}$. It has a convergent subsequence $(F_n^1(1))_{n\in\mathbb{N}}$. The sequence $(F_n^1(\frac{1}{2}))_{n\in\mathbb{N}}$ has a convergent subsequence $(F_n^2(\frac{1}{2}))_{n\in\mathbb{N}}$, etc. Thus we arrive at sequences $(F_n^m)_{n\in\mathbb{N}} \subset K$ such that $(F_n^m)_{n\in\mathbb{N}} \subset (F_n^\ell)_{n\in\mathbb{N}}$, $m > \ell$, and $(F_n^m(\frac{1}{m}))_{n\in\mathbb{N}}$ converges to an element of $\{F(\frac{1}{m}) \mid F \in K\}$. Now the diagonal sequence $(F_n^n)_{n\in\mathbb{N}}$ converges in K. □

(I.2.13) Proposition

A sequence $(F_n)_{n\in\mathbb{N}} \subset \mathcal{T}_{X,A}$ is a Cauchy sequence iff there exists $\psi \in B_+(\mathbb{R})$ such that $(F_n)_{n\in\mathbb{N}}$ is a Cauchy sequence in the Hilbert space $\psi(A)X$.

Finally, we mention some topological properties of $\mathcal{T}_{X,A}$.

(I.2.14) Theorem

(i) $\mathcal{T}_{X,A}$ is complete.

(ii) $\mathcal{T}_{X,A}$ is barreled.

(iii) $\mathcal{T}_{X,A}$ is bornological.

(iv) $\mathcal{T}_{X,A}$ is Montel iff for all $t > 0$ the operator e^{-tA} is compact.

(v) $\mathcal{T}_{X,A}$ is nuclear iff for all $t > 0$ the operator e^{-tA} is Hilbert-Schmidt.

Proof. Being a Frechet space, $\mathcal{T}_{X,A}$ is complete, barreled and bornological.

(iv) We prove that $\mathcal{T}_{X,A}$ is Montel iff there exists $\psi \in B_+(\mathbb{R})$ such that the operator $\psi(A)^{-1}$ is compact on X. It is not hard to see that the latter assertion is equivalent with the assertion that the operator e^{-tA} is compact for all $t > 0$.

⇐) Consider the following diagram

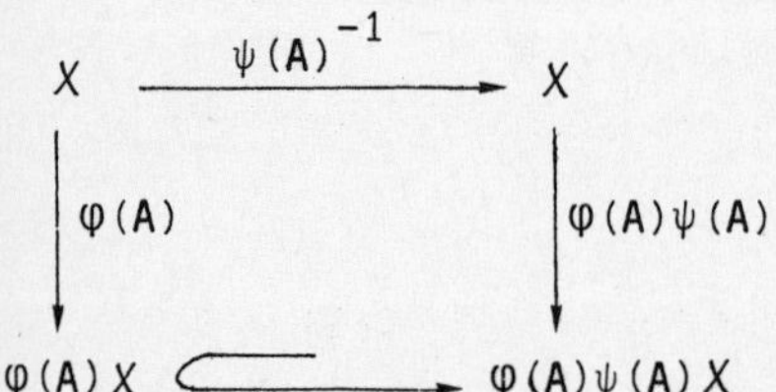

Let B be closed and bounded in $\mathcal{T}_{X,A}$. Then B is closed and bounded in $\varphi(A)X$ for some $\varphi \in B_+(\mathbb{R})$. From the above diagram we derive that B is a compact subset of $\varphi(A)\psi(A)X$.

⇒) Let $(x_n)_{n\in\mathbb{N}}$ be a bounded subset in X. Then the set $\overline{\{x_n \mid n \in \mathbb{N}\}}$ is compact in $\mathcal{T}_{X,A}$. So for all $t > 0$ the sequence $(e^{-tA}x_n)_{n\in\mathbb{N}}$ contains a convergent subsequence in X. It follows that e^{-tA} is a compact operator. So A has a discrete spectrum without accumulation points. Put $\psi(\lambda) = 1 + \lambda^2$, $\lambda \in \mathbb{R}$. Then $\psi \in B_+(\mathbb{R})$ and $\psi(A)^{-1}$ is compact.

(v) $\Leftarrow$) It follows that A has a discrete spectrum with eigenvalues $(\lambda_k)_{k\in\mathbb{N}}$ and orthonormal eigenvectors $(v_k)_{k\in\mathbb{N}}$. Let $t > 0$. Let J denote the embedding $\hat{T}_t \hookrightarrow \hat{T}_{2t}$, where $\hat{T}_\tau$, $\tau > 0$, denotes the completion of $T_{X,A}$ with respect to the norm q_τ. Since J can be written as

$$JG = \sum_{k=1}^{\infty} e^{-\lambda_k t} (G(t), v_k)(e^{2\lambda_k t} v_k) , \qquad G \in T_{X,A} ,$$

J is a nuclear map.

$\Rightarrow$) For each $\tau > 0$ the embedding $I: X \hookrightarrow \hat{T}_\tau$ is bounded. Let $t > 0$. Then there exists $\tau > 0$ such that the mapping

$$J_{t,\tau}: \hat{T}_\tau \hookrightarrow \hat{T}_t$$

is nuclear. So the injection $X \hookrightarrow \hat{T}_t$ is nuclear. It follows that e^{-tA} is a trace class operator on X. □

I.3. Pairing and duality of $S_{X,A}$ and $T_{X,A}$

In this section we turn $S_{X,A}$ and $T_{X,A}$ into a dual pair. Therefore, we introduce the sesquilinear form $\langle\, ,\, \rangle$ on $S_{X,A} \times T_{X,A}$ by

(I.3.1) $\langle w,F\rangle = (e^{tA}w, F(t))$, $\quad w \in S_{X,A}$, $\quad F \in T_{X,A}$.

Here $t > 0$ has to be chosen so small that $w \in e^{-tA}(X)$. Definition (I.3.1) does not depend on the choice of t. To show this, let $u \in e^{-tA}(X)$. Then for any τ with $0 < \tau < t$

$$(e^{\tau A}u, F(\tau)) = (e^{-(t-\tau)A} e^{\tau A}u, F(\tau)) = (e^{tA}u, F(t)) .$$

Further, the sesquilinear form is nondegenerate: If $\langle w,F\rangle = 0$ for all $F \in T_{X,A}$ then $w = 0$ because $X \subset T_{X,A}$. Conversely, if $\langle w,F\rangle = 0$ for all $w \in S_{X,A}$, then $F(t) = 0$ for all $t > 0$ and so $F = 0$. The following theorem is the extension of the Riesz-Fischer representation theorem to our kind of spaces.

(I.3.2) <u>Theorem</u>

(i) A linear functional ℓ on $S_{X,A}$ is continuous in the topology σ_{ind} for $S_{X,A}$ iff there exists $G \in T_{X,A}$ such that

$$\ell(u) = \langle u,G\rangle \,, \qquad u \in S_{X,A} \,.$$

(ii) A linear functional m on $T_{X,A}$ is continuous in the topology τ_{proj} iff there exists $w \in S_{X,A}$ such that

$$m(F) = \overline{\langle w,F\rangle} \,, \qquad F \in T_{X,A} \,.$$

Hence $T_{X,A}$ is a representation of the dual $S'_{X,A}$ of $S_{X,A}$ and $S_{X,A}$ is a representation of the dual $T'_{X,A}$ of $T_{X,A}$.

Proof

(i) $\Leftarrow$) Let $G \in T_{X,A}$. Then $G = \psi(A)x$ for some $x \in X$ and $\psi \in \mathbf{B}_+(\mathbb{R})$. Hence

$$|\langle w,G\rangle| = |(\psi(A)w,x)| \leq \|\psi(A)w\|\,\|x\| \,.$$

Since $\sigma_{\text{ind}} = \sigma_{\text{proj}}$ the functional $w \mapsto \langle w,G\rangle$ is continuous.

$\Rightarrow$) Let ℓ be a continuous linear functional on $S_{X,A}$. Then $\ell \circ e^{-tA}$ is a continuous linear functional on X for all $t > 0$. So there exists $x_t \in X$ such that $(\ell \circ e^{-tA})(x) = (x,x_t)$, $x \in X$. Put $G\colon t \mapsto x_t$, $t > 0$. Then it is easy to see that $G(t+\tau) \equiv e^{-\tau A}\,G(t)$, for all $t,\tau > 0$. Hence $G \in T_{X,A}$. Moreover,

$$\ell(u) = \ell(e^{-tA}x) = (x,x_t) = \langle u,G\rangle \,.$$

(ii) $\Leftarrow$) Let $w = e^{-tA}x$ with $x \in X$ and $t > 0$. Then we have

$$|\langle w,F\rangle| \leq \|x\|\,\|F(t)\| \,.$$

So the linear functional $F \mapsto \langle w,F\rangle$ is τ_{proj}-continuous.

$\Rightarrow$) Let m be a continuous linear functional on $T_{X,A}$. Then for each $\psi \in \mathbf{B}_+(\mathbb{R})$ the functional $m \circ \psi(A)$ extends to a continuous linear functional on the whole of X. So there exists $x_\psi \in X$ such that $(m \circ \psi(A))(x) = (x,x_\psi)$, $x \in \mathcal{D}(\psi(A))$. Put $w_\psi = \psi(A)^{-1}x_\psi$. Then it follows that $\varphi(A)x_\psi = \psi(A)x_\varphi$ for all $\varphi,\psi \in \mathbf{B}_+(\mathbb{R})$. Hence $w_\psi = w_\varphi$ and $w = w_\psi$ is well defined. Further, $w \in \bigcap_{\psi \in \mathbf{B}_+(\mathbb{R})} \mathcal{D}(\psi(A)) = S_{X,A}$ and $m(F) = m(\varphi(A)x) = \overline{(x_\varphi,x)} = \langle w,F\rangle$. □

We note that it would be incorrect to write $S'_{X,A} = T_{X,A}$ since $T_{X,A}$ is only a representation of the dual of $S_{X,A}$. Similarly, we do not write $T'_{X,A} = S_{X,A}$.

It makes sense to introduce the weak topology $\sigma(S_{X,A}, T_{X,A})$ for $S_{X,A}$, and $\sigma(T_{X,A}, S_{X,A})$ for $T_{X,A}$. It is a natural question whether weak boundedness of a subset of $S_{X,A}$ ($T_{X,A}$) implies its boundedness.

(I.3.3) Theorem (Banach-Steinhaus)

(i) Each weakly bounded set B in $S_{X,A}$ is bounded

(ii) Each weakly bounded set V in $T_{X,A}$ is bounded.

Proof

(i) Let B be a weakly bounded subset of $S_{X,A}$, i.e., for each $G \in T_{X,A}$ there exists $M_G > 0$ such that

$$\forall_{w\in B}: |\langle w,G\rangle| \leq M_G .$$

Let $\psi \in \mathbf{B}_+(\mathbb{R})$. Then for all $x \in X$ we have

$$\forall_{w\in B}: |(\psi(A)w,x)| \leq M_{\psi(A)x} .$$

So the set $\{\psi(A)w \mid w \in B\}$ is weakly bounded in X, and hence norm bounded. It means that there exists $M_\psi > 0$ such that

$$\forall_{w\in B}: \|\psi(A)w\| \leq M_\psi .$$

Since $\psi \in \mathbf{B}_+(\mathbb{R})$ is arbitrary, the set B is bounded.

(ii) Let V be a weakly bounded subset of $T_{X,A}$. Let $t > 0$. Then for each $x \in X$ there exists $M_{x,t} > 0$ such that

$$\forall_{G\in V}: |(x,G(t))| \leq M_{x,t} .$$

So there exists $M_t > 0$ such that

$$\forall_{G\in V}: \|G(t)\| \leq M_t . \qquad \square$$

From the characterization of the bounded subsets of $S_{X,A}$ and $T_{X,A}$ given in Sections I.1 and I.2, we obtain that the topology σ_{ind} is equivalent to the Mackey topology $\tau(S_{X,A}, T_{X,A})$ for $S_{X,A}$ with respect to the duality $\langle S_{X,A}, T_{X,A}\rangle$, and, similarly, the topology τ_{proj} is equivalent to the Mackey topology $\tau(T_{X,A}, S_{X,A})$ for $T_{X,A}$. For the terminology and definitions we refer to [SCH], p. 131.

In Theorem (I.1.10) and Theorem (I.2.11) we stated that bounded subsets of $S_{X,A}$ and $T_{X,A}$, respectively, are homeomorphic to bounded subsets of the Hilbert space X. Now we show that bounded subsets of $S_{X,A}$ and $T_{X,A}$ are also weakly homeomorphic to bounded subsets of X.

(I.3.4) Theorem

(i) Let B be a (weakly) bounded subset of $S_{X,A}$. Then there exists $t > 0$ such that $B \subset e^{-tA}(X)$, and such that each weakly convergent net $(w_\alpha) \subset B$ with weak limit $w \in B$ is convergent in the weak topology of the Hilbert space $e^{-tA}(X)$.

(ii) Let V be a (weakly) bounded subset of $T_{X,A}$. Then there exists $\psi \in \mathbf{B}_+(\mathbb{R})$ such that $V \subset \psi(A)X$, and such that each weakly convergent net $(F_\alpha) \subset V$ with weak limit $F \in V$ is convergent in the weak topology of the Hilbert space $\psi(A)X$.

Proof

(i) From Lemma (I.1.8) we obtain $t' > 0$ and a bounded subset B_0 of X such that $B = e^{-t'A}(B_0)$. Let $0 < t < t'$ be fixed. Then $B = e^{-tA}(e^{-(t'-t)A}B_0)$. Next, consider the net (w_α) in B which converges weakly to $w \in B$. Then the net $(w_\alpha - w)$ converges weakly to 0. Let $K := \sup_{u\in B} \|e^{t'A}u\|$. Then for all α

$$\|e^{t'A}(w_\alpha - w)\| \leq 2K .$$

Let $\varepsilon > 0$. There is $L > 0$ such that $\exp(-L(t'-t)) < \varepsilon$. So we obtain for all $\xi \in X$

$$(*) \qquad \left|\int_L^\infty e^{\lambda t}\, d(E_\lambda(w_\alpha - w),\xi)\right| < 2e^{-L(t'-t)}K\|x\| = 2\varepsilon K\|\xi\| .$$

Next, observe that the function $\lambda \mapsto \chi_{[0,L)}(\lambda)e^{\lambda t}$ belongs to $\mathbf{B}(\mathbb{R})$. So by assumption we find that for each $x \in X$ there exists α_0 such that

$$(**) \qquad |(e^{tA}\chi_{[0,L)}(A)(w_\alpha - w),x)| < \varepsilon .$$

Combination of (*) and (**) yields the wanted result.

(ii) From Theorem (I.2.8) we obtain a function $\psi \in \mathbf{B}_+(\mathbb{R})$ and a bounded set $V_1 \subset X$ such that $V = \psi(A)V_1$. Put $\varphi(\lambda) = (1 + \lambda^2)\psi(\lambda)$, $\lambda \geq 0$. Then

$\varphi \in B_+(\mathbb{R})$, and $V = \varphi(A)V_0$, where $V_0 = (I + A^2)^{-1} V_1$. Put $K = \sup \{\|b\| \mid b \in V_1\}$.

Let (F_α) be a net in V with weak limit $F \in V$. Then there exists x_α, $x \in V_0$ and y_α, $y \in V$, such that $F_\alpha = \varphi(A)x_\alpha = \psi(A)y_\alpha$ and $F = \varphi(A)x = = \psi(A)y$. Let $\varepsilon > 0$. Then we can take $L > 0$ so large that $(1+L^2)^{-1} < \varepsilon$. It follows that for all $\xi \in X$

$$(*) \qquad |(\chi_{[L,\infty)}(A)(x_\alpha - x),\xi)| \leq (1 + L^2)^{-1}(y_\alpha - y,\xi) < 2\varepsilon K\|\xi\| \ .$$

Let $t > 0$ be fixed. Then for each $\xi \in X$ there exists α_0 such that

$$|(\xi,F_\alpha(t) - F(t))| < \varepsilon \ , \qquad \alpha \geq \alpha_0$$

and hence

$$(**) \qquad |(\xi,\chi_{[0,L)}(A)(x_\alpha - x))| \leq [\sup_{\lambda\in[0,L)} (e^{\lambda t}\,\varphi(\lambda)^{-1})]\,|(\xi,F_\alpha(t) - F(t))| \leq$$

$$\leq \varepsilon e^{tL} \ .$$

So from (*) and (**) we derive that for all $\alpha \geq \alpha_0$

$$|(\xi,x_\alpha - x)| \leq \varepsilon(e^{tL} + 2K\|\xi\|) \ . \qquad \square$$

(I.3.5) <u>Corollary</u>

(i) Let $(w_n)_{n\in\mathbb{N}}$ be a weakly convergent sequence in $S_{X,A}$. Then there exists $t > 0$ such that the sequence $(e^{tA}w_n)_{n\in\mathbb{N}}$ is weakly convergent in the Hilbert space X.

(ii) Let $(G_n)_{n\in\mathbb{N}}$ be a weakly convergent sequence in $T_{X,A}$. Then there exists $\psi \in B_+(\mathbb{R})$ and a weakly convergent sequence $(x_n)_{n\in\mathbb{N}}$ in X such that $G_n = \psi(A)x_n$. We note that $\text{w-lim}\, G_n = \psi(A)(\text{w-lim}\, x_n)$.

(I.3.6) <u>Corollary</u>

(i) Any bounded sequence $(w_n)_{n\in\mathbb{N}}$ in $S_{X,A}$ contains a weakly convergent subsequence.

(ii) Any bounded sequence $(F_n)_{n\in\mathbb{N}}$ in $T_{X,A}$ contains a weakly convergent subsequence.

The final theorem of this section deals with the following question: When does weak convergence of a sequence imply its strong convergence? The following result is quite standard.

(I.3.7) Theorem

(i) Each weakly convergent sequence in $S_{X,A}$ is also σ_{ind}-convergent iff $S_{X,A}$ is a Montel space.

(ii) Each weakly convergent sequence in $T_{X,A}$ is also τ_{proj}-convergent iff $T_{X,A}$ is a Montel space.

Proof

(i) $\Leftarrow$) A weakly convergent sequence in $S_{X,A}$ converges weakly in $e^{-tA}(X)$ for some $t > 0$. Since $S_{X,A}$ is Montel, the injection $X_t \subset X_\alpha$ is compact for all α with $0 < \alpha < t$. So the sequence converges strongly in $e^{-\alpha A}(X)$.

$\Rightarrow$) Let $t > 0$, and let $(x_n)_{n\in\mathbb{N}}$ be a weakly convergent sequence in X. Then $(e^{-tA}x_n)_{n\in\mathbb{N}}$ is a weakly convergent sequence in $S_{X,A}$. So $(e^{-tA}x_n)_{n\in\mathbb{N}}$ is convergent with respect to σ_{ind}. In particular it follows that the sequence $(e^{-tA}x_n)_{n\in\mathbb{N}}$ converges strongly in X. So e^{-tA} is a compact operator. Since $t > 0$ has been taken arbitrarily it follows from Theorem (I.1.13) that $S_{X,A}$ is a Montel space.

(ii) $\Leftarrow$) Let $(F_n)_{n\in\mathbb{N}}$ be a sequence in $T_{X,A}$ which converges weakly to $F \in T_{X,A}$, i.e., $F_n(t) \to F(t)$ weakly in X for all $t > 0$. Since $T_{X,A}$ is Montel, it follows from Theorem (I.2.14) that the operator $e^{-\tau A}$ is compact for all $\tau > 0$. Hence $\|F_n(t-\tau) - F(t-\tau)\| \to 0$ for all τ with $0 < \tau < t$. But then $F_n \to F$ in τ_{proj}-sense.

$\Rightarrow$) Let $(x_n)_{n\in\mathbb{N}}$ be a weakly convergent sequence in X. Then $(x_n)_{n\in\mathbb{N}}$ is also a weakly convergent sequence in $T_{X,A}$, and hence it is τ_{proj}-convergent. It follows that for all $t > 0$ the sequence $(e^{-tA}x_n)_{n\in\mathbb{N}}$ is strongly convergent in X. Since the sequence $(x_n)_{n\in\mathbb{N}}$ is arbitrarily taken, all operators e^{-tA}, $t > 0$, are compact. □

One might wonder whether the Mackey and the weak topology coincide if all operators e^{-tA}, $t > 0$, are compact. They indeed do when restricted to bounded subsets of $S_{X,A}$ ($T_{X,A}$). However, it cannot be true on $S_{X,A}$ ($T_{X,A}$) as a whole, since a set which belongs to a weak basis of 0-neighbourhoods always contains a closed subspace of finite co-dimension. Such 'large' sets do not fit in σ_{ind}- (τ_{proj}-) open null-neighbourhoods.

I.4. Continuous linear mappings between analyticity spaces and trajectory spaces

Let B be a nonnegative self-adjoint operator in a separable Hilbert space Y. As in the previous sections A is a nonnegative self-adjoint operator in the separable Hilbert space X. In this chapter we give conditions implying the continuity of linear mappings: $S_{X,A} \to S_{Y,B}$, $S_{X,A} \to T_{Y,B}$, $T_{X,A} \to S_{Y,B}$, $T_{X,A} \to T_{Y,B}$. Also, we look for conditions which imply that a linear operator from X into Y gives rise to a continuous linear mapping of one of the previous types. It will be clear to the reader that in this connection the description of analyticity spaces and trajectory spaces both as inductive and as projective limits is of great use.

Further, we mention that we also investigate a fifth type of linear mappings, viz. continuous linear mappings from $S_{X,A}$ into $S_{Y,B}$ which have an extension to a continuous linear mapping from $T_{X,A}$ into $T_{Y,B}$. We call them the extendible mappings. The next theorem is an immediate consequence of the definition of the topology σ_{ind} for $S_{X,A}$ and the fact that $S_{X,A}$ is bornological.

(I.4.1) Theorem

A linear mapping $L: S_{X,A} \to V$, where V is an arbitrary locally convex topological vector space, is continuous

(i) iff for each $t > 0$ the mapping $Le^{-tA}: X \to V$ is continuous;

(ii) iff for each null sequence $(u_n)_{n\in\mathbb{N}} \subset S_{X,A}$, the sequence $(Lu_n)_{n\in\mathbb{N}}$ is a null sequence in V.

Proof

(i) $\Rightarrow$) e^{-tA} is an isomorphism from X to $e^{-tA}(X)$. By the definition of the topology σ_{ind} the injection $e^{-tA}(X) \hookrightarrow S_{X,A}$ is continuous. So, if L is continuous, it follows that Le^{-tA} is continuous on X.

$\Leftarrow$) Let L_t denote the restriction of L to $e^{-tA}(X)$. Then by assumption L_t is continuous on $e^{-tA}(X)$. Now let Ω be an open null-neighbourhood in V. Then for each $t > 0$, $L^{-1}(\Omega) \cap e^{-tA}(X) = L_t^{-1}(\Omega)$ is an open neighbourhood in $e^{-tA}(X)$. Therefore, $L^{-1}(\Omega)$ is σ_{ind}-open.

(ii) The proof of (ii) follows from (i) because null sequences in $S_{X,A}$ are always null sequences in some $e^{-tA}(X)$, $t > 0$, and vice versa. □

In the next theorems we characterize continuous linear mappings from $S_{X,A}$ into $S_{Y,B}$ and from $S_{X,A}$ into $T_{Y,B}$.

(I.4.2) Theorem

Suppose $P: S_{X,A} \to S_{Y,B}$ is a linear mapping. The following seven conditions are equivalent:

(i) P is continuous as a linear mapping from $(S_{X,A},\sigma_{ind})$ into $(S_{Y,B},\sigma_{ind})$.

(ii) $u_n \to 0$ in $(S_{X,A},\sigma_{ind})$ implies $Pu_n \to 0$ in $(S_{Y,B},\sigma_{ind})$.

(iii) For each $t > 0$ the operator $Pe^{-tA}: X \to Y$ is bounded.

(iv) For each $t > 0$, and each $\psi \in B_+(\mathbb{R})$, $Pe^{-tA}(X) \subset \mathcal{D}(\psi(B))$ and $\psi(B)Pe^{-tA}: X \to Y$ is bounded.

(v) For each $t > 0$ there exists $\tau > 0$ such that $Pe^{-tA}(X) \subset e^{-\tau B}(Y)$ and $e^{\tau B}\, Pe^{-tA}: X \to Y$ is bounded.

(vi) There exists a dense linear subspace $\mathcal{D} \subset Y$ such that for each fixed $y \in \mathcal{D}$ the linear functional $\ell_{P,y}(w) = (Pw,y)$, $w \in S_{X,A}$, is continuous on $S_{X,A}$.

(vii) For each $t > 0$ the operator $(Pe^{-tA})^*: Y \to X$ is bounded.

Proof

(i) ⇔ (ii) See Theorem (I.4.1).

(i) ⇒ (iii) For each $t > 0$, the operator $e^{-tA}: X \to S_{X,A}$ is continuous. Hence $Pe^{-tA}: X \to S_{Y,B}$ is continuous. In particular, $Pe^{-tA}: X \to Y$ is bounded.

(iii) ⇒ (iv) The operator $\psi(B)Pe^{-tA}$ is closed and defined on the whole of X. Therefore it is bounded.

(iv) ⇒ (v) Let E denote the unit ball in X. Because of (iv) the set $Pe^{-tA}(E)$ is bounded in $S_{Y,B}$. Now apply Lemma (I.1.8).

(iii) ⇒ (vi) We can take $\mathcal{D} = Y$. Take $y \in Y$ fixed. For each $x \in X$ and each $t > 0$ we have

$$|(\ell_{P,y} \circ e^{-tA})(x)| = |(Pe^{-tA}x,y)| \leq c_{t,y}\|x\| .$$

Following Theorem (I.4.1), $\ell_{P,y}$ is continuous on $S_{X,A}$.

(vi) ⇒ (vii) The linear functional $\ell_{P,y}$ on $S_{X,A}$ can be represented by

(*) $\ell_{P,y}(w) = \langle w, F_y \rangle_X$

where $F_y \in T_{X,A}$. Observe that $F_y(t) \in S_{X,A}$ for all $t > 0$. Let $t > 0$. From (*) we get $F_y(t) = (Pe^{-tA})^* y$. So $\mathcal{D} \subset \mathcal{D}((Pe^{-tA})^*)$ and $\mathcal{D}((Pe^{-tA})^*)$ is dense in Y. Since Pe^{-tA} is defined on the whole X and closable, it is bounded. Hence, $(Pe^{-tA})^*$ is bounded. Repeating the argument with arbitrary $y \in Y$ shows that $(Pe^{-tA})^* y \in S_{X,A}$.

(viii) ⇒ (iii) Pe^{-tA} is bounded because $(Pe^{-tA})^*$ is bounded.

(v) ⇒ (ii) Trivial. □

The next corollaries are important for applications.

(I.4.3) Corollary

Let $Q: X \to Y$ be a densely defined operator, with $\mathcal{D}(Q) \supset S_{X,A}$. Suppose the operator Qe^{-tA} is bounded, and for all $\psi \in B_+(\mathbb{R})$, $Qe^{-tA}(X) \subset \mathcal{D}(\psi(B))$. Then Q maps $S_{X,A}$ continuously into $S_{Y,B}$. (It follows that $\forall_{t>0} \exists_{s>0}$: $e^{sB} Q e^{-tA}$ is bounded on X.)

Proof. Let $w \in S_{X,A}$. Then $w = e^{-tA}x$ for some $t > 0$ and $x \in X$. It follows that

$$Qe^{-tA}x \in \bigcap_{\psi \in B_+(\mathbb{R})} \mathcal{D}(\psi(B)) = S_{Y,B} .$$

So Q maps $S_{X,A}$ into $S_{Y,B}$. The result follows from Theorem (I.4.2),(iii). □

(I.4.4) Corollary

Suppose $Q: X \to Y$ is a densely defined closable operator. Let $\mathcal{D}(Q) \supset S_{X,A}$ and $Q(S_{X,A}) \subset S_{Y,B}$. Then Q maps $S_{X,A}$ continuously into $S_{Y,B}$.

Proof. Q is closable iff $\mathcal{D}(Q^*)$ is dense in Y. The operator $Qe^{-tA}: X \to Y$ is defined on the whole of X. Since $(Qe^{-tA})^* y = e^{-tA} Q^* y$ for all $y \in \mathcal{D}(Q^*)$, the operator $(Qe^{-tA})^*$ is densely defined and hence Qe^{-tA} is closable. Thus the boundedness of Qe^{-tA} follows. Application of Theorem (I.4.2),(iii), yields the wanted result. □

(I.4.5) Theorem

Let $K: S_{X,A} \to T_{Y,B}$ be a linear mapping. The following four conditions are equivalent:

(i) K is continuous as a linear mapping from $(S_{X,A}, \sigma_{ind})$ into $(T_{Y,B}, \tau_{proj})$.

(ii) For each $t > 0$, $s > 0$, the operator $e^{-sB} K e^{-tA}: X \to Y$ is bounded.

(iii) For each $t > 0$ there exists $\psi \in B_+(\mathbb{R})$ such that the operator $\psi(B)^{-1} K e^{-tA}: X \to Y$ is bounded.

(iv) For each $s > 0$ the mapping $e^{-sB} K: S_{X,A} \to S_{Y,B}$ is continuous.

Proof

(i) ⇒ (ii) Let $t > 0$ and $s > 0$. Then the mapping $Ke^{-tA}: X \to T_{Y,B}$ is continuous, and also the mapping $e^{-sB}: T_{Y,B} \to Y$. Hence $e^{-sB} K e^{-tA}$ is a continuous linear operator from X into Y.

(ii) ⇒ (iii) Let $t > 0$. Let E denote the unit ball in X. Then the set $Ke^{-tA}(E)$ is bounded in $T_{Y,B}$. By Lemma (I.2.10) there exists $\psi \in B_+(\mathbb{R})$ such that $\psi(B)^{-1} K e^{-tA}(E)$ is a bounded subset of X.

(iii) ⇒ (ii) Trivial.

(ii) ⇒ (iv) Let $s > 0$. Since $e^{-sB} K e^{-tA}: X \to Y$ is bounded, io follows from Theorem (I.4.2) that $e^{-sB} K: S_{X,A} \to S_{Y,B}$ is continuous.

(iv) ⇒ (i) Let $s > 0$, and let $f \in S_{X,A}$. Then $(Kf)(s) = e^{-sB} Kf$, and by assumption there exists $\varphi \in B_+(\mathbb{R})$ such that

$$\|(Kf)(s)\|_Y \leq \|\varphi(A)f\|_X \, .$$

□

The following theorem is similar to Theorem (I.4.1). We use the equivalence of the topologies τ_{proj} and τ_{ind}.

(I.4.6) Theorem

Let V be a locally convex topological vector space. A linear mapping $R: T_{X,A} \to V$ is continuous

(i) iff for each $\varphi \in B_+(\mathbb{R})$ the mapping $R_r\, \varphi(A): X \supset \mathcal{D}(\varphi(A)) \to V$ is continuous (here R_r denotes the restriction from R to X);

(ii) iff for each null sequence $(F_n)_{n\in\mathbb{N}} \subset T_{X,A}$, the sequence $(RF_n)_{n\in\mathbb{N}} \subset V$ is a null sequence in V.

Proof. The proof is very similar to the proof of Theorem (I.4.1) and is omitted. □

(I.4.7) Theorem

Let $M: T_{X,A} \to S_{Y,B}$ be a linear mapping. Let $M_r: X \to Y$ denote the restriction of M to X. The following four conditions are equivalent:

(i) M is continuous as a linear mapping from $(T_{X,A}, \tau_{proj})$ into $(S_{Y,B}, \sigma_{ind})$.

(ii) For all $\varphi, \psi \in B_+(\mathbb{R})$ the operator $\psi(B) M_r\, \varphi(A): \mathcal{D}(\varphi(A)) \to Y$ is bounded.

(iii) There exist $t, s > 0$ such that the operator $e^{sB} M_r\, e^{tA}: X \supset e^{-tA}(X) \to Y$ is bounded.

(iv) There exists $t > 0$ such that the operator $M_r\, e^{tA}: X \supset e^{-tA}(X) \to Y$ is bounded.

Proof

(i) ⇒ (ii) Let $\varphi, \psi \in B_+(\mathbb{R})$. Then from Theorem (I.4.5) we obtain that $M_r\, \varphi(A): X \supset \mathcal{D}(\varphi(A)) \to S_{Y,B}$ is continuous. Moreover, $\psi(B)$ maps $S_{Y,B}$ continuously into Y. So the operator $\psi(B) M_r\, \varphi(A) : X \supset \mathcal{D}(\varphi(A)) \to S_{Y,B}$ is bounded.

(ii) ⇒ (iii) Let $\varphi \in B_+(\mathbb{R})$. Then for all $\psi \in B_+(\mathbb{R})$ the operator $\psi(B) M_r\, \varphi(A)$ is bounded. So there exists $s > 0$, such that the operator $e^{sB} M_r \varphi(A)$ is bounded (cf. Corollary (I.4.3)). Now for all $\varphi \in B_+(\mathbb{R})$ the operator $\varphi(A)(e^{sB} M_r)^* = (e^{sB} M_r\, \varphi(A))^*$ is bounded, since $e^{sB} M_r$ is a bounded operator. A similar argument yields $t > 0$ such that $e^{tA}\,(e^{sB} M_r)^*$ is a bounded operator. Hence $e^{sB} M_r\, e^{tA}: X \supset e^{-tA}(X) \to Y$ is bounded.

(iii) ⇒ (iv) Trivial.

(iv) ⇒ (i) Note first that M_r is a bounded operator. Let Q denote the bounded extension of $M_r\, e^{tA}$. For each $\tau > 0$, $Qe^{-\tau A}$ is the bounded extension of $M_r\, e^{(t-\tau)A}$. So by Theorem (I.4.2) Q maps $S_{X,A}$ continuously into $S_{Y,B}$. We have $M = Qe^{-tA}$. Since $e^{-tA}: T_{X,A} \to S_{X,A}$ continuously, the proof is complete. □

(I.4.8) Theorem

Let $W: T_{X,A} \to T_{Y,B}$ be a linear mapping. Let $W_r: X \to T_{Y,B}$ denote the restriction of W to X. The following five conditions are equivalent:

(i) W is continuous as a linear mapping from $(T_{X,A}, \tau_{proj})$ into $(T_{Y,B}, \tau_{proj})$.

(ii) For each $\varphi \in B_+(\mathbb{R})$ and each $s > 0$ the operator $e^{-sB} W_r \varphi(A)$: $X \supset \mathcal{D}(\varphi(A)) \to Y$ is bounded.

(iii) For each $s > 0$ there exists $t > 0$ such that the operator $e^{-sB} W_r e^{tA}$: $X \supset e^{-tA}(X) \to Y$ is bounded.

(iv) For each $s > 0$, $(e^{-sB} W_r)^*(Y) \subset S_{X,A}$.

(v) For each $g \in S_{Y,B}$, the functional $F \mapsto \overline{\langle g, WF\rangle}$, $F \in T_{X,A}$, is continuous.

<u>Proof</u>

(i) ⇒ (ii) Let $s > 0$ and $\varphi \in B_+(\mathbb{R})$. From Theorem (I.4.6) we derive that $W_r\, \varphi(A)$ is continuous as a mapping from X into $T_{Y,B}$. Moreover, $e^{-sB}: T_{Y,B} \to Y$ is continuous. Hence $e^{-sB} W_r\, \varphi(A)$ is a bounded operator from $\mathcal{D}(\varphi(A))$ into Y.

(ii) ⇒ (i) It follows from Theorem (I.4.6) that $e^{-sB} W: T_{X,A} \to Y$ is continuous for all $s > 0$. Hence for all $s > 0$ there exists $t > 0$ and a constant $c > 0$ such that

$$\forall_{F \in T_{X,A}} : \|(WF)(s)\|_Y \le c\|F(t)\|_X .$$

Thus the continuity of W follows.

(iii) ⇒ (iv) Trivial.

(iv) ⇒ (v) Let $g \in S_{Y,B}$. Then $\langle g, WF\rangle_Y = (e^{sB} g, (WF)(s))_Y$ for $s > 0$ sufficiently small. Hence we get

$$(e^{sB} g, (WF)(s)) = (e^{tA}(e^{-sB} W_r)^* e^{sB} g, F(t))$$

for $t > 0$ sufficiently small. So the linear functional

$$F \mapsto \overline{\langle g, WF\rangle_Y} = \overline{\langle (e^{-sB} W_r)^* e^{sB} g, F\rangle_X}$$

is continuous on $T_{X,A}$.

(v) ⇒ (ii) Let $s > 0$ and let $\varphi \in B_+(\mathbb{R})$. For each $g \in e^{-sB}(Y)$ the linear functional

$$x \mapsto \overline{\langle g, W\varphi(A)x\rangle}$$

is continuous on X. It follows that the operator $(e^{-sB} W_r\, \varphi(A))^*$ is densely defined. Hence $e^{-sB} W_r\, \varphi(A)$ is bounded on $\mathcal{D}(\varphi(A))$. □

(I.4.9) Theorem

Let the linear mappings

$$L: \mathcal{S}_{X,A} \to \mathcal{S}_{Y,B}\,, \qquad M: \mathcal{T}_{X,A} \to \mathcal{S}_{Y,B}\,,$$

$$K: \mathcal{S}_{X,A} \to \mathcal{T}_{Y,B}\,, \qquad W: \mathcal{T}_{X,A} \to \mathcal{T}_{Y,B}$$

be continuous with respect to the topologies σ_{ind} and τ_{proj}. Then the dual linear mappings

$$L': \mathcal{T}_{Y,B} \to \mathcal{T}_{X,A}\,, \qquad M': \mathcal{T}_{Y,B} \to \mathcal{S}_{X,A}\,,$$

$$K': \mathcal{S}_{Y,B} \to \mathcal{T}_{X,A}\,, \qquad W': \mathcal{S}_{Y,B} \to \mathcal{S}_{X,A}$$

are also continuous with respect to the topologies σ_{ind} and τ_{proj}.

Proof. Compare Theorem (I.4.2),(v) with Theorem (I.4.8),(iii).
Look at Theorem (I.4.5),(ii).
Look at Theorem (I.4.7),(iii). □

It is an interesting question whether a densely defined (possibly unbounded) operator from X into Y can be extended to a continuous linear mapping from $\mathcal{T}_{X,A}$ into $\mathcal{T}_{Y,B}$.

(I.4.10) Theorem

Let E be a densely defined linear operator from X into Y. Then E can be extended to a continuous linear mapping $\bar{E}: \mathcal{T}_{X,A} \to \mathcal{T}_{Y,B}$ iff $\mathcal{S}_{Y,B} \subset \mathcal{D}(E^*)$ and $E^*(\mathcal{S}_{Y,B}) \subset \mathcal{S}_{X,A}$.

Proof

$\Leftarrow$) If $\bar{E}$ exists as a continuous linear mapping, its dual mapping $\bar{E}': \mathcal{S}_{Y,B} \to \mathcal{S}_{X,A}$ is continuous. For each $x \in \mathcal{D}(E)$ and $g \in \mathcal{S}_{Y,B}$ we have $(g,Ex)_Y = \langle g,Ex\rangle_Y = (\bar{E}'g,x)_X$. Hence $E^* \supset \bar{E}'$ and $E^*(\mathcal{S}_{Y,B}) \subset \mathcal{S}_{X,A}$.

$\Rightarrow$) From Corollary (I.4.3) it follows that E^* maps $\mathcal{S}_{Y,B}$ continuously into $\mathcal{S}_{X,A}$. By the previous theorem, its dual $(E^*)'$ maps $\mathcal{T}_{X,A}$ continuously into $\mathcal{T}_{Y,B}$, and $(E^*)'$ is an extension of E. □

(I.4.11) Corollary

A continuous linear mapping Q: $\mathcal{S}_{X,A} \to \mathcal{S}_{Y,B}$ can be extended to a continuous linear mapping $\bar{Q}$: $\mathcal{T}_{X,A} \to \mathcal{T}_{Y,B}$ iff Q has an adjoint $Q^*: \mathcal{Y} \to \mathcal{X}$ with $\mathcal{D}(Q^*) \supset \mathcal{S}_{Y,B}$ and $Q^*(\mathcal{S}_{Y,B}) \subset \mathcal{S}_{X,A}$.

CHAPTER II

Illustrative examples of analyticity spaces

Introduction

If we consider a nonnegative self-adjoint operator A in the Hilbert space $X = \mathbf{L}_2(M,\mu)$, where M is a measure space and μ a σ-finite measure, then $S_{X,\mathsf{A}}$ represents a space of test 'functions' on M and $T_{X,\mathsf{A}}$ a space of generalized functions on M. We remark that for a topological measure space M with metrizable topology, there exist conditions (Cf. [Fe]) which indicate a certain compatibility between measure and topology, such that any nuclear analyticity space $S_{X,\mathsf{A}}$ can be canonically represented by a space of genuine test functions. These functions obey a measure theoretical Sobolev lemma. Thus $S_{\mathbf{L}_2(M),\mathsf{A}}$ can be made into a genuine test function space if M is e.g. (an open set in) $\mathbb{R}^n$, a smooth manifold, or a denumerable disjoint union of these. For some results in this direction, see Chapter V and [EG 7].
In the previous chapter we have presented the functional analytic features of the type of distribution theories based on the Gelfand triple $S_{X,\mathsf{A}} \subset X \subset T_{X,\mathsf{A}}$. However, an important aspect of our research is the construction and study of test function-spaces which can be characterized as analyticity spaces. On one hand examples give the theory a wider scope, on the other hand they contribute in deriving new ideas. This chapter is composed of material on this theme gathered in the last few years.

First we consider analyticity spaces based on the Laplacian operator. We study spaces S_{X,Δ^ν}, where $X = \mathbf{L}_2([-\pi,\pi])$, $\mathbf{L}_2([0,\pi])$, $\mathbf{L}_2((0,\infty))$ and $\mathbf{L}_2((-\infty,\infty))$, and where Δ denotes the differential operator $- d^2/dx^2$. Also their higher dimensional analogues will have our attention. We note that the spaces $S_{X,\Delta}$ are related to the heat equation $\partial\psi/\partial t = -\Delta\psi$. In this connection we also study the spaces $S_{\mathbf{L}_2(S^{n-1}),(\Delta_{LB})^{\frac{1}{2}}}$ where S^{n-1} denotes the unit sphere and Δ_{LB} the Laplace-Beltrami operator for S^{n-1}. We observe that the Laplacian $\Delta^{(n)}$ for $\mathbb{R}^n$ satisfies

$$\Delta^{(n)} = -\frac{d^2}{dr^2} + \frac{n-1}{r}\frac{d}{dr} + \frac{1}{r^2}\Delta_{LB}$$

in spherical coordinates. The elements of $S_{L_2(S^{n-1}),(\Delta_{LB})^{\frac{1}{2}}}$ are restrictions to S^{n-1} of functions which are harmonic on open sets in $\mathbb{R}^n$ which contain the unit ball. The corresponding generalized functions appear to be 'boundary values' of harmonic functions.

Second we consider the Gelfand-Shilov spaces S_α^β. They are (countable) inductive limits of Frechet spaces. So it seems an appropriate question whether these spaces are analyticity spaces. The answer is positive if we can find a positive self-adjoint operator A in a Hilbert space X such that $S_\alpha^\beta = S_{X,A}$. For the spaces $S_{1/k+1}^{k/k+1}$, $k \in \mathbb{N}$, and S_α^α, $\alpha \geq \frac{1}{2}$, such an operator has been given. For the spaces S_α^β with $\alpha = \frac{p}{p+q}\rho$ and $\beta = \frac{q}{p+q}\rho$ with $p,q \in \mathbb{N}$ and $\rho \geq 1$, we present a serious candidate.

In Section II.3 we discuss analyticity spaces which are related to classical polynomials. In this connection three differential operators are considered.

- $$H = \tfrac{1}{2}\left(-\frac{d^2}{dx} + x^2 + 1\right)$$

- $$L_\alpha = -\frac{d^2}{dx^2} + (2\alpha+1)\frac{1}{x}\frac{d}{dx} + x^2 , \qquad \alpha > -1$$

- $$J_{\alpha,\beta} = -(1-x^2)\frac{d}{dx^2} + (\alpha+\beta+2)x\frac{d}{dx} + (\alpha-\beta)\frac{d}{dx} , \qquad \alpha,\beta > -1 .$$

The operators H, L_α and $J_{\alpha,\beta}$ are nonnegative and self-adjoint in the Hilbert spaces $L_2(\mathbb{R})$, $X_\alpha = L_2((0,\infty), x^{2\alpha+1}dx)$ and $X_{\alpha,\beta} = L_2([-1,1],(1-x)^\alpha(1+x)^\beta dx)$, respectively. The operators have a discrete spectrum. The eigenfunctions of H are related to the Hermite polynomials. The eigenfunctions of L_α to the Laguerre polynomials. The eigenfunctions of $J_{\alpha,\beta}$ are the Jacobi polynomials.

The central problem in the research of the spaces $S_{L_2(\mathbb{R}),H^\nu}$, $S_{X_\alpha,(L_\alpha)^\nu}$ and $S_{X_{\alpha,\beta},(A_{\alpha,\beta})^\nu}$, $\nu > 0$, is the characterization of their elements in classical analytic terms. In short, we outline this problem here. Let X denote the Hilbert space $L_2((a,b),w(x)dx)$, $-\infty \leq a < b \leq \infty$, where $w(x)$ denotes a positive weight function. Further, let $(p_k)_{k\in\mathbb{N}}$ be an orthonormal basis in X. Assume that each function p_k is extendible to an entire function and also that the functions p_k are the eigenfunctions of the positive self-adjoint operator A in X with eigenvalues λ_k. The functional analytic characterization of the elements of the space $S_{X,A}$ is simple, i.e.

$$f \in S_{X,A} \;:\Leftrightarrow\; \exists_{t>0}: \sum_{k=1}^{\infty} e^{\lambda_k t} |(f,p_k)|^2 < \infty \,.$$

In order to obtain the classical analytic characterization of the elements of $S_{X,A}$, we have to find necessary and sufficient conditions in terms of classical analysis such that any function satisfying these conditions, can be expanded in a series

$$f: \lambda \mapsto \sum a_k \, p_k(\lambda) \;,$$

if and only if $\sum e^{\lambda_k t} |a_k|^2 < \infty$ for some $t > 0$.

In Section 4 we connect the theory of analyticity spaces and trajectory spaces to Lie group representation theory. Let G be a semi-simple Lie group and U a unitary representation of G into $U(X)$, the group of unitary operators on a Hilbert space X. Then the set of analytic vectors in X belonging to this unitary representation establishes an analyticity space. We discuss results of Nelson [Ne 1], Goodman [Go 1-2] and Nagel [Na].

II.1. Analyticity spaces based on the Laplacian operator

The diffusion equation

$$\dagger \qquad \frac{\partial u}{\partial t} = -\Delta^{\nu} u \;, \qquad \nu \geq \tfrac{1}{2}$$

and the Gelfand triple $S_{X,A} \subset X \subset T_{X,A}$ corresponding to it, can be considered on any Riemannian manifold. Here we restrict ourselves to $\mathbb{R}$, $(0,\infty)$, $[-\pi,\pi]$, $[0,\pi]$, and for $\nu = \frac{1}{2}$ to the unit sphere S^{q-1} in $\mathbb{R}^q$.

Case (II.a) $X = L_2(\mathbb{R})$, $A = \Delta^{\nu} = \left(-\dfrac{d^2}{dx^2}\right)^{\nu}$, $\nu > \frac{1}{2}$.

The operator Δ has a continuous spectrum. For an initial condition $u_0 \in L_2(\mathbb{R})$ the equation ($\dagger$) can be solved by means of the Fourier transformation

$$u(x,t) = \frac{1}{\sqrt{2\pi}} \int_{-\infty}^{\infty} \hat{u}_0(\xi) e^{-t|\xi|^{2\nu}} e^{i\xi x} \, d\xi$$

where $\hat{u}_0$ denotes the Fourier transform of u_0. For fixed $t > 0$ and fixed $\nu > \frac{1}{2}$, $u(x,t)$ can be extended to an entire analytic function $u(z,t)$ which obeys the estimate

$$\exists_{A>0} \forall_{y \in \mathbb{R}} : \int_{-\infty}^{\infty} |u(x+iy,t)|^2 \, dx \leq A \exp\left(2B \, |y|^{\frac{2\nu}{2\nu-1}}\right), \tag{$*$}$$

with

$$B = (2\nu-1)\left(\frac{1}{t(2\nu)^{2\nu}}\right)^{\frac{1}{2\nu-1}}.$$

Conversely, if an entire function $v(z)$ satisfies an estimate of type $(*)$, then for each $t < (2\nu)^{-2\nu}\left(\frac{2\nu-1}{B}\right)^{2\nu-1}$ there exists $h \in \mathbf{L}_2(\mathbb{R})$ such that $v = e^{-t\Delta^\nu} h$. Proofs of these statements can be given in the spirit of our paper [EG 3], Ch. 3.

Thus we come to the following conclusion:

(II.1.1) <u>Theorem</u>

The space $S_{\mathbf{L}_2(\mathbb{R}),\Delta^\nu}$, $\nu > \frac{1}{2}$, contains precisely all functions f with the properties

(i) f can be extended to an entire analytic function;

(ii) $\exists_{A>0} \exists_{B>0} \forall_{y \in \mathbb{R}} : \int_{-\infty}^{\infty} |f(x+iy)|^2 \, dx \leq A \exp\left(2B \, |y|^{\frac{2\nu}{2\nu-1}}\right)$.

The space $S_{\mathbf{L}_2(\mathbb{R}),\Delta^\nu}$ cannot be nuclear since Δ^ν has continuous spectrum. (See Theorem (I.1.13).)

Examples of continuous linear functionals on $S_{\mathbf{L}_2(\mathbb{R}),\Delta^\nu}$ are δ-functions with complex argument $\delta_a^k(f) = f^{(k)}(a)$, $a \in \mathbb{C}$, $k \in \mathbb{N} \cup \{0\}$. More general functionals are given by

$$f \mapsto \int_C \varphi(z) f(z) \, dz$$

where C is a Jordan curve and φ a continuous bounded function on C.

Examples of continuous linear mappings on $S_{\mathbf{L}_2(\mathbb{R}),\Delta^\nu}$, $\nu > \frac{1}{2}$, are

$e^{z\Delta^\alpha}$, $z \in \mathbb{C},\ 0 \leqq \alpha < \nu$ $\qquad$ $e^{z\Delta^\nu}$, $z \in \mathbb{C}$, $\mathrm{Re}\, z \leqq 0$

R_a , $a \in \mathbb{R}$ $\qquad$ T_b , $b \in \mathbb{C}$

Z_λ , $\lambda \in \mathbb{R},\ \lambda \neq 0$ $\qquad$ D

M_ψ

Here

$$(R_a f)(x) = e^{iax} f(x) \ , \quad (T_b f)(x) = f(x+b) \ , \quad (Z_\lambda f)(x) = f(\lambda x) \ ,$$

$$(Df)(x) = \frac{df}{dy}(x) \quad , \quad (M_\psi f)(x) = \psi(x) f(x) \ ,$$

where ψ is a 'multiplyer', i.e. an entire function with the property

$$\sup_{x \in \mathbb{R}} |\psi(x+iy)| \leqq A \exp\left(B\ |y|^{\frac{2\nu}{2\nu-1}}\right) , \quad A > 0,\ B > 0 \ .$$

It follows from simple analytic and geometric considerations that the mentioned operators map $S_{\mathbf{L}_2(\mathbb{R}),\Delta^\nu}$ into itself. Moreover, their $\mathbf{L}_2(\mathbb{R})$-adjoints have the same property. Then the continuity and extendibility of these operators follow from Corollary (I.4.4) and Theorem (I.4.10).

Finally, we want to show that certain strongly divergent Fourier integrals can be interpreted as elements of $T_{\mathbf{L}_2(\mathbb{R}),\Delta^\nu}$. Our interpretation as trajectories is closely related to the Gauss-Weierstrass summation method. Let g be a measurable function on $\mathbb{R}$ such that for each $\varepsilon > 0$ the function $y \mapsto g(y)e^{-\varepsilon|y|^{2\nu}}$ is in $\mathbf{L}_2(\mathbb{R})$. The possibly divergent Fourier integral $\int_{-\infty}^{\infty} g(y)e^{iyx}\, dy$ can be considered as an element of $T_{\mathbf{L}_2(\mathbb{R}),\Delta^\nu}$. Its trajectory is given by

$$G(x,t) = \int_{-\infty}^{\infty} g(y)e^{-t|y|^{2\nu}}\, e^{iyx}\, dy \ .$$

This first illustration of our theory certainly has some elegant features, but in some respects the illustration is too simple. Since $e^{-t\Delta^\nu}$ is not a Hilbert-Schmidt operator there are no kernel theorems in this case. That is to say, there exist continuous linear mappings from $S_{\mathbf{L}_2(\mathbb{R}),\Delta^\nu}$ into $T_{\mathbf{L}_2(\mathbb{R}),\Delta^\nu}$ which do not arise from a trajectory of the diffusion equation in $\mathbb{R}^2$.

<u>Case</u> (II.b) $X = \mathbf{L}_2(\mathbb{R})$, $A = \Delta^{\frac{1}{2}} = \left(-\frac{d^2}{dx^2}\right)^{\frac{1}{2}}$.

We have

$$(e^{-t\Delta^{\frac{1}{2}}}u)(x) = \int_{-\infty} K_t(x,y)u(y)dy$$

with

$$K_t(x,y) = \frac{1}{\pi}\int_0^\infty \exp(-|k|t)\cos k(x-y)dk = \frac{1}{\pi}\frac{1}{t^2+(x-y)^2},$$

which is just the Poisson kernel for solving the Dirichlet problem on the halfplane $t \geqq 0$. This is not surprising, since, at least formally,

$$\left(\frac{\partial^2}{\partial t^2}+\frac{\partial^2}{\partial x^2}\right) = \left(\frac{\partial}{\partial t}-\sqrt{-\frac{\partial^2}{\partial x^2}}\right)\left(\frac{\partial}{\partial t}+\sqrt{-\frac{\partial^2}{\partial x^2}}\right).$$

$S_{\mathbf{L}_2(\mathbb{R}),\Delta^{\frac{1}{2}}}$ consists of functions f which are analytic on a strip around the real axis and which satisfy

$$\sup_{-h<y<h}\int_{-\infty}^{\infty}|f(x+iy)|^2\,dx < \infty$$

for h sufficiently small. The width of the strip depends on f.

Examples of extendible linear operators on $S_{\mathbf{L}_2(\mathbb{R}),\Delta^{\frac{1}{2}}}$ are

$e^{z\Delta^\alpha}$,	$z \in \mathbb{C}$, $0 \leqq \alpha < \frac{1}{2}$,	$e^{z\Delta^{\frac{1}{2}}}$,	$\operatorname{Re} z \leqq 0$
R_a,	$a \in \mathbb{R}$	T_b,	$b \in \mathbb{R}$
Z_λ,	$\lambda \in \mathbb{R}, \lambda \neq 0$	D	
M_ψ			

Here ψ is a function which is analytic and uniformly bounded on a strip around the real axis. Most remarks of Case (II.a) also apply here.

<u>Case</u> (II.c) $X = \mathbf{L}_2((0,\infty))$, $A = \left(-\frac{d^2}{dx^2}\right)^\nu$, $\nu \geqq \frac{1}{2}$.

In order to make A into a self-adjoint operator, boundary conditions at

$x = 0$ have to be prescribed. Therefore, define Δ_0 by the differential operator $-d^2/dx^2$ on the domain

$$\mathcal{D}(\Delta_0) = \{u \mid u \in \mathbf{H}^2((0,\infty)),\ u(0) = 0\}\ ,\quad \mathbf{H}^2((0,\infty)) \text{ a Sobolev space.}$$

Then we get

$S_{\mathbf{L}_2(0,\infty),\Delta_0^\nu}$, $\nu \geq \frac{1}{2}$, contains precisely the *odd* analytic functions in $S_{\mathbf{L}_2(\mathbb{R}),\Delta^\nu}$, $\nu \geq \frac{1}{2}$.

On the other hand, define Δ_1 by $-d^2/dx^2$ on the domain

$$\mathcal{D}(\Delta_1) = \{u \mid u \in \mathbf{H}^2((0,\infty)),\ u'(0) = 0\}\ .$$

Then we get

$S_{\mathbf{L}_2((0,\infty)),\Delta_1^\nu}$, $\nu \geq \frac{1}{2}$, contains precisely the *even* analytic functions in $S_{\mathbf{L}_2(\mathbb{R}),\Delta^\nu}$, $\nu \geq \frac{1}{2}$.

It is an easy exercise to construct examples of extendible continuous linear mappings and continuous linear functionals in the spirit of Case(II.a-b).

<u>Case</u> (II.d) $X = \mathbf{L}_2([-\pi,\pi])$, $\mathbf{A} = \Delta_p^\nu = \left(-\dfrac{d^2}{dx^2}\right)^\nu$, $\nu \geq \frac{1}{2}$.

We consider the operator Δ with periodic boundary conditions

$$\mathcal{D}(\Delta_p) = \{u \mid u \in \mathbf{H}^2([-\pi,\pi]),\ u(-\pi) = u(\pi),\ u'(-\pi) = u'(\pi)\}\ .$$

In [EG 3], Theorem 3.10, we have shown that $S_{\mathbf{L}_2([-\pi,\pi]),\Delta_p^\nu}$, $\nu > \frac{1}{2}$, consists precisely of those 2π-periodic entire functions f which satisfy

$$\sup_{x\in\mathbb{R}} |f(x+iy)| \leq A \exp\left(B\,|y|^{\frac{2\nu}{2\nu-1}}\right)$$

with $A > 0$, $B > 0$, dependent on f.

In [EG 3], Theorem (3.1.3), we have shown that $S_{\mathbf{L}_2([-\pi,\pi]),\Delta_p^{1/2}}$ consists of all functions f which can be extended to a 2π-periodic analytic function on a neighbourhood of the real axis.

Examples of continuous linear mappings on $S_{\mathbf{L}_2([-\pi,\pi]),\Delta_p^\nu}$, $\nu \geq \frac{1}{2}$, are

$e^{z\Delta_p^\alpha}$, $z \in \mathbb{C},\ 0 \leq \alpha < \nu$, $\quad e^{z\Delta_p^\nu}$, $\mathrm{Re}\, z \leq 0$

R_n , $n \in \mathbb{Z}$ $\quad T_b$, $b \in \mathbb{R}$

Z_λ , $\lambda \in \mathbb{Z}$ $\quad D$

M_ψ , $\psi \in S_{L_2([-\pi,\pi]),\Delta_p^\nu}$

If $\nu > \frac{1}{2}$ we also have T_b with $b \in \mathbb{C}$.
The operator Δ_p^ν has discrete spectrum. Its eigenvectors are the functions $x \mapsto e^{inx}$, $n \in \mathbb{Z}$, with eigenvalues $|n|^{2\nu}$. All operators $e^{-t\Delta_p^\nu}$, $t > 0$, $\nu \geq \frac{1}{2}$, are Hilbert-Schmidt and therefore all spaces $S_{L_2([-\pi,\pi]),\Delta_p^\nu}$ are nuclear. These spaces are also commutative rings with identity and no zero divisors. An interesting class of continuous linear functionals on $S_{L_2([-\pi,\pi]),\Delta_p^\nu}$, $\nu > \frac{1}{2}$, is given by

$$\ell_\mu(f) = \int_{-\pi}^{\pi} \int_{-\infty}^{\infty} f(x+iy)\, d\mu(x,y)$$

with $\mu(x,y)$ a suitable Borel measure on $[-\pi,\pi] \times i\mathbb{R}$. Special cases are the δ-functions with complex argument δ_z and their derivatives. The continuous linear functionals on $S_{L_2([-\pi,\pi]),\Delta_p^{1/2}}$ are the 2π-periodic *hyperfunctions*. For a detailed discussion of all those subjects, see [EG3], Ch. 4.
If a sequence of complex numbers $(c_n)_{n\in\mathbb{Z}}$ is such that $(c_n \exp - \varepsilon|n|^{2\nu})$, $n \in \mathbb{Z}$, is a bounded sequence for each $\varepsilon > 0$, then $\sum_{n=-\infty}^{\infty} c_n e^{inx}$ can be seen as a generalized function with corresponding trajectory in $T_{L_2([-\pi,\pi]),\Delta_p^\nu}$ given by

$$G(x,t) = \sum_{n=-\infty}^{\infty} c_n \exp(-|n|^{2\nu} t + inx)\ , \quad x \in [-\pi,\pi]\ , \quad t > 0\ .$$

In fact all trajectories arise in this way.

Case (II.e) $X = L_2([0,\pi])$, $A = \left(-\frac{d^2}{dx^2}\right)^\nu$, $\nu \geq \frac{1}{2}$.

Analogously to Case (II.c) the closed subspace of odd functions in $S_{L_2([-\pi,\pi]);\Delta_p^\nu}$ is equal to the space $S_{L_2([0,\pi]);\Delta_0^\nu}$. Here Δ_0 equals $-d^2/dx^2$ with domain

$$\mathcal{D}(\Delta_0) = \{u \mid u \in H^2([0,\pi]),\ u(0) = u(\pi) = 0\}\ .$$

Similarly, the closed subspace of even functions in $S_{\mathbf{L}_2([-\pi,\pi]),\Delta_p^\nu}$ can be written as $S_{\mathbf{L}_2([0,\pi]),\Delta_e^\nu}$. Then Δ_e equals $-d^2/dx^2$ on the domain

$$\mathcal{D}(\Delta_e) = \{v \mid v \in \mathbf{H}^2([0,\pi]),\ v'(0) = v'(\pi) = 0\} .$$

Examples of continuous linear mappings can be found by taking suitable combinations of the operators mentioned in Case (II.d).
We note that in [EG 3] the spaces $S_{\mathbf{L}_2([0,\pi]),\Delta_e^\nu}$, $\nu \geq \frac{1}{2}$, play an important role in the investigation of test spaces based on Jacobi polynomials. See also the third section of this chapter.

<u>Case</u> (II.f) $X = \mathbf{L}_2(S^{n-1})$, $A = \Delta_{LB}^{\frac{1}{2}}$.

Here S^{n-1} denotes the unit sphere in $\mathbb{R}^n$ and Δ_{LB} denotes the (positive) Lapace-Beltrami operator on S^{n-1}.
Having introduced orthogonal spherical coordinates $x_j = F_j(r,\theta_1,\ldots,\theta_{n-1})$, $1 \leq j \leq n$, in $\mathbb{R}^n$ we obtain for the Laplacian on $\mathbb{R}^n$

$$\ddagger \qquad \Delta = -\sum_{j=1}^{n} \frac{\partial^2}{\partial x_j^2} = -\frac{\partial^2}{\partial r^2} - \frac{n-1}{r}\frac{\partial}{\partial r} + \frac{1}{r^2}\Delta_{LB} .$$

From this it follows that any m-th order spherical harmonic is an eigenvector of Δ_{LB} with eigenvalue $m(m+n-2)$. (Cf. [Mü], p. 4.) The number of mutually orthogonal spherical harmonics of order m is of the growth order m^{n-2}. Therefore, the operators $e^{-t\Delta_{LB}^{\frac{1}{2}}}$, $t > 0$, are all Hilbert-Schmidt and the space $S_{\mathbf{L}_2(S^{n-1}),\Delta_{LB}^{\frac{1}{2}}}$ is nuclear.

Next, consider the operator

$$\widetilde{A} = -\tfrac{1}{2}(n-1)I + \sqrt{\tfrac{1}{4}(n-1)^2 I + \Delta_{LB}} .$$

A simple calculation shows that a sperical harmonic of order m is an eigenvector of $\widetilde{A}$ with eigenvalue m. For this reason $e^{-tA}\, e^{t\widetilde{A}}$ is a bounded operator for each $t > 0$ and we have

$$S_{\mathbf{L}_2(S^{n-1}),\Delta_{LB}^{\frac{1}{2}}} = S_{\mathbf{L}_2(S^{n-1}),\widetilde{A}} \, .$$

In [G 5] the following result has been proved:

$f \in S_{L_2(S^{n-1}),\Delta_{LB}^{\frac{1}{2}}}$ iff f can be extended to a harmonic function on an open neighbourhood of the closed unit ball

$$B^{n-1} = \{\underline{x} \in \mathbb{R}^n \mid \|\underline{x}\|_n \leq 1\} .$$

There are some heuristic arguments for this result. Introduction of $r = e^{-t}$ in ‡ transforms the Laplacian into

$$e^{2t}\left\{-\frac{\partial^2}{\partial t^2} + (n-1)\frac{\partial}{\partial t} + \Delta_{LB}\right\} .$$

The expression between { } can be factored into two evolution equations as follows:

$$- e^{2t}\{\frac{\partial}{\partial t} - \tfrac{1}{2}(n-1)I - \{\tfrac{1}{4}(n-1)^2 I + \Delta_{LB}\}^{\frac{1}{2}}\} \cdot$$

$$\cdot \{\frac{\partial}{\partial t} - \tfrac{1}{2}(n-1)I + \{\tfrac{1}{4}(n-1)^2 I + \Delta_{LB}\}^{\frac{1}{2}}\} .$$

The second factor can be written as $\partial/\partial t + \tilde{A}$.
So a solution u of $\partial u/\partial t = -\tilde{A}u$, $t \geq 0$, corresponds to a harmonic function on the unit ball B^{n-1} and vice versa.
Similar to Case (II.d) certain strongly divergent series of spherical harmonics can be viewed as generalized functions on S^{n-1}, i.e. as elements of $T_{L_2(S^{n-1}),\Delta_{LB}^{\frac{1}{2}}}$.

We conclude this section with the description of some classes of linear operators on $S_{L_2(S^{n-1}),\Delta_{LB}^{\frac{1}{2}}}$ which are based on a combination of analytical and geometrical considerations. They are taken from [G 5].

- Multipliers

Let $h \in S_{L_2(S^{n-1}),\Delta_{LB}^{\frac{1}{2}}}$ be fixed. Consider the mapping $f \to M_h f = h\cdot f$, where the product is pointwise on S^{n-1}. The mapping M_h acts continuously from $S_{L_2(S^{n-1}),\Delta_{LB}^{\frac{1}{2}}}$ into itself. Since the product of two harmonic functions is not harmonic in general, it is not trivial at all that M_h is well defined. Actually, it must be shown that the restriction to S^{n-1} of the product $h\cdot f$ of two harmonic functions h and f can be extended to a harmonic function. We note further that M_h is extendible.

- Differentiation operators

Let $\underline{a} \in \mathbb{R}^n$. The operator $f \mapsto (\underline{a},\underline{\nabla})f$ is defined as follows. First extend f to a harmonic function. Then calculate $a_1 \frac{\partial f}{\partial x_1} + \dots + a_n \frac{\partial f}{\partial x_n}$. Finally, restrict this to S^{n-1}.
Instead of the constants a_j we can also take multipliers, thus getting differential operators with variable coefficients. An interesting subclass of this kind is obtained in the following way: Take a matrix $A \in \mathbb{R}^{n,n}$. The operator $f \mapsto (\underline{x},A\underline{\nabla})f$ maps $S_{L_2(S^{n-1}),\Delta_{LB}^{\frac{1}{2}}}$ into itself. If $A = I$, then $(\underline{x},A\nabla)$ indicates the normal derivative. If A is antisymmetric, $A^T = -A$, then the vector fields $(\underline{x},A\underline{\nabla})$ are tangent to S^{n-1}. They are linear combinations of the moment of momentum operators in quantum mechanics. All mentioned differentiation operators are extendible.

- Harmonic contractions

Take a matrix $A \in \mathbb{R}^{n,n}$ with (sup norm) $\|A\| \leq 1$. Define $(L_A f)(\xi) = f(A\xi)$, $\xi \in S^{n-1}$. Then $L_A f$ can be extended to a harmonic function on a neighbourhood of the closed unit ball. For the proof, see [G 5]. Therefore, L_A maps $S_{L_2(S^{n-1}),\Delta_{LB}^{\frac{1}{2}}}$ into itself. Note that $L_{AB} \neq L_A \circ L_B$ in general. If $\|A\| < 1$, the L_A is smoothing, i.e. L_A can be extended to an operator from $T_{L_2(S^{n-1}),\Delta_{LB}^{\frac{1}{2}}}$ into $S_{L_2(S^{n-1}),\Delta_{LB}^{\frac{1}{2}}}$.

II.2. The Gelfand-Shilov spaces S_α^β

In [GS2] the S_α^β-spaces are introduced. They consist of infinitely differentiable functions f on $\mathbb{R}$ for which there are constants $A,B,C > 0$ such that

$$\text{(II.2.1)} \quad \forall_{k,\ell \in \mathbb{N}\cup\{0\}} : \sup_{x\in\mathbb{R}} |x^k f^{(\ell)}(x)| \leq C A^k B^\ell k^{\alpha k} \ell^{\beta\ell} .$$

If the constant β is not greater than 1 every function $\varphi \in S_\alpha^\beta$ has an analytic continuation into the complex plane. In particular for $\beta < 1$, the space S_α^β consists of precisely all entire functions f with the following growth behaviour:

$$\text{(II.2.2)} \quad \exists_{a>0} \exists_{b>0} : |f(x+iy)| \leq C \exp(-a|x|^{1/\alpha} + b|y|^{1/1-\beta}) .$$

The above inequality imposes such strong constraints on the function φ that

for $\alpha+\beta < 1$, the space S_α^β is trivial, i.e. it contains only the function which is identically zero.
The space S_α^β is a union of Frechet spaces, i.e.

$$S_\alpha^\beta = \bigcup_{A,B>0} S_{\alpha,A}^{\beta,B} ,$$

where $S_{\alpha,A}^{\beta,B}$ denote the spaces of the functions $\varphi \in S_\alpha^\beta$ with the property that for all $\bar{A} > A$ and all $\bar{B} > B$ the inequality (II.2.1) is satisfied with A replaced by $\bar{A}$ and B by $\bar{B}$. Now the topology for $S_{\alpha,A}^{\beta,B}$ is generated by the countable family of seminorms:

$$\|f\|_{\delta\rho} = \sup_{x,k,\ell} \left[\frac{|x^k f^{(\ell)}(x)|}{(A+\delta)^k (B+\rho)^\ell k^{\alpha k} \ell^{\beta\ell}}\right] , \qquad \delta,\rho = 1,\tfrac{1}{2},\ldots .$$

The topology for S_α^β is the locally convex inductive limit topology induced by the Frechet spaces $S_{\alpha,A}^{\beta,B}$, $A,B > 0$.

We list some properties of the spaces S_α^β.
Each space S_α^β is contained in the space of infinitely differentiable functions of rapid decrease S. The embedding $S_\alpha^\beta \subset S$ is continuous. So each tempered distribution acts as a continuous linear functional on S_α^β. A natural class of continuous linear mappings on S_α^β are the multipliers M_φ (cf. Section II.1). In [GS2] it is proved that any function φ with

$$\forall_{\varepsilon>0} \exists_{c_\varepsilon} \forall_{\ell\in\mathbb{N}}: \sup_{x\in\mathbb{R}} |\varphi^{(\ell)}(x)| \leq c_\varepsilon \varepsilon^q \ell^{\beta\ell} \exp(\varepsilon |x|^{1/\alpha})$$

gives rise to a multiplier on S_α^β.
Other examples of continuous linear mappings on S_α^β are the translation operators T_a, $a \in \mathbb{R}$, phase shift operators R_b, $b \in \mathbb{R}$, dilation operators Z_λ, $\lambda \in \mathbb{R}$, and the differential operator D. Further, we mention the Fourier transformation which maps the spaces S_α^β bijectively onto S_β^α. So the spaces S_α^α, $\alpha \geq \tfrac{1}{2}$, are invariant under Fourier transformation.
A number of S_α^β-spaces are of type $S_{X,A}$. In this section, we present a candidate $A_{\alpha,\beta}$ for $\alpha = q/(p+q)\rho$ and $\beta = p/(p+q)\rho$ with $p,q \in \mathbb{N}$ and $0 < \rho \leq 1$.
To this end, consider the differential operator $B_{p,q}$:

$$\text{(II.2.3)} \quad B_{p,q} = \left(-\frac{d^2}{dx^2}\right)^q + (x^2)^p , \qquad p,q \in \mathbb{N} .$$

The operator $B_{p,q}$ is positive and self-adjoint in the Hilbert space $L_2(\mathbb{R})$. Its spectrum is discrete (cf. [Mü Pf]). Let λ_k, $k \in \mathbb{N}$, denote its eigenvalues with $0 < \lambda_1^{(p,q)} < \lambda_2^{(p,q)} < \dots$. Moreover, let $e_k^{p,q}$, $k \in \mathbb{N}$, denote the normalized eigenfunctions belonging to the eigenvalue $\lambda_k^{p,q}$. We have

$$\forall_{t>0}: \sum_{k=1}^{\infty} \exp(-\lambda_k^{p,q} t) < \infty .$$

In the definition (II.2.1) of the spaces S_α^β we may as well take the $L_2(\mathbb{R})$-norm $\|\cdot\|_2$ instead of the sup norm $\|\cdot\|_\infty$. The following result is taken from [EGP].

(II.2.4) Lemma

Let f be an infinitely differentiable function which satisfies the following inequalities for fixed $A,B,C > 0$ and fixed $\alpha,\beta > 0$, $\alpha+\beta \geqq 1$,

$$\|x^k D^\ell f\|_2 \leqq C A^k B^\ell k^{\alpha k} \ell^{\beta \ell} , \qquad k,\ell = 0,1,2,\dots .$$

Then for each $n \in \mathbb{N}$ and each $r,s \in (\mathbb{N} \cup \{0\})^n$

$$\|D^{r_1} x^{s_1} D^{r_2} \dots x^{s_{n-1}} D^{r_n} x^{s_n} f\|_2 \leqq C_1 A_1^{|s|} B_1^{|r|} |s|^{\alpha|s|} |r|^{\beta|r|}$$

where

$$C_1 = C , \quad A_1 = 2^{\sigma\beta+1} e^{\sigma\alpha} A , \quad B_1 = 2^{\sigma\alpha+1} e^{\sigma\beta} B , \quad \sigma = (\alpha+\beta)^{-1} .$$

We put $|s| = s_1 + \dots + s_n$, $|r| = r_1 + \dots + r_n$.

This lemma leads to the following result:

(II.2.5) Lemma

Let $0 < \rho \leqq 1$ and let $k,\ell \in \mathbb{N}$. Put $\alpha = q/(p+q)\rho$ and $\beta = p/(p+q)\rho$. Let $f \in S_\alpha^\beta$. Then for all $n \in \mathbb{N}$

$$\|(B_{p,q})^n f\|_2 \leqq KN^n \exp\left[\frac{2npq}{(p+q)\rho} \log n\right] .$$

Here the constants $K,N > 0$ depend on the choice of f.

Proof. By assumption there exist positive constants A, B, C such that

$$\|x^k D^\ell f\|_2 \leqq C A^k B^\ell k^{\alpha k} \ell^{\beta \ell} , \qquad k,\ell = 0,1,2,\dots .$$

Let $n \in \mathbb{N}$. Then

$$(B_{p,q})^n = \sum_{t=0}^{n} v_t(D^{2q},x^{2p})$$

where $v_t(D^{2q},x^{2p})$ is a sum of $\binom{n}{t}$ combinations of the form

$$(D^{2q})^{r_1}(x^{2p})^{s_1} \ \dots \ (D^{2q})^{r_b}(x^{2p})^{s_b}$$

where $r_1,\dots,r_b \in \mathbb{N} \cup \{0\}$, $s_1,\dots,s_b \in \mathbb{N} \cup \{0\}$, $r_1 + \dots + r_b = t$ and $s_1 + \dots + s_b = n-t$. By Lemma (II.2.4) we get

$$\| v_t(D^{2q},x^{2p})f\|_2 \leq \binom{n}{t} C\, A_1^{2p(n-1)}\, B_1^{2qt}(2p(n-t))^{2\alpha p(n-t)}(2qt)^{2\beta qt} \ .$$

It leads to the following estimation:

$$\|(B_{p,q})^n f\|_2 \leq C \sum_{t=0}^{n} \binom{n}{t} A_2^{n-t} B_2^{t}(n-t)^{2\alpha p(n-t)} t^{2\beta qt}$$

$$\leq C \sum_{t=0}^{n} \binom{n}{t} A_2^{n-t} B_2^{t}\, n^{2\alpha p(n-t)} n^{2\beta qt}$$

$$= C(A_2 + B_2)^n \exp\left(\frac{2pqn}{(p+q)\rho} \log n\right) \ .$$

Here we have put $A_2 = ((2p)^{\alpha} A_1)^{2p}$, $B_2 = ((2q)^{\beta} B_1)^{2q}$. □

We have the following inclusion result.

(II.2.6) Theorem

Put $\alpha_{p,q}(\rho) = q/(p+q)\rho$, $\beta_{p,q}(\rho) = p/(p+q)\rho$ and $\varepsilon_{p,q}(\rho) = (p+q)\rho/2pq$. Then

$$S_{\alpha_{p,q}(\rho)}^{\beta_{p,q}(\rho)} \subset S_{L_2(\mathbb{R}),(B_{p,q})}^{\varepsilon_{p,q}(\rho)} \ .$$

(Note that $(B_{p,q})^{\varepsilon_{p,q}(\rho)} = \left(\left(-\frac{d^2}{dx^2}\right)^p + (x^2)^q\right)^{(p+q)\rho/2pq}$.)

Proof. Let $f \in S^{\beta_{p,q}(\rho)}_{\alpha_{p,q}(\rho)}$. We put

$$a_k = (f, e_k^{p,q})_{\mathbf{L}_2(\mathbb{R})} = (\lambda_k^{p,q})^{-n}((\mathbf{B}_{p,q})^n f, e_k^{p,q})_{\mathbf{L}_2(\mathbb{R})} .$$

Then by Lemma (II.2.5) we derive for all $n \in \mathbb{N}$

$$|a_k| \leq K(\lambda_k^{p,q})^{-n} N^n \exp\left[\frac{n}{\varepsilon_{p,q}(\rho)} \log n\right] .$$

Taking the infimum of the right hand side with respect to n it follows that

$$|a_k| \leq D \exp(-\tau(\lambda_k^{p,q})^{\varepsilon_{p,q}(\rho)})$$

for some $D,\tau > 0$. Now for all $\tau' < \tau$ the sequence

$$(\exp(-\tau'(\lambda_k^{p,q})^{\varepsilon_{p,q}(\rho)}))_{k\in\mathbb{N}}$$

belongs to ℓ_2, and hence $f \in \exp(-\alpha(\mathbf{B}_{p,q})^{\varepsilon_{p,q}(\rho)})(\mathbf{L}_2(\mathbb{R}))$. □

(II.2.7) Conjecture

$$S_{\mathbf{L}_2(\mathbb{R}),(\mathbf{B}_{p,q})^{\varepsilon_{p,q}(\rho)}} = S^{\beta_{p,q}(\rho)}_{\alpha_{p,q}(\rho)} .$$

In the following cases Conjecture (II.2.7) has been proved.

Case (II.g) $S_{\mathbf{L}_2(\mathbb{R}),(\frac{1}{2}(-\frac{d^2}{dx^2}+x^2+1))^{1/2\alpha}} = S^\alpha_\alpha$, $\alpha \geq \frac{1}{2}$.

This result can be derived from a paper by Ging-Zong [GZ].

Case (II.h) $S_{\mathbf{L}_2(\mathbb{R}),(-\frac{d^2}{dx^2}+x^{2k})^{\frac{1}{2}}} = S^1_{1/k}$, $k \in \mathbb{N}$.

The proof of this assertion can be achieved from the paper [Go 2] of Goodman. Taking Fourier transformation, we get

$$S_{\mathbf{L}_2(\mathbb{R}),((-\frac{d^2}{dx^2})^k+x^2)^{\frac{1}{2}}} = S_1^{1/k} , \quad k \in \mathbb{N} .$$

<u>Case</u> (II.i) $S_{L_2(\mathbb{R}),(-\frac{d^2}{dx^2}+x^{2k})^{k+1/2k}} = S^{k/k+1}_{1/k+1}$, $k \in \mathbb{N}$.

This result has been established by the authors in cooperation with R.S. Pathak. Proofs can be found in [EGP]. Also we have

$$S_{L_2(\mathbb{R}),((-\frac{d^2}{dx^2})^k+x^2)^{k+1/2k}} = S^{1/k+1}_{k/k+1} , \qquad k \in \mathbb{N} .$$

II.3. Analyticity spaces related to classical polynomials

A very transparent way to introduce generalized functions is by means of formal series expansions with respect to some orthonormal basis in a Hilbert space. We clarify this statement a bit more. Let $(\lambda_n)_{n\in\mathbb{N}}$ be a sequence of real numbers with $0 \leq \lambda_1 \leq \lambda_2 \leq \ldots$ and $\sum_{n=1}^{\infty} e^{-\lambda_n t} < \infty$ for all $t > 0$. We consider a vector space T of formal series $\sum_{n=1}^{\infty} a_n v_n$ where the coefficients $(a_n)_{n\in\mathbb{N}}$ satisfy

$$\forall_{t>0}: \sup_{n\in\mathbb{N}} |a_n| e^{-\lambda_n t} < \infty .$$

A candidate for the dual of T can be obtained by considering all sequences $(b_n)_{n\in\mathbb{N}}$ for which $\sum_{n=1}^{\infty} |a_n b_n| < \infty$ for all elements of T. Then a representation of the dual is given by the space S which consists of the elements in X represented by $\sum_{n=1}^{\infty} b_n v_n$. We have $S \subset X \subset T$. It is not hard to prove that each such sequence $(b_n)_{n\in\mathbb{N}}$ has to satisfy

$$\exists_{\tau>0}: \sup_{n\in\mathbb{N}} |b_n| e^{\lambda_n \tau} < \infty .$$

Now we introduce the operator A as follows:

$$D(A) = \left\{ f \mid \sum_{n=1}^{\infty} \lambda_n^2 |(f,v_n)|^2 < \infty \right\} ,$$

$$Af = \sum_{n=1}^{\infty} \lambda_n (f,v_n) v_n , \qquad f \in D(A) .$$

Then A is a nonnegative self-adjoint operator in the Hilbert space X. Moreover, $S = S_{X,A}$ and $T = T_{X,A}$.

In this section we consider orthonormal bases $(v_n)_{n\in\mathbb{N}}$ based on classical polynomials.

Case (II.j) Analyticity spaces related to the Hermite polynomials.

The Hermite polynomials H_n, $n \in \mathbb{N} \cup \{0\}$, are defined by

$$H_n(x) = (-1)^n\, e^{x^2}\, D^n(e^{-x^2}), \qquad x \in \mathbb{R} .$$

Related to the Hermite polynomials are the normalized Hermite functions ψ_n defined by

$$\psi_n(x) = (\pi^{\frac{1}{2}}\, 2^n\, n!)^{-\frac{1}{2}}\, e^{-\frac{1}{2}x^2}\, H_n(x), \qquad x \in \mathbb{R} .$$

The functions ψ_n, $n \in \mathbb{N} \cup \{0\}$, constitute an orthonormal basis for the Hilbert space $L_2(\mathbb{R})$. They are the eigenfunctions of the differential operator H,

$$H = \tfrac{1}{2}\left(-\frac{d^2}{dx^2} + x^2 + 1\right)$$

with the natural numbers as eigenvalues, i.e. $H\psi_n = (n+1)\psi_n$. As already mentioned in the previous section, the space $S_{L_2(\mathbb{R}),H^{1/2\alpha}}$ equals the space S_α^α for all $\alpha \geq \frac{1}{2}$. Thus we get the following characterization.

(II.3.1) Theorem

Let f be an infinite differentiable function on $\mathbb{R}$. Then $f \in S_\alpha^\alpha$, $\alpha \geq \frac{1}{2}$, iff there exists $t > 0$ such that $(f,\psi_n) = \mathcal{O}(\exp(-tn^{1/2\alpha}))$. The series $\sum_{n=0}^\infty (f,\psi_n)\psi_n(x)$, $x \in \mathbb{R}$, converges uniformly on $\mathbb{R}$.

The spaces $S_{\frac{1}{2}}^{\frac{1}{2}}$ and S_1^1 seem the most interesting, and we devote some extra attention to them.

In 1974, De Bruijn published a paper on "a new theory of generalized function ..." which is quoted as [B]. In [B] the space $S_{\frac{1}{2}}^{\frac{1}{2}}$ has been taken as a test space for a new distribution theory. De Bruijn proved that any element $f \in S_{\frac{1}{2}}^{\frac{1}{2}}$ can be written as

$$f(z) = \frac{1}{\sqrt{2\pi \sinh\alpha}} \int_{-\infty}^{\infty} g(y)\, \exp\left(-(z^2+y^2)\coth\alpha + \frac{2zy}{\sinh\alpha}\right) dy$$

where $g \in S_{\frac{1}{2}}^{\frac{1}{2}}$ and $\alpha > 0$ sufficiently small. Since

$$K_\alpha(x,y) = \frac{1}{\sqrt{2\pi \sinh \alpha}} \exp\left(-(x^2+y^2)\coth \alpha + \frac{1}{\sinh \alpha} xy\right)$$

is the Hilbert-Schmidt kernel of the operator $\exp(-\alpha H)$, his result can be rephrased as

$$S_{\frac{1}{2}}^{\frac{1}{2}} = \bigcup_{\alpha>0} \exp(-\alpha H)(S_{\frac{1}{2}}^{\frac{1}{2}}) .$$

Now the elements of the dual can be described as mappings F from the open interval $(0,\infty)$ into $S_{\frac{1}{2}}^{\frac{1}{2}}$ satisfying

$$F(\alpha+\beta) = \exp(-\alpha H)F(\beta) , \quad \alpha,\beta > 0 .$$

In [B] the mappings F are called 'traces'.
The set-up in [B] which in fact only concerns the space $S_{\frac{1}{2}}^{\frac{1}{2}}$, inspired De Graaf to the general functional analytic construction of the Gelfand triple $S_{X,A} \subset X \subset T_{X,A}$.
For reasons which will become clearer in Section II.4 the Space S_1^1 is also an inspiring example. The elements of S_1^1 are infinitely differentiable functions f which have an analytic extension to a strip $|\mathrm{Im}\, z| < r$ in the complex plane and on this strip satisfy

$$\sup_{|\mathrm{Im}\, z|<r} |e^{r|x|} f(x+iy)| < \infty , \quad z = x + iy .$$

Here the width r of the strip depends on f.
An interesting class of continuous linear functionals on S_1^1 can be obtained in the following way. Let Ω be a function which is analytic on $\mathbb{C} \setminus \mathbb{R}$, and for which the functions $x \mapsto \Omega(x+iy)$, $y \in \mathbb{R} \setminus \{0\}$, do not increase faster than polynomially. Then we define the functional ℓ_Ω by

$$\ell_\Omega(f) = \left(\int_{\infty-i\varepsilon}^{-\infty-i\varepsilon} + \int_{-\infty+i\varepsilon}^{\infty+i\varepsilon} \right) f(z)\Omega(z)dz$$

where $\varepsilon > 0$ must be taken so small that f is analytic on $|\mathrm{Im}\, z| < \varepsilon$ and satisfies the mentioned estimation.
Examples of continuous linear mappings are the operators T_a, $a \in \mathbb{R}$, Z_λ, $\lambda \in \mathbb{R}$, D, R_b, $b \in \mathbb{R}$, and the Fourier transformation.

The space $S_{\frac{1}{2}}^{\frac{1}{2}}$ is continuously embedded in S_1^1. So all continuous linear functionals on S_1^1 are automatically continuous on $S_{\frac{1}{2}}^{\frac{1}{2}}$. It is worth mentioning that all evaluation functionals δ_z, $z \in \mathbb{C}$, are continuous on $S_{\frac{1}{2}}^{\frac{1}{2}}$. Consequently, the operator $\mathcal{Q}$ defined by

$$\mathcal{Q}f: x \mapsto xf(x) \ , \quad x \in \mathbb{R} \ ,$$

has the whole complex plane as its spectrum. The operators T_a, $a \in \mathbb{C}$, R_b, $b \in \mathbb{C}$, D, Z_λ, $\lambda \in \mathbb{R}$, map $S_{\frac{1}{2}}^{\frac{1}{2}}$ continuously into $S_{\frac{1}{2}}^{\frac{1}{2}}$.

Finally, we devote some attention the a closely related subject: the Bargmann-Fock space.

The lineair operator $\mathbb{A}$ on $\mathbf{L}_2(\mathbb{R})$ is defined as follows:

$$(\mathbb{A}f)(z) = \int_{\mathbb{R}} \mathbb{A}(z,q)f(q)dq \ , \quad z \in \mathbb{C}, \quad f \in \mathbf{L}_2(\mathbb{R})$$

where

$$\mathbb{A}(z,q) = \pi^{-\frac{1}{4}} \exp\left(-\tfrac{1}{2}(z^2+q^2) + \sqrt{2}\, zq\right) \ .$$

For each $f \in \mathbf{L}_2(\mathbb{R})$ the function Af is entire analytic, and it satisfies

$$\int_{\mathbb{R}^2} |(\mathbb{A}f)(x+iy)|^2 \, e^{-(x^2+y^2)} dxdy < \infty \ .$$

In a natural way the transformation $\mathbb{A}$ leads to the introduction of the Bargmann-Fock space F. The Hilbert space F is the closed subspace of $\mathbf{L}_2(\mathbb{R}^2, e^{-(x^2+y^2)}dxdy)$ which consists of all functions $(x,y) \to u(x+iy)$ with u an entire function. The operator $\mathbb{A}$ maps $\mathbf{L}_2(\mathbb{R})$ unitarily onto F. In particular, $\mathbb{A}\psi_n = u_n$, $n \in \mathbb{N} \cup \{0\}$ with $u_n(z) = (1/\sqrt{n!})z^n$, $z \in \mathbb{C}$.

Bargmann has extensively studied the Hilbert space F. Cf. [Ba 1-2]. In [Ba 2] the operator $\mathbb{A}$ is extended to the space of tempered distributions. Similarly, $\mathbb{A}$ can be restricted to $S_{\mathbf{L}_2(\mathbb{R}),H}$ and extended to $T_{\mathbf{L}_2(\mathbb{R}),H}$. Putting $\hat{H} = \mathbb{A}H\mathbb{A}^*$ yields

$$\mathbb{A}(S_{\mathbf{L}_2(\mathbb{R}),H}) = S_{F,\hat{H}} \quad \text{and} \quad \mathbb{A}(T_{\mathbf{L}_2(\mathbb{R}),H}) = T_{F,\hat{H}} \ .$$

We have $\hat{H} = z\frac{d}{dz} + 1$. The space $S_{F,\hat{H}}$ consists of all entire functions for which there are constants $c < \frac{1}{2}$ and $D > 0$ such that $|h(z)| < D \exp(c|z|^2)$. The space $T_{F,\hat{H}}$ consists also of entire functions. They are characterized as follows.

$$\Phi \in T_{F,A} \Leftrightarrow \Phi \text{ is an entire function for which}$$

$$\forall_{c>\frac{1}{2}} \exists_{D>0} : |\Phi(z)| < D \exp(c|z|^2) .$$

One of the merits of the Bargmann-Fock space is the very simple representation of the coherent states. Each coherent state in $\mathbf{L}_2(\mathbb{R})$ is represented by the exponential function $e_w : z \mapsto e^{\bar{w}z}$, $w,z \in \mathbb{C}$. In the paper [G4] this property of the space F has been exploited.

Case (II.k) Analyticity spaces related to the Laguerre polynomials.

The Laguerre polynomials $L_n^{(\alpha)}$, $\alpha > -1$, are introduced as follows:

$$L_n^{(\alpha)}(x) = \frac{1}{n!} e^x x^{-\alpha} D^n(e^{-x} x^{n+\alpha}) .$$

Closely connected to these polynomials $L_n^{(\alpha)}$ are the functions $\mathcal{L}_n^{(\alpha)}$:

$$\mathcal{L}_n^{(\alpha)}(x) = \left\{\frac{2n!}{\Gamma(n+\alpha+1)}\right\}^{\frac{1}{2}} e^{-\frac{1}{2}x^2} L_n^{(\alpha)}(x^2) , \quad x \in \mathbb{R} .$$

For each $\alpha > -1$ the functions $\mathcal{L}_n^{(\alpha)}$ constitute an orthonormal basis in the Hilbert space $X_\alpha = \mathbf{L}_2((0,\infty), x^{2\alpha+1}\,dx)$. The functions $\mathcal{L}_n^{(\alpha)}$ are related to the Hermite functions ψ_n in the following way:

$$\text{(II.3.2)} \quad \psi_{2n}(x) = (-1)^m \mathcal{L}_m^{-\frac{1}{2}}(x^2) \; ; \quad \psi_{2m+1}(x) = (-1)^m x\, \mathcal{L}_m^{(\frac{1}{2})}(x^2) .$$

Let $\alpha > -1$. The orthonormal basis $(\mathcal{L}_n^{(\alpha)})$ consists of eigenvectors of the positive self-adjoint operator

$$A_\alpha = -\frac{d^2}{dx^2} + \frac{2\alpha+1}{x}\frac{d}{dx} + x^2 .$$

We have $A_\alpha \mathcal{L}_n^{(\alpha)} = (4n+2\alpha+2)\mathcal{L}_n^{(\alpha)}$.

In our papers [EG 1-2] we have proved that the spaces $S_{X_\alpha,(A_\alpha)^\nu}$, with $\alpha > -1$ and $\frac{1}{2} \leq \nu \leq 1$, are spaces of analytic functions. We give a short description of the method that has been employed in the mentioned papers.
First observe that a precise characterization of the spaces $S_{X_{-\frac{1}{2}},(A_{-\frac{1}{2}})^\nu}$ can be derived from the relation (II.3.2) and Case (II.j).

(II.3.3) Lemma

$f \in S_{X_{-\frac12},(A_{-\frac12})^\nu}$ iff f is an even function on $\mathbb{R}$ which is extendible to an analytic function belonging to the Gelfand-Shilov space $S^{1/2\nu}_{1/2\nu}$.

The second observation concerns the particular form of the operator A_α. Each A_α can be written as $A_\alpha = A_{-\frac12} + P_\alpha$ with $P_\alpha = + (2\alpha+1)\frac{1}{x}D$. In [EG 2] the general case of two analyticity spaces $S_{X,A}$ and $S_{Y,B}$ is considered, where B is a perturbation of A, i.e. $B = A + P$. Conditions on the pair (Y,P) are given such that $S_{X,A^\nu} = S_{Y,B^\nu}$, $\nu > 0$, as topological vector spaces. Due to these results, we have the following classification.

(II.3.4) Lemma

Let $\alpha > -1$ and let $\frac12 \leq \nu \leq 1$. Then we have

$$S_{X_\alpha,(A_\alpha)^\nu} = S_{X_{-\frac12},(A_{-\frac12})^\nu}\ .$$

Lemma (II.3.3) and Lemma (II.3.4) establish the classical analytic characterization of the spaces $S_{X_\alpha,(A_\alpha)^\nu}$, $\alpha > -1$, $\frac12 \leq \nu \leq 1$.

Closely related to the functions $L_n^{(\alpha)}$ are the functions $\tilde{L}_n^{(\alpha)}$:

$$\tilde{L}_n^{(\alpha)}(x) = x^{\alpha+\frac12}\, L_n^{(\alpha)}(x)\ , \qquad x > 0\ .$$

For each $\alpha > -1$, the set $\{\tilde{L}_n^{(\alpha)} \mid n \in \mathbb{N}\}$ is an orthonormal basis for the Hilbert space $L_2((0,\infty))$. Put

$$\tilde{A}_\alpha = -\frac{d^2}{dx^2} + x^2 + \frac{\alpha^2 - \frac14}{x^2}\ .$$

Then $\tilde{A}_\alpha \tilde{L}_n^{(\alpha)} = (4n + 2\alpha + 2)\tilde{L}_n^{(\alpha)}$. The description of the spaces $S_{L_2((0,\infty)),(\tilde{A}_\alpha)^\nu}$ is easily obtained from the preceding lemmas.

(II.3.5) Lemma

Let $\frac12 \leq \nu \leq 1$ and let $\alpha > -1$. Then $f \in S_{L_2((0,\infty)),(\tilde{A}_\alpha)^\nu}$ iff the function $x \mapsto x^{-(\alpha+\frac12)} f(x)$, $x \in \mathbb{R}^+$, is extendible to an even analytic function in the space $S^{1/2\nu}_{1/2\nu}$.

The Hankel transformation $\mathbb{H}_\alpha$ is defined by

$$(\mathbb{H}_\alpha f)(x) = \int_0^\infty f(y)\, \sqrt{xy}\, J_\alpha(xy)\,dy \;.$$

From [MOS], p. 242, the following relation can be derived:

$$\mathbb{H}_\alpha \tilde{\mathcal{L}}_n^{(\alpha)} = (-1)^n \tilde{\mathcal{L}}_n^{(\alpha)} \;.$$

So $\mathbb{H}_\alpha$ extends to a unitary operator on $\mathbf{L}_2((0,\infty))$, which commutes with A_α on $S_{\mathbf{L}_2((0,\infty)),\tilde{A}_\alpha}$. Therefore, $\mathbb{H}_\alpha$ maps $S_{\mathbf{L}_2((0,\infty)),(\tilde{A}_\alpha)^\nu}$ continuously and bijectively onto itself. Thus $S_{\mathbf{L}_2((0,\infty)),(\tilde{A}_\alpha)^\nu}$ is a Hankel invariant test space and hence $T_{\mathbf{L}_2((0,\infty)),(\tilde{A}_\alpha)^\nu}$ a Hankel invariant distribution space.

(II.3.6) Corollary

Let $\nu \in [\frac{1}{2},1]$. Then the even part of $S_{1/2\nu}^{1/2\nu}$ is invariant under the transformations $\mathbf{Q}^{-(\alpha+\frac{1}{2})}\, \mathbb{H}_\alpha\, \mathbf{Q}^{(\alpha+\frac{1}{2})}$, $\alpha > -1$. Here $\mathbf{Q}$ denotes the operator of multiplication by the identity function.

Finally, we list the following continuous linear functionals and continuous linear mappings on the space $S_{\mathbf{L}_2((0,\infty)),(\tilde{A}_\alpha)^\nu}$, $\frac{1}{2} \leq \nu \leq 1$.

- The linear functionals $\delta_x^{(\alpha)}: f \mapsto f(x)$ with $x > 0$, and for $\frac{1}{2} < \nu \leq 1$, the lineair functionals $\delta_z^{(\alpha)}$, $|\arg(z)| < \pi$.
- The linear functionals $J_\alpha(x) = \delta_x^{(\alpha)} \circ \mathbb{H}_\alpha$, $x > 0$. For $\frac{1}{2} < \nu \leq 1$ also the linear functionals $J_\alpha(z)$, $|\arg(z)| < \pi$.
- The linear operator $\mathbf{Q}^2$. We have $\mathbf{Q}^2\, \delta_z^{(\alpha)} = z^2\, \delta_z^{(\alpha)}$.
- The linear operator $\mathbf{B}_\alpha = -\frac{d^2}{dx^2} + \frac{\alpha^2 - \frac{1}{4}}{x^2}$. We have $\mathbf{B}_\alpha = \mathbb{H}_\alpha\, \mathbf{Q}^2\, \mathbb{H}_\alpha$ and $\mathbf{B}_\alpha(J_\alpha(z)) = z^2\, J_\alpha(z)$.
- The linear operator $\mathbf{Q}^\delta$, $\delta > 0$, which maps $S_{\mathbf{L}_2((0,\infty)),\tilde{A}_\alpha}$ bijectively onto $S_{\mathbf{L}_2((0,\infty)),\tilde{A}_{\alpha+\delta}}$.
- The linear operator $\mathbf{D}$ which maps $S_{\mathbf{L}_2((0,\infty)),\tilde{A}_{\alpha+1}}$ into $S_{\mathbf{L}_2((0,\infty)),\tilde{A}_\alpha}$.

Case (II.1) Analyticity spaces related to Jacobi polynomials.

Let $\alpha,\beta > -1$. Consider the Hilbert space

$$X_{\alpha,\beta} = \mathbf{L}_2([-1,1],(1-x)^{\alpha}(1+x)^{\beta}\,dx)$$

and the nonnegative self-adjoint operator $A_{\alpha,\beta}$ in $X_{\alpha,\beta}$

$$A_{\alpha,\beta} = -(1-x^2)\frac{d^2}{dx^2} - ((\beta-\alpha) - (\alpha+\beta+2)x)\frac{d}{dx}$$

The operator $A_{\alpha,\beta}$ has the discrete spectrum $\{n(n+\alpha+\beta+1) \mid n \in \mathbb{N} \cup \{0\}\}$. Its normalized eigenvectors are the normalized Jacobi polynomials $R_n^{(\alpha,\beta)}$

$$R_n^{(\alpha,\beta)} = \left(\frac{2n+\alpha+\beta+1}{2^{\alpha+\beta+1}}\,\frac{\Gamma(n+1)\Gamma(n+\alpha+\beta+1)}{\Gamma(n+\alpha+1)\Gamma(n+\beta+1)}\right)^{\frac{1}{2}} P_n^{(\alpha,\beta)}$$

where

$$P_n^{(\alpha,\beta)}(x) = \frac{(-1)^n}{n!\,2^n}\,\frac{1}{(1-x)^{\alpha}(1+x)^{\beta}}\,\left(\frac{d}{dx}\right)^n\,[(1-x)^{\alpha+n}(1+x)^{\beta+n}]\,.$$

A special type of Jacobi polynomials are the Legendre polynomials $P_n^{(0,0)}$, $n \in \mathbb{N} \cup \{0\}$, and the Chebyshev polynomials $P_n^{(-\frac{1}{2},-\frac{1}{2})}$, $n \in \mathbb{N} \cup \{0\}$.

With the tranformation $x = \cos\Theta$ we get the Hilbert spaces

$$Y_{\alpha,\beta} = \mathbf{L}_2([0,\pi],(1-\cos\Theta)^{\alpha}(1+\cos\Theta)^{\beta}\sin\Theta\,d\Theta)$$

and the operators

$$C_{\alpha,\beta} = -\frac{d^2}{d\Theta^2} + (\beta-\alpha)\,\frac{1}{\sin\Theta}\,\frac{d}{d\Theta} - (\alpha+\beta+1)\cot\Theta\,\frac{d}{d\Theta}$$

In [EG 3] the following program has been carried out:

- Classification of the spaces $S_{X_{\alpha,\beta},(A_{\alpha,\beta})^{\nu}}$ where $\nu \geq \frac{1}{2}$ and $\alpha,\beta > -1$. We have proved that for all admissible pairs (α,β) and (γ,δ), and for all $\nu \geq \frac{1}{2}$

$$S_{X_{\alpha,\beta},(A_{\alpha,\beta})^{\nu}} = S_{X_{\gamma,\delta},(A_{\gamma,\delta})^{\nu}}\,.$$

  In this connection new results have been derived with respect to classification of analyticity spaces, in general. Besides the method which has been developed for the classification of Laguerre analyticity spaces, a classification methode has been invented which is based on diagonalization of upper triangular infinite matrices. We note that we also have the equality

$$S_{Y_{\alpha,\beta},(C_{\alpha,\beta})^{\nu}} = S_{Y_{\gamma,\delta},(C_{\gamma,\delta})^{\nu}} .$$

- Characterization of the spaces $S_{Y_{-\frac{1}{2},-\frac{1}{2}},(C_{-\frac{1}{2},-\frac{1}{2}})^{\nu}}$, $\nu \geqq \frac{1}{2}$. The Chebyshev polynomials have the following typifying property

$$R_n^{(-\frac{1}{2},-\frac{1}{2})}(\cos\Theta) = \sqrt{\frac{2}{\pi}}\cos n\Theta$$

and

$$R_0^{(-\frac{1}{2},-\frac{1}{2})}(\cos\Theta) = \sqrt{\frac{1}{\pi}} .$$

So the space $S_{Y_{-\frac{1}{2},-\frac{1}{2}},(C_{-\frac{1}{2},-\frac{1}{2}})^{\nu}}$ equals the space $S_{L_2([0,\pi]),\Delta_e^{\nu}}$ which is described in Case (II.e).

Consider the conformal mapping w = cos z. This mapping describes the complex analytic relation between the spaces $S_{Y_{-\frac{1}{2},-\frac{1}{2}},(C_{-\frac{1}{2},-\frac{1}{2}})^{\nu}}$ and $S_{X_{-\frac{1}{2},-\frac{1}{2}},(A_{-\frac{1}{2},-\frac{1}{2}})^{\nu}}$.

- Characterization of the spaces $S_{X_{-\frac{1}{2},-\frac{1}{2}},(A_{-\frac{1}{2},-\frac{1}{2}})^{\nu}}$, $\nu \geqq \frac{1}{2}$.

(i) $h \in S_{X_{-\frac{1}{2},-\frac{1}{2}},(A_{-\frac{1}{2},-\frac{1}{2}})^{\frac{1}{2}}}$ if and only if $h \in X_{-\frac{1}{2},-\frac{1}{2}}$ can be extended to an analytic function on an open neighbourhood of the interval [-1,1].

(ii) Let $\nu \geqq \frac{1}{2}$. Then $h \in S_{X_{-\frac{1}{2},-\frac{1}{2}},(A_{-\frac{1}{2},-\frac{1}{2}})^{\nu}}$ if and only if $h \in X_{-\frac{1}{2},-\frac{1}{2}}$ can be extended to an entire analytic function for which there exist A,B > 0 such that

$$|h(w)| < A\exp(B(\log|w|)^{2\nu/2\nu-1}) , \quad |w| \geqq 1 .$$

Also sequential convergence in $S_{X_{-\frac{1}{2},-\frac{1}{2}},(A_{-\frac{1}{2},-\frac{1}{2}})^{\nu}}$ has been described in classical analytic terms.
We mention the following examples of continuous linear functionals and continuous linear mappings on $S_{X_{-\frac{1}{2},-\frac{1}{2}},(A_{-\frac{1}{2},-\frac{1}{2}})^{\nu}}$, $\nu > \frac{1}{2}$.

- The linear functionals $\delta_z^{(k)}$ with $z \in \mathbb{C}$ and $k \in \mathbb{N} \cup \{0\}$. A wider class is obtained as follows. Consider a Borel measure μ on $\mathbb{R}^2$ with the property

$$\forall_{B>0}: \iint_{\mathbb{R}^2} \exp B(\log(1+x^2+y^2))^{2\nu/2\nu-1}\, d|\mu(x,y)| < \infty .$$

Then define ℓ_μ by

$$\ell_\mu(f) = \iint_{\mathbb{R}^2} f(x+iy)\,d\mu(x,y) .$$

- The linear operators M_ψ, $\psi \in S_{X_{-\frac{1}{2},-\frac{1}{2}},(A_{-\frac{1}{2},-\frac{1}{2}})^\nu}$, T_a, $a \in \mathbb{C}$, Z_λ, $\lambda \in \mathbb{C}$, D.

For $\nu = \frac{1}{2}$ we have

- The linear functionals $\delta_x^{(k)}$, $x \in [-1,1]$, $k \in \mathbb{N} \cup \{0\}$. At the end of this section we shall relate the dual of the space $S_{X_{-\frac{1}{2},-\frac{1}{2}},(A_{-\frac{1}{2},-\frac{1}{2}})^{\frac{1}{2}}}$ to a space of hyperfunctions.
- The linear operators, M_ψ, $\psi \in S_{X_{-\frac{1}{2},-\frac{1}{2}},(A_{-\frac{1}{2},-\frac{1}{2}})^{\frac{1}{2}}}$, Z_λ, $\lambda \in \mathbb{R}$, with $|\lambda| \leq 1$, D.

The spaces $T_{Y_{-\frac{1}{2},-\frac{1}{2}},(C_{-\frac{1}{2},-\frac{1}{2}})^{\frac{1}{2}}}$ and $T_{X_{0,0},(A_{0,0})^{\frac{1}{2}}}$ can be considered as spaces of hyperfunctions in the following way.
Consider the function

$$\Theta_z : x \mapsto \frac{1}{2\pi i} \frac{\sin \bar{z}}{\cos x - \cos \bar{z}} , \qquad \operatorname{Im} z \neq 0 .$$

For each $z \in \mathbb{C} \setminus \mathbb{R}$, the function Θ_z belongs to $S_{Y_{-\frac{1}{2},-\frac{1}{2}},(C_{-\frac{1}{2},-\frac{1}{2}})^{\frac{1}{2}}}$. Let $G \in T_{Y_{-\frac{1}{2},-\frac{1}{2}},(C_{-\frac{1}{2},-\frac{1}{2}})^{\frac{1}{2}}}$. Then $H_G: z \mapsto \overline{\langle\Theta_z, G\rangle}_{X_{-\frac{1}{2},-\frac{1}{2}}}$ is an analytic 2π-periodic odd function on $\mathbb{C} \setminus \mathbb{R}$ with $H_G(\infty) = 0$. Conversely, if the function Φ is 2π-periodic odd and analytic on $\mathbb{C} \setminus \mathbb{R}$ with normalization $\Phi(\infty) = 0$, then $\Phi = H_G$ for some $G \in T_{Y_{-\frac{1}{2},-\frac{1}{2}},(C_{-\frac{1}{2},-\frac{1}{2}})^{\frac{1}{2}}}$.
The hyperfunction corresponding to H_G is the distributional limit

$$\lim_{\varepsilon \downarrow 0} [H_G(x+i\varepsilon) - H_G(x-i\varepsilon)] .$$

For each $f \in S_{Y_{-\frac{1}{2},-\frac{1}{2}},(C_{-\frac{1}{2},-\frac{1}{2}})^{\frac{1}{2}}}$ we have

$$\overline{\langle f,G\rangle} = \int_{C_t} \overline{f(\bar{z})} H_G(z)\,dz\ , \qquad t > 0 \text{ sufficiently small,}$$

where C_t consists of the two directed line segments

$$\{z \in \mathbb{C} \mid \operatorname{Im} z = t,\ 0 \leq \operatorname{Re} z \leq \pi\} \cup \{z \in \mathbb{C} \mid \operatorname{Im} z = -t,\ \pi \geq \operatorname{Re} z \geq 0\}\ .$$

If $x \mapsto \lim_{\varepsilon \downarrow 0} (H_G(x+i\varepsilon) - H_G(x-i\varepsilon))$ is an L_1-function h, then

$$\overline{\langle f,G\rangle} = \int_0^\pi \overline{f(x)}\, h(x)\,dx\ .$$

Similarly, the space $T_{X_{0,0},(A_{0,0})^{\frac{1}{2}}}$ can be considered as a space of hyper-function on the interval $[-1,1]$. In this case we consider the functions

$$\tau_w : x \to \frac{1}{2\pi i}\,\frac{1}{y - \bar{w}}\ , \qquad w \notin [-1,1]\ .$$

Each function τ_w is analytic in an open neighbourhood of $[-1,1]$. So τ_w, $w \notin [-1,1]$, belongs to $S_{X_{0,0},(A_{0,0})^{\frac{1}{2}}}$. Let $F \in T_{X_{0,0},(A_{0,0})^{\frac{1}{2}}}$. Then we define the function H_F

$$H_F : w \to \overline{\langle \tau_w, F\rangle}\ , \qquad w \notin [-1,1]\ .$$

The function H_F is analytic outside the interval $[-1,1]$ and $H_F(\infty) = 0$. Also the converse is valid: Any function Ω which is analytic on $\mathbb{C} \setminus [-1,1]$ with $\Omega(\infty) = 0$, can be written as $\Omega = H_F$. For each $h \in S_{X_{0,0},(A_{0,0})^{\frac{1}{2}}}$ we have

$$\overline{\langle h,F\rangle} = \oint_{E_t} \overline{h(\bar{w})}\, H_F(w)\,dw\ , \qquad t > 0 \text{ sufficiently small,}$$

where the contour E_t is given by

$$E_t = \{w \in \mathbb{C} \mid w = \cos x \cosh t + i \sin x \sinh t,\ 0 \leq x < 2\pi\}$$

and E_t is run in clockwise fashion.
The distributional limit $x \mapsto \lim_{\varepsilon \downarrow 0} [H_F(x+i\varepsilon) - H(x-i\varepsilon)]$, $x \in [-1,1]$, is the hyperfunction corresponding to F.

Let $\alpha,\beta > -1$. From [Sz] we derive that each function H_F, $F \in \mathcal{T}_{X_{0,0},(A_{0,0})^{\frac{1}{2}}}$ can be expanded in a series with respect to the function $\tilde{Q}_n^{(\alpha,\beta)}$. These functions are related to the Jacobi functions of the second kind in the following way:

$$\tilde{Q}_n^{(\alpha,\beta)}(w) = \left[\frac{2n+\alpha+\beta+1}{2^{\alpha+\beta+1}} \frac{\Gamma(n+1)\Gamma(n+\alpha+\beta+1)}{\Gamma(n+\alpha+1)\Gamma(n+\beta+1)}\right]^{\frac{1}{2}} (1-w)^{\alpha}(1+w)^{\beta}\, Q_n^{(\alpha,\beta)}(w) .$$

They are analytic outside the interval $[-1,1]$.
For each $F \in \mathcal{T}_{X_{0,0},(A_{0,0})^{\frac{1}{2}}}$ we have

$$H_F(w) = \sum_{n=0}^{\infty} \langle \overline{R_n^{(\alpha,\beta)}}, F\rangle\, \tilde{Q}_n^{(\alpha,\beta)}(w) , \qquad w \notin [-1,1] .$$

The above series converges uniformly outside each open neighbourhood of the interval $[-1,1]$.
Although the proof of some results can be taken from Szegö's book [Sz], our approach is different because it is based on purely functional analytic considerations.

II.4. Analyticity spaces related to unitary representations of Lie groups

In this section we devote some attention to the notion of analytic vector for unitary representations of Lie groups and its relation to analyticity spaces.
In [Ne 1], Nelson introduced the notion of analytic vector for a self-adjoint operator P in a separable Hilbert space X. We recall that a vector $w \in \mathcal{D}^{\infty}(P)$ is analytic for P if

$$\exists_{a>0}\, \exists_{b>0}\, \forall_{n\in\mathbb{N}}: \|P^n w\| < ab^n\, n! .$$

The analyticity domain, i.e. the set of analytic vectors, for P is denoted by $\mathcal{D}^{\omega}(P)$. It can easily be checked that $C^{\omega}(P) = S_{X,|P|}$ with $|P| = (P^2)^{\frac{1}{2}}$.
The notion of analytic vector has been introduced also for unitary representations of Lie groups. Here we present a short discussion.
Let G be a finite dimensional Lie group. A unitary representation U of G is a homomorphism

$$g \mapsto U(g) , \qquad g \in G$$

from G into the group of unitary operators on a Hilbert space X. A vector $w \in X$ is called an analytic vector for the representation U if the mapping

$$g \mapsto U(g)w$$

is analytic on G. Cf. [Wa], p. 276. We denote the set of analytic vectors for U by $\mathcal{D}^{\omega}(U)$. The set $\mathcal{D}^{\omega}(U)$ can be considered as an analyticity space in the following way. Let $A(G)$ denote the Lie algebra corresponding to the Lie group G and let $\{p_1,\dots,p_d\}$ be a basis for $A(G)$. Then for every $p \in A(G)$, the operator valued function

$$s \mapsto U(\exp(sp))$$

is a one-parameter group of unitary operators on X. By Stone's theorem its infinitesimal generator denoted by $\partial U(p)$, is skew-adjoint. Thus the infinitesimal representation ∂U of the Lie algebra $A(G)$ links a skew-adjoint operator $\partial U(p)$ to any element $p \in A(G)$. Now put

$$\Delta = I - \sum_{k=1}^{d} (\partial U(p_k))^2 .$$

Then it has been proved in Nelson's paper [Ne 1], that the operator Δ is positive and esssentially self-adjoint on $\mathcal{D}^{\omega}(U)$. We have the following results.

(II.4.1) <u>Theorem</u>

The set of analytic vectors $\mathcal{D}^{\omega}(U)$ equals the space $S_{X,\Delta^{\frac{1}{2}}}$.

<u>Proof</u>. Cf. [Ne 1] or [Go 1]. □

The following result says something about the action of G and $A(G)$ on the space $S_{X,\Delta^{\frac{1}{2}}}$.

(II.4.2) <u>Theorem</u>

(i) The unitary operators $U(g)$, $g \in G$, map $S_{X,\Delta^{\frac{1}{2}}}$ continuously into $S_{X,\Delta^{\frac{1}{2}}}$; they are extendible.

(ii) The skew-adjoint operators $\partial U(p)$, $p \in A(G)$, map $S_{X,\Delta^{\frac{1}{2}}}$ continuously into $S_{X,\Delta^{\frac{1}{2}}}$; they are extendible.

Proof

(i) Let $g \in \mathbf{G}$. Then $U(g)$ maps $S_{X,\Delta^{\frac{1}{2}}}$ onto itself. Since $U(g)$ is unitary also $U(g)^*$ maps $S_{X,\Delta^{\frac{1}{2}}}$ onto itself. By Corollary (I.4.4), the mapping $U(g)$ is continuous and by Corollary (I.4.11) it is extendible.

(ii) Let $p \in \mathbf{A}(\mathbf{G})$. Then the function $t \mapsto U(\exp(tp))w$ is analytic in a neighbourhood of zero for any $w \in S_{X,\Delta^{\frac{1}{2}}}$. Hence by taking its derivative, it follows that $\partial U(p)w \in S_{X,\Delta^{\frac{1}{2}}}$ for any $w \in S_{X,\Delta^{\frac{1}{2}}}$. Since $\partial U(p)$ is skew-adjoint, it maps $S_{X,\Delta^{\frac{1}{2}}}$ continuously into $S_{X,\Delta^{\frac{1}{2}}}$. It also follows that $\partial U(p)$ is extendible. □

It is an interesting problem under which conditions $S_{X,\Delta^{\frac{1}{2}}}$ is a nuclear space. In Warner's book [Wa] there can be found some indications which can help to resolve this problem. From [Na] we obtain the following cases.

(II.4.3) Theorem

Let U be an irreducible unitary representation of a Lie group $\mathbf{G}$. Then $S_{X,\Delta^{\frac{1}{2}}}$ is nuclear in each of the following cases:

(i) $\mathbf{G}$ is semi-simple with finite center.

(ii) $\mathbf{G}$ is the semi-direct product $A \otimes K$ where A is an abelian invariant subgroup and K is a compact subgroup, e.g. the Euclidean groups.

(iii) $\mathbf{G}$ is nilpotent.

In the remaining part of this section we work out some examples.

Case (II.m) The Heisenberg group

The Heisenberg group $\mathbf{G}_H$ is a simply connected three dimensional Lie group in which the multiplication operation $\circ$ is defined by

$$(a,b,c) \circ (a',b',c') = (a+a',b+b',c+c'+ab') .$$

Its Lie algebra is spanned by p_1, p_2 and p_3 which satisfy the bracket relations $[p_1,p_2] = p_3$, p_3 central.
The irreducible unitary representation U of $\mathbf{G}_H$ on $\mathbf{L}_2(\mathbb{R})$ is defined by

$$U(a,b,c)f : x \mapsto \exp(ibx+ic)f(x+a) .$$

Correspondingly, $\partial U(p_1) = D$, $\partial U(p_2) = iQ$ and $\partial U(p_3) = iI$.

We have $T_a = U(a,0,0) = \exp(aD)$, $R_b = U(0,b,0) = \exp(ibQ)$ and $U(0,0,c) = e^{ic}I$. The operator

$$\Delta_H = I - \sum_{k=1}^{3} (\partial U(p_k))^2$$

equals the self-adjoint operator $-d^2/dx^2 + x^2$. So the set of analytic vectors for the unitary representation U is given by

$$S_{L_2(\mathbb{R}),(-\frac{d^2}{dx^2}+x^2)^{\frac{1}{2}}} = S_1^1 \ .$$

This space has been described in Case (II.j).
An extension of the Heisenberg group can be derived by taking a semi-direct product of the Heisenberg group G_H and the group $SL(2,\mathbb{R})$. It results in the so-called Schrödinger group, the symmetry group of the Schrödinger equation (cf. [Mi]). Since the unitary representation of the Schrödinger group G_S on $L_2(\mathbb{R})$ is rather complicated, we do not present it here. The corresponding infinitesimal representation of the Lie algebra $A(G_S)$ is spanned by the differential operators

$$i \ , \quad ix \ , \quad \frac{d}{dx} \ , \quad ix^2 \ , \quad x\frac{d}{dx} + \frac{d}{dx}x \ , \quad -i\frac{d^2}{dx^2}$$

It can be proved that the corresponding analyticity space $S_{L_2(\mathbb{R}),\Delta_S^{\frac{1}{2}}}$ equals the space $S_{\frac{1}{2}}^{\frac{1}{2}}$, cf. Case (II.j).

Another extension of G_H can be found in [Go 2]. Here we give the infinitesimal representation of the Lie algebra belonging to this extension. It is spanned by the differential operators

$$i \ , \quad ix \ , \quad -\frac{d}{dx} \ , \quad -i\frac{d^2}{dx^2} \ , \quad \ldots \ , \quad i(i\frac{d}{dx})^n \ .$$

The corresponding analyticity domain is given by

$$S_{L_2(\mathbb{R}),((-\frac{d^2}{dx^2})^k+x^2)^{\frac{1}{2}}} = S_1^{1/k} \ .$$

See also Case (II.h).

<u>Case</u> (II.n) The Euclidean group $\mathbf{E}_2$

The Euclidean group $\mathbf{E}_2$ is a three parameter group (Θ,a,b) with $\Theta \in [0,2\pi]$ and $a,b \in \mathbb{R}$. The group multiplication is defined by

$$(\Theta,a,b) \circ (\Theta',a',b') =$$

$$= (\Theta + \Theta', a \cos\Theta' + b \sin\Theta' + a', - a \sin\Theta' + b \cos\Theta' + b') .$$

A well-known realization of $\mathbf{E}_2$ is the group of 3×3-matrices

$$(\Theta,a,b) \mapsto \begin{pmatrix} \cos\Theta & -\sin\Theta & 0 \\ \sin\Theta & \cos\Theta & 0 \\ a & b & 1 \end{pmatrix} .$$

$\mathbf{E}_2$ acts as a transformation group in the plane (rotations and translations). The point (x,y) is mapped into $(a + x\cos\Theta + y \sin\Theta, b - x \sin\Theta + y\cos\Theta)$. The Lie algebra $\mathbf{A}(\mathbf{E}_2)$ is spanned by p_1, p_2 and m, which satisfy the following bracket relations:

$$[p_1,p_2] = 0 , \quad [m,p_1] = p_2 , \quad [m,p_2] = - p_1 .$$

Let S^1 denote the unit sphere in $\mathbb{R}^2$. Then an irreducible unitary representation U of $\mathbf{E}_2$ on $\mathbf{L}_2(S^1)$ is given by

$$U(\Theta,r,\alpha)f : \varphi \mapsto e^{ir\cos(\varphi-\alpha)} f(\varphi-\Theta) .$$

Here we have used polar coordinates (r,α) instead of cartesian coordinates (a,b). The operators $U(\Theta,r,\alpha)$ are unitary. For the corresponding infinitesimal representation we have

$$\partial U(p_1) = i\cos\varphi , \quad \partial U(p_2) = i \sin\varphi , \quad \partial U(p_3) = \frac{d}{d\varphi} .$$

So the operator $\Delta = 2 - d^2/d\varphi^2$, and the set of analytic vectors is

$$S_{\mathbf{L}_2(S^1),(2-\frac{d^2}{d\varphi^2})^{\frac{1}{2}}} = S_{\mathbf{L}_2(S^1),(-\frac{d^2}{d\varphi^2})^{\frac{1}{2}}} .$$

This space has been described in Case (II.d).

<u>Case</u> (II.o) The rotation group $\mathbf{O}_3$

The rotation group $\mathbf{O}_3$ is the group of all real 3×3-matrices with the property that $RR^T = I$ and $\det R = 1$. Here R^T denotes the 3×3 identity matrix. O_3 is a real 3 parameter compact Lie group. The real Lie algebra $A(O_3)$ is spanned by j_1, j_2, j_3 with bracket relations

$$[j_1,j_2] = j_3 \ , \quad [j_3,j_1] = j_2 \ , \quad [j_2,j_3] = j_1 \ .$$

Let S^2 denote the unit sphere in $\mathbb{R}^3$. Then a reducible unitary representation of O_3 on $L_2(S^2)$ is obtained as follows:

$$U(R)f : \underline{\xi} \to f(\underline{\xi}R) \ , \qquad \underline{\xi} \in S^2 \ .$$

If we express the elements of S^2 by means of spherical coordinates (Θ,φ), $0 \leqq \Theta < \pi$, $0 \leqq \varphi < 2\pi$, then the skew-adjoint operator $\partial U(j_k)$, $k = 1,2,3$, are of the form

$$\partial U(j_1) = \sin\varphi \frac{\partial}{\partial\Theta} + \cos\varphi \cot\Theta \frac{\partial}{\partial\varphi}$$

$$\partial U(j_2) = -\cos\varphi \frac{\partial}{\partial\Theta} + \sin\varphi \cot\Theta \frac{\partial}{\partial\varphi}$$

$$\partial U(j_3) = -\frac{\partial}{\partial\varphi} \ .$$

The analyticity domain of this unitary representation is given by

$$S_{L_2(S^2),\Delta^{\frac{1}{2}}}$$

where Δ is just the Laplace-Beltrami operator for S^2. In spherical coordinates we have

$$\Delta = -\frac{1}{\sin\Theta}\frac{\partial}{\partial\Theta}\sin\Theta\frac{\partial}{\partial\Theta} - \frac{1}{\sin^2\Theta}\frac{\partial^2}{\partial\varphi^2} \ .$$

It is well-known that the spherical harmonics $Y_\ell^{(k)}$,

$$Y_\ell^{(k)}(\Theta,\varphi) = e^{ik\varphi} P_\ell^{(k)}(\cos\Theta) \ , \qquad \ell \in \mathbb{N} \cup \{0\} \ , \quad k = -\ell,-\ell+1,\ldots,\ell$$

are the eigenfunctions of the operator Δ with

$$\Delta Y_\ell^{(k)} = \ell(\ell+1)Y_\ell^{(k)} \ .$$

It follows that $S_{L_2(S^2),\Delta^{\frac{1}{2}}}$ is nuclear. This space is also one of the spaces presented in Case (II.f).

CHAPTER III

Compound spaces, tensor products and kernel theorems

Introduction

In Chapter I we have introduced the space $T_{X,A}$ as the space which consists of all mappings (trajectories) F from the open interval $(0,\infty)$ into X with the property that $F(t+\tau) = e^{-\tau A} F(t)$, $t,\tau > 0$. The spaces $S_{X,A}$ and X are embedded in $T_{X,A}$. So $\mathrm{emb}(S_{X,A})$ and $\mathrm{emb}(X)$ are spaces of trajectories. The trajectories in $\mathrm{emb}(X)$ are precisely the trajectories in $T_{X,A}$ for which $\lim_{t\downarrow 0} F(t)$ exists in X-sense. The trajectories in $\mathrm{emb}(S_{X,A})$ are precisely the trajectories of $T_{X,A}$ which have an extension to some interval $[-\varepsilon,\infty)$, $\varepsilon > 0$, with

$$\mathrm{emb}(w)(t) \in X \ , \qquad t \geqq -\varepsilon \ ,$$

$$\mathrm{emb}(w)(t+\tau) = e^{-\tau A}\, \mathrm{emb}(w)(t) \ , \qquad \tau \geqq -\varepsilon \ , \quad t \geqq 0 \ .$$

In this chapter we consider a pair of nonnegative self-adjoint operators C and D in a separabele Hilbert space Z, which commute strongly, i.e. their spectral projections commute; we discuss trajectories of two variables.

The vector space $TT_{Z;C,D}$ consists of all mappings F from $(0,\infty) \times (0,\infty)$ into Z for which

$$F(t+\tau,s+\sigma) = e^{-\tau C} e^{-\sigma D} F(t,s) \ .$$

We note that for fixed $t > 0$ the mapping $s \to F(t,s)$ is a trajectory in $T_{Z,D}$ and similarly for fixed $s > 0$ the mapping $t \to F(t,s)$ belongs to $T_{Z,C}$. The vector space $SS_{Z;C,D}$ can be considered as the subspace of all elements W in $TT_{Z;C,D}$ which have an extension to the region $[-\varepsilon_1,\infty) \times [-\varepsilon_2,\infty)$ for certain $\varepsilon_1,\varepsilon_2 > 0$ in the sense that

$$W(t,s) \in Z \quad \text{for } t \geqq -\varepsilon_1 \text{ and } s \geqq -\varepsilon_2 \ ,$$

$$W(t+\tau,s+\sigma) = e^{-\tau C} e^{-\sigma D} W(t,s) \quad \text{for } \tau \geqq -\varepsilon_1,\ \sigma \geqq -\varepsilon_2 \text{ and } t,s \geqq 0 \ .$$

We note that for each fixed $t \geqq -\varepsilon_1$ the trajectory $s \mapsto W(t,s)$ belongs to $\mathrm{emb}(S_{Z,D})$, and for each fixed $s \geqq -\varepsilon_2$ the trajectory $t \mapsto W(t,s)$ belongs to $\mathrm{emb}(S_{Z,C})$.

Besides the spaces $SS_{Z;C,D}$ and $TT_{Z;C,D}$ we shall introduce the so called 'compound' spaces $ST_{Z;C,D}$ and $TS_{Z;C,D}$ as subspaces of $TT_{Z;C,D}$.
In Section (III.1) we give the precise definition of all four spaces $SS_{Z;C,D}$, $ST_{Z;C,D}$, $TS_{Z;C,D}$ and $TT_{Z;C,D}$. We describe their relation with the ordinary analyticity and trajectory spaces.
In sections (III.2) and (III.3) we discuss the features of the topological structure of $ST_{Z;C,D}$ and $TS_{Z;C,D}$. We give a necessary and sufficient condition on the pair (C,D) such that the space $ST_{Z;C,D}$ is complete. In many cases the space $ST_{Z;C,D}$ is not complete, thus leading to examples of non-strict incomplete inductive limits of Frechet (Hilbert) spaces.
In Section III.4 we discuss the duality between $ST_{Z;C,D}$ and $TS_{Z;C,D}$.
In Section III.5 we consider sums and intersections of all introduced spaces. They are again of one of the mentioned types. Moreover, we introduce an inclusion diagram in which all spaces occur. For suitably chosen C, D and Z the compound spaces $TS_{Z;C,D}$ and $ST_{Z;C,D}$ describe topological tensor products of analyticity spaces and trajectory spaces; the spaces $SS_{Z;C,D}$ and $TT_{Z;C,D}$ describe topological tensor products of two analyticity spaces and of two trajectory spaces, respectively. The description of these tensor products and the resulting kernel theorems are given in the final section.

III.1. Compound spaces

Let C and D denote two strongly commuting nonnegative self-adjoint operators in a separable Hilbert space Z. We take them fixed throughout the Sections III.1 - III.5. The spectral resolutions of the identity belonging to C and D are denoted by $(G_\lambda)_{\lambda\in\mathbb{R}}$ and $(H_\mu)_{\mu\in\mathbb{R}}$, respectively. So for every pair of Borel sets Δ_1, Δ_2 in $\mathbb{R}$

$$G(\Delta_1)H(\Delta_2) = H(\Delta_2)G(\Delta_1) .$$

The mapping $\Delta_1 \times \Delta_2 \to G(\Delta_1)H(\Delta_2)$ can be extended to an operator valued measure on the Borel subsets of $\mathbb{R}^2$. So for any everywhere finite Borel function θ on $\mathbb{R}^2$ the operator

$$\theta(C,D) = \iint_{\mathbb{R}^2} \theta(\lambda,\mu)\, d\, G_\lambda\, H_\mu$$

is a well-defined normal operator on Z with domain

$$\mathcal{D}(\theta(C,D)) = \{x \in Z \mid \iint_{\mathbb{R}^2} |\theta(\lambda,\mu)|^2 \, d(G_\lambda H_\mu x,x) < \infty\} .$$

We introduce the following convention. By the expression

$$\int_a^b \int_c^d \theta(\lambda,\mu) \, d\, G_\lambda H_\mu$$

we mean

$$\iint_{\mathbb{R}^2} \chi_{a,b}(\lambda)\chi_{c,d}(\mu) \, d\, G_\lambda H_\mu$$

with

$$\chi_{a,b}(\lambda) = \begin{cases} 1 & \text{on } [a,b) \\ 0 & \text{elsewhere,} \end{cases}$$

and similarly $\chi_{c,d}$.

(III.1.1) Definition

The space $SS_{Z;C,D}$ is defined by

$$SS_{Z;C,D} = \bigcup_{s>0} \bigcup_{t>0} e^{-sD} e^{-tC}(Z) .$$

The topology on $SS_{Z;C,D}$ is the locally convex inductive limit topology generated by the Hilbert spaces $e^{-sD} e^{-tC}(Z)$, $s,t > 0$. Here the inner product in $e^{-sD} e^{-tC}(Z)$ is given by $(\cdot,\cdot)_{s,t} = (e^{sD} e^{tC} \cdot, e^{sD} e^{tC} \cdot)_Z$. We observe that

$$SS_{Z;C,D} = \bigcup_{t>0} e^{-tD} e^{-tC}(Z) = \bigcup_{t>0} e^{-t(D+C)}(Z) .$$

Moreover, each Hilbert space $e^{-sD} e^{-tC}(Z)$ is continuously embedded in $e^{-\tau(C+D)}(Z)$, where $\tau = \min\{t,s\}$. So the locally convex topological vector $SS_{Z;C,D}$ is equal to the analyticity space $S_{Z,C+D}$.

(III.1.2) Definition

The space $TT_{Z;C,D}$ consists of all mappings F from $(0,\infty) \times (0,\infty)$ into Z with the property

$$F(t+\tau,s+\sigma) = e^{-\tau C}\, e^{-\sigma D}\, F(t,s)\ , \qquad t,\tau,s,\sigma > 0\ .$$

We topologize the space $TT_{Z;C,D}$ by means of the seminorms

$$q_{t,s}(F) = \|F(t,s)\|\ , \qquad t,s > 0\ .$$

The space $TT_{Z;C,D}$ is in one-to-one correspondence with the trajectory space $T_{Z,C+D}$. To show this, let $F \in TT_{Z;C,D}$. Define the reduction red: $TT_{Z;C,D} \to T_{Z,C+D}$

$$\mathrm{red}(F): t \mapsto F(t,t)\ , \qquad t > 0\ .$$

The mapping red is a linear homeomorphism. Its inverse red^{-1} on $T_{Z;C+D}$ is given by

$$\mathrm{red}^{-1}(H): (t,s) \mapsto \begin{cases} e^{-(t-s)C}\, H(s)\ , & t \geq s > 0 \\ e^{-(s-t)D}\, H(t)\ , & s \geq t > 0\ . \end{cases}$$

We identified $SS_{Z;C,D}$ and $S_{Z,C+D}$. Similarly, if no confusion is likely to arise, we identify the spaces $TT_{Z;C,D}$ and $T_{Z,C+D}$ where the identification is established by the mapping red.

We now arrive at the definition of genuinely compound spaces $ST_{Z;C,D}$ and $TS_{Z;C,D}$.

(III.1.3) Definition

Let $r > 0$. By $e^{-rD}(T_{Z,C})$ we denote the subspace of $TT_{Z;C,D}$ which consists of all mappings F from $(0,\infty) \times [-r,\infty) \to Z$ with the property $F(t+\tau,s+\sigma) = e^{-\tau C}\, e^{-\sigma D}\, F(t,s)$ for all $t,\tau > 0$ and $s > -r$, $s+\sigma > -r$.

We endow the space $e^{-rD}(T_{Z,C})$ with the topology generated by the seminorms

$$q_t^{(r)}(F) = \|F(t,-r)\|_Z\ , \qquad t > 0\ .$$

(III.1.4) Lemma

The space $e^{-rD}(T_{Z,C})$ is linearly homeomorphic to the trajectory space $T_{e^{-rD}(Z),C}$.

Proof. Define the natural linear mapping h on $e^{-rD}(T_{Z,C})$ by

$$h(F): t \mapsto F(t,-r) \ , \quad F \in e^{-rD}(T_{Z,C}) \ .$$

Then h maps $e^{-rD}(T_{Z,C})$ bijectively onto $T_{e^{-rD}(Z),C}$. Further,

$$\|h(F)(t)\|_Z = \|F(t,-r)\|_Z = q_t^{(r)}(F) \ , \quad t > 0 \ .$$

So h is a homeomorphism. □

It follows that $e^{-rD}(T_{Z,C})$ is a Frechet space.

(III.1.5) Definition

The space $ST_{Z;C,D}$ is introduced as the union

$$ST_{Z;C,D} = \bigcup_{r>0} e^{-rD}(T_{Z,C})$$

The topology $\sigma\tau_{ind}$ on $ST_{Z;C,D}$ is the inductive limit topology brought about by the Frechet spaces $e^{-rD}(T_{Z,C})$, i.e. the finest locally convex topology on $ST_{Z;C,D}$ for which all injections

$$i_r: e^{-rD}(T_{Z,C}) \subset ST_{Z;C,D}$$

are continuous. The spaces of type $ST_{Z;C,D}$ are called analyticity-trajectory spaces.

In Section III.2 we shall introduce a set of seminorms on $ST_{Z;C,D}$ which induces the locally convex topology $\sigma\tau_{ind}$.

(II.1.6) Definition

The space $TS_{Z;C,D}$ is the subspace of $TT_{Z;C,D}$ which consists of all mappings F from $[0,\infty) \times (0,\infty)$ into Z with the property that

$$\forall_{t \geq 0} \ \forall_{s>0} : F(t,s) \in S_{Z,C} \ .$$

The topology $\tau\sigma_{proj}$ on $TS_{Z;C,D}$ is generated by the uncountable family of seminorms

$$p_{\psi,s}: F \mapsto \|\psi(C)F(0,s)\| , \qquad F \in TS_{Z;C,D}$$

where $\psi \in \mathbf{B}_+(\mathbb{R})$ and $s > 0$. The spaces of type $TS_{Z;C,D}$ are called trajectory-analyticity spaces.

Let $F \in TS_{Z;C,D}$, and let $s > 0$ be fixed. Then the trajectory

$$t \mapsto F(t,s) , \qquad t \geq 0$$

belongs to $\mathrm{emb}(S_{Z,C}) \subset T_{Z,C}$. It means that for $t_s > 0$ sufficiently small, $t \mapsto F(t-t_s,s)$, $t \geq 0$, belongs to $\mathrm{emb}(Z)$. So we have

$$\forall_{s>0} \exists_{t_s>0}: F(-t_s,s) \in Z .$$

Put differently, for each $F \in TS_{Z;C,D}$ the mapping

$$s \mapsto F(0,s)$$

is an element of $T_{Z,D}$ with values in $S_{Z,C}$.

III.2. The analyticity-trajectory space $ST_{Z;C,D}$

The locally convex topological vector space $ST_{Z;C,D}$ has been introduced in Definition (III.1.5). The topology $\sigma\tau_{ind}$ for $ST_{Z;C,D}$ is the inductive limit topology brought about by the Frechet spaces $e^{-rD}(T_{Z,C})$. The topology σ_{ind} for $S_{X,A}$ is generated by an explicit system of seminorms, which arises from the algebra of Borel functions $\mathbf{B}_+(\mathbb{R})$. Similarly, we introduce the algebra of Borel functions $\mathbf{F}_+(\mathbb{R}^2)$, which induce a set of seminorms that describes the topology $\sigma\tau_{ind}$ for $ST_{Z;C,D}$.

(III.2.1) <u>Definition</u>

Let θ be an everywhere finite Borel function on $\mathbb{R}^+ \times \mathbb{R}^+$. Then $\theta \in \mathbf{F}(\mathbb{R}^2)$ if and only if

$$\forall_{s>0} \exists_{t>0}: \sup_{\lambda,\mu \geq 0} (|\theta(\lambda,\mu)|\, e^{\lambda t}\, e^{-\mu s}) < \infty .$$

By $\mathbf{F}_+(\mathbb{R}^2)$ we denote the subset of all positive functions in $\mathbf{F}(\mathbb{R}^2)$ with the property: $\forall_{T>0} \exists_{\delta>0} \forall_{\lambda,\mu \in [0,T]}: \theta(\lambda,\mu) > \delta$.

For each $\theta \in \mathbf{F}(\mathbb{R}^2)$ the operator $\theta(C,D)$ is defined by

$$\theta(C,D) = \iint_{\mathbb{R}^2} \theta(\lambda,\mu)\, dG_\lambda H_\mu \quad .$$

The operators $\theta(C,D)$ extend to linear mappings from $ST_{Z;C,D}$ into Z. This can be seen as follows. Let $h \in ST_{Z;C,D}$. Then there exists $s_0 > 0$ such that for all $t > 0$, $h(t,-s_0) \in Z$. Put

(III.2.2) $\theta(C,D)h = (e^{t_0 C}\, \theta(C,D) e^{-s_0 D})h(t_0,-s_0)$

where $t_0 > 0$ must be chosen so small that

$$\sup_{\lambda,\mu \geq 0} (e^{t_0\lambda}\, e^{-s_0\mu} |\theta(\lambda,\mu)|) < \infty ,$$

i.e., $e^{t_0 C}\, \theta(C,D) e^{-s_0 D}$ is a bounded operator on Z.

The following definition makes sense.

(III.2.3) <u>Definition</u>

For each $\theta \in \mathbf{F}_+(\mathbb{R}^2)$ the seminorm p_θ is defined by

$$p_\theta(h) = \|\theta(C,D)h\|_Z \, , \qquad h \in ST_{Z;C,D}$$

Further, the set $U_{\theta,\varepsilon}$, $\varepsilon > 0$, is defined by

$$U_{\theta,\varepsilon} = \{h \in ST_{Z;C,D} \mid p_\theta(h) < \varepsilon\} \, .$$

The next theorem is the generalization of Theorem (I.1.4) for analyticity spaces to analyticity-trajectory spaces.

(III.2.4) <u>Theorem</u>

(i) For each $\theta \in \mathbf{F}_+(\mathbb{R}^2)$ the seminorm p_θ is continuous with respect to the topology $\sigma\tau_{ind}$ for $ST_{Z;C,D}$.

(ii) Let a convex set $\Omega \subset ST_{Z;C,D}$ have the property that for each $s > 0$ the set $\Omega \cap e^{-sD}(T_{Z,C})$ contains a neighbourhood of 0 in the Frechet space $e^{-sD}(T_{Z,C})$. Then there exist $\theta \in \mathbf{F}_+(\mathbb{R}^2)$ and $\varepsilon > 0$ such that Ω contains a set $U_{\theta,\varepsilon}$.

Hence the topology $\sigma\tau_{ind}$ is generated by the seminorms p_θ.

<u>Proof</u>

(i) We have to show that p_θ is continuous on each space $e^{-sD}(T_{Z,C})$, $s > 0$. So let $s > 0$ and let $h \in e^{-sD}(T_{Z,C})$. Then we have

$$p_\theta(h) = \|\theta(C,D)h\| \leq \|e^{tC}\,\theta(C,D)e^{-sD}\|\,\|h(t,-s)\| \ .$$

Here $t > 0$ has to be chosen so small that $e^{tC}\,\theta(C,D)e^{-sD}$ is a bounded operator on Z. The result follows from the definition of $\sigma\tau_{ind}$ in Definition (III.1.3).

(ii) Introduce the orthogonal projections P_{nm}, $n,m \in \mathbb{N}$,

$$P_{nm} = \int_{n-1}^{n}\int_{m-1}^{m} d\,G_\lambda\,H_\mu \ .$$

Since the function $(\lambda,\mu) \mapsto \chi_{n-1,n}(\lambda)\chi_{m-1,m}(\mu)$ belongs to $\mathbf{F}(\mathbb{R}^2)$ we have $P_{nm}(ST_{Z;C,D}) \subset P_{nm}(Z)$.
The set $P_{nm}(\Omega)$ contains an open neighbourhood of 0 in $P_{nm}(Z)$. Define the numbers r_{nm}, $n,m \in \mathbb{N}$, by

$$r_{nm} = \sup\,\{\rho \mid (h \in P_{nm}(Z) \wedge \|P_{nm}h\| < \rho) \Rightarrow h \in \Omega\}$$

and the function θ by

$$\theta(\lambda,\mu) = \frac{n^2\,m^2}{r_{nm}}\ , \qquad \lambda \in [n-1,n)\ , \quad \mu \in [m-1,m)\ .$$

We prove that $\theta \in \mathbf{F}(\mathbb{R}^2)$.
Let $s > 0$. The set $\Omega \cap e^{-\frac{1}{2}sD}(T_{Z,C})$ contains an open neighbourhood of 0 in $e^{-\frac{1}{2}sD}(T_{Z,C})$. It means that there are $t > 0$ and $\varepsilon > 0$ such that

$$\{h \in e^{-\frac{1}{2}sD}(T_{Z,C}) \mid \|h(t,-\tfrac{1}{2}s)\| < \varepsilon\} \subset \Omega \cap e^{-\frac{1}{2}sD}(T_{Z,C}) \ .$$

So we derive

$$r_{nm} > e^{(n-1)t}\,e^{-\frac{1}{2}ms}\ , \qquad n,m \in \mathbb{N} \ .$$

Take $\lambda \in [n-1,n)$ and $\mu \in [m-1,m)$. Then we have

$$\theta(\lambda,\mu)e^{\frac{1}{2}\lambda t}\ e^{-\mu s} \leq \frac{n^2\ m^2}{r_{nm}}\ e^{\frac{1}{2}nt}\ e^{-(m-1)s}$$

$$\leq \frac{n^2\ m^2}{\varepsilon}\ e^{-\frac{1}{2}(n-1)t}\ e^{-\frac{1}{2}(m-1)s}\ e^{\frac{1}{2}(t+s)}\ .$$

So

$$\sup_{\lambda,\mu\geqq 0}\ (e^{\frac{1}{2}\lambda t}\ e^{-\mu s}\ \theta(\lambda,\mu)) < \infty\ .$$

Next we claim that $U_{\theta,1} \subset \Omega$, i.e.

(*) $\quad \|\theta(C,D)h\| < 1 \Rightarrow h \in \Omega\ .$

Suppose $h \in e^{-sD}(T_{Z,C})$ for some $s > 0$. Then for alle $t > 0$

$$\sum_{n,m\in\mathbb{N}} \|P_{nm}(h(t,-s))\|^2 < \infty$$

and for a fixed σ with $0 < \sigma < s$ and for each $\tau > t > 0$

(**) $\quad \|P_{nm}(h(\tau,-\sigma))\| \leqq e^{-(m-1)(s-\sigma)}\ e^{-(n-1)(\tau-t)}\ \|P_{nm}(h(t,-s))\|\ .$

Further, because we assume that $h \in U_{\theta,1}$, we have

$$\|P_{nm}h\| < \frac{r_{nm}}{n^2\ m^2}\ .$$

Hence $n^2 m^2 P_{nm} h \in \Omega$ for every $n,m \in \mathbb{N}$.
Now for each $\tau > 0$, $h(\tau,-\sigma)$ can be written as

(***) $\quad h(\tau,-\sigma) = \sum_{n,m}^{N,M} \frac{1}{n^2\ m^2}\ (n^2 m^2 P_{nm}\ h) + \left(\sum_{\substack{(n>N)\vee\\(m>M)}} \frac{1}{n^2\ m^2}\right) h_{NM}(\tau,-\sigma)$

where h_{NM} is defined by

$$h_{NM} = \left(\sum_{\substack{(j>N)\vee\\(i>M)}} i^{-2}\ j^{-2}\right)^{-1}\left(\sum_{\substack{(n>N)\vee\\(m>M)}} P_{nm}\ h\right)\ .$$

With (**) we calculate

$$\|h_{NM}(\tau,-\sigma)\|^2 \leqq \left(N^4 \sum_{n=N+1}^{\infty}\sum_{m=0}^{\infty} + M^4 \sum_{m=M+1}^{\infty}\sum_{n=0}^{\infty}\right) \|P_{nm}(h(\tau,-\sigma))\|^2$$

$$\leqq (N^4\ e^{-2N(\tau-t)} + M^4\ e^{-2M(s-\sigma)})\,\|h(t,-s)\|^2\ .$$

Because both $t > 0$ and $\tau > t$ are taken arbitrarily, $h_{NM} \to 0$ in $e^{-\sigma D}(\mathcal{T}_{Z,C})$. It follows that $h_{NM} \in \Omega \cap e^{-\sigma D}(\mathcal{T}_{Z,C})$ for sufficiently large N and M. In (***), h is expressed as a subconvex combination of elements in the convex set Ω. Therefore, h belongs to Ω. □

In the next lemma we characterize the elements of the analyticity-trajectory space $\mathcal{ST}_{Z;C,D}$.

(III.2.5) <u>Lemma</u>

Let $h \in \mathcal{TT}_{Z;C,D}$. Then $h \in \mathcal{ST}_{Z;C,D}$ iff there are $\psi \in \mathbf{B}_+(\mathbb{R})$, $x \in Z$ and $s_0 > 0$ such that

$$h(t,s) = e^{-(s+s_0)D}\,\psi(C)e^{-tC}\,x\,, \qquad t > 0\,,\quad s \geqq -s_0\,.$$

<u>Proof</u>. The proof follows from the characterization of the elements of $\mathcal{T}_{Z,C}$ given in Proposition (I.2.5). □

We observe that $\mathcal{ST}_{Z;C,D}$ can be introduced as the union

$$\mathcal{ST}_{Z;C,D} = \bigcup_{\psi\in\mathbf{B}_+(\mathbb{R})}\ \bigcup_{s>0}\ \psi(C)e^{-sD}\cdot Z$$

where $\psi(C)e^{-sD}\cdot Z$ becomes a Hilbert subspace of $\mathcal{TT}_{Z;C,D}$ in the usual manner.

Similar to Theorem (I.1.14) it can be proved that $\mathcal{ST}_{Z;C,D}$ is barreled and bornological.

In order to be able to proceed along similar lines as in Section (I.1) we need a result for the function algebra $\mathbf{F}_+(\mathbb{R}^2)$ which is similar to the result of Lemma (I.1.5) for $\mathbf{B}_+(\mathbb{R})$. More precisely, we should be able to prove the following

(III.2.6) <u>Conjecture</u>

Let η be an everywhere finite Borel function on $\mathbb{R}^2$ with the property that for all $\theta \in \mathbf{F}_+(\mathbb{R}^2)$

$$\sup_{\lambda\geqq 0,\mu\geqq 0} |\eta(\lambda,\mu)\theta(\lambda,\mu)| < \infty\,.$$

Then η satisfies

$$\exists_{s>0}\ \forall_{t>0}:\ \sup_{\lambda\geqq 0,\mu\geqq 0} (|\eta(\lambda,\mu)|\ e^{\mu s}\ e^{-\lambda t}) < \infty$$

(or similarly

$$\exists_{s>0}\ \exists_{\psi\in \mathbf{B}_+(\mathbb{R})}:\ |\eta(\lambda,\mu)| \leqq \psi(\lambda)e^{-\mu s}\ ).$$

However, the conjecture (III.2.6) is false. A counterexample has been constructed by De Bruijn (Cf. [B 1]). We discuss it here in full detail.

(III.2.7) Lemma

There exists a bounded Borel function on $\mathbb{R}^2$ with the property that for all $\theta \in \mathbf{F}_+(\mathbb{R}^2)$

(i) $$\sup_{\lambda\geqq 0,\mu\geqq 0} (|\eta(\lambda,\mu)|\theta(\lambda,\mu)) < \infty$$

and

(ii) $$\forall_{s>0}\ \exists_{t>0}:\ \sup_{\lambda\geqq 0,\mu\geqq 0} |\eta(\lambda,\mu)|\ e^{\mu s}\ e^{-\lambda t} = \infty\ .$$

Proof. Put

$$\eta(\lambda,\mu) = \exp\ (-\ \mu(\frac{\mu}{\lambda+\mu})^2)\ ,\qquad \lambda,\mu \in \mathbb{R}^+\ .$$

Then $0 < \eta \leqq 1$. First we prove assertion (ii).
Let $s > 0$. Choose a so large that $(\frac{1}{1+a})^2 < \frac{1}{2}s$. Next, let $b \geqq a$ and take $t = \frac{1}{8b}s$. Then for all $\lambda,\mu \in \mathbb{R}^+$ with $b \geqq \lambda/\mu \geqq a$ we get

$$\eta(\lambda,\mu)e^{\mu s}\ e^{-\lambda t} = \exp \mu\left[-\frac{1}{(\frac{\lambda}{\mu}+1)^2} + s - \frac{2\lambda}{\mu}\,t\right] \exp(\lambda t)$$

$$\geqq \exp(\mu[-\ (\frac{1}{a+1})^2 + s - \tfrac{1}{4}s])\ \exp(\lambda t)$$

$$\geqq \exp(\tfrac{1}{4}\mu s)\ \exp(\lambda t)\ .$$

It follows that the function

$$(\lambda,\mu) \mapsto \eta(\lambda,\mu)e^{\mu s}\ e^{-\lambda t}$$

is unbounded on $\mathbb{R}^+ \times \mathbb{R}^+$.

Next we prove assertion (i).

Let $\theta \in \mathbf{F}_+(\mathbb{R}^2)$. For $s = 1$ there exists $t_1 > 0$ such that

$$\sup_{\lambda\geq 0,\mu\geq 0} (\theta(\lambda,\mu)e^{-\mu+\lambda t_1}) < \infty .$$

Put $p = \frac{1}{2}\frac{t_1}{1+t_1}$, and write $\nu(\lambda,\mu) = \frac{\mu}{\lambda+\mu}$. Then $\frac{\lambda}{\mu} = \frac{1}{\nu(\lambda,\mu)} - 1$.

First we consider $(\lambda,\mu) \in \mathbb{R}^+ \times \mathbb{R}^+$ with $0 < \nu(\lambda,\mu) < p$. It follows that $\frac{1}{2}t_1(\lambda+\mu) < -\mu+t_1\lambda$. Since $0 < \eta \leq 1$, we get

$$(*) \qquad \sup_{\substack{\lambda\geq 0,\mu\geq 0,\\ 0<\nu(\lambda,\mu)<p}} (\eta(\lambda,\mu)\theta(\lambda,\mu)e^{\frac{1}{2}t_1(\lambda+\mu)}) < \infty .$$

Next we investigate the region $\nu(\lambda,\mu) \geq p$.

Take ε with $0 < \varepsilon < p^2$ and put $q = p(p^2-\varepsilon)$. From the definition of $\mathbf{F}_+(\mathbb{R}^2)$ we obtain

$$\sup_{\lambda\geq 0,\mu\geq 0} \theta(\lambda,\mu)e^{-\varepsilon\mu} < \infty .$$

Since $p \leq \nu(\lambda,\mu) \leq 1$ we obtain the inequality

$$\frac{q}{\nu(\lambda,\mu)} - \nu(\lambda,\mu)^2 \leq \frac{q}{p} - \nu(\lambda,\mu)^2 = (p^2-\varepsilon) - \nu(\lambda,\mu)^2 \leq -\varepsilon$$

and hence

$$q(\lambda+\mu) - \mu\left(\frac{\mu}{\lambda+\mu}\right)^2 \leq -\varepsilon\mu .$$

It leads to the estimation

$$(**) \qquad \sup_{\substack{\lambda\geq 0,\mu\geq 0,\\ \nu(\lambda,\mu)\geq p}} e^{q(\mu+\lambda)} \eta(\lambda,\mu)\theta(\lambda,\mu) \leq \sup_{\lambda\geq 0,\mu\geq 0} e^{-\varepsilon\mu}\,\theta(\lambda,\mu) < \infty .$$

From (*) and (**) we derive

$$\sup_{\lambda\geq 0,\mu\geq 0} (\eta(\lambda,\mu)\theta(\lambda,\mu)) < \infty . \qquad \square$$

The preceding lemma has a number of interesting consequences. The space $ST_{Z;C,D}$ is a non-strict inductive limit of Frechet spaces. So it is not surprising that the characterization of bounded subsets and null sequences in

$ST_{Z;C,D}$ is not the same as in the case of strict inductive limits. In fact is is surprising indeed that we derived characterizations of bounded subsets, compact subsets and null sequences in the nonstrict inductive limit $S_{X,A}$ which are so very close to the corresponding characterizations for strict inductive limits. The main role in these characterizations is played by Lemma (I.1.5). However, from the preceding lemma it follows that the corresponding result for $ST_{Z;C,D}$ is false. We shall prove that only in the trivial case that $ST_{Z;C,D} = S_{Z,D}$ or $ST_{Z;C,D} = T_{Z,C}$, or a combination hereof, we get a description which is very close to strict inductive limit results; only, in this case, the space $ST_{Z;C,D}$ is (sequentially) complete. To this end we first give necessary and sufficient conditions on the pair (C,D) such that either $ST_{Z;C,D} = S_{Z,D}$ or $ST_{Z;C,D} = T_{Z,C}$.

(III.2.8) <u>Lemma</u>

$ST_{Z;C,D} \triangleq S_{Z,D}$ ($\triangleq$: "can be identified with") iff there exist constants $\gamma,\kappa > 0$ such that $C \leqq \kappa I + \gamma D$, i.e. for all $s > 0$ there exists $t > 0$ such that the operator $e^{tC}\, e^{-sD}$ is bounded.

<u>Proof</u>

$\Leftarrow$) From the characterization of the elements of $ST_{Z;C,D}$ it follows that for each $h \in ST_{Z;C,D}$ there are $\psi \in B_+(\mathbb{R})$, $x \in Z$ and $s > 0$, such that $w = e^{-sD}\,\psi(C)x$. Because of the assumption $C \leqq \kappa I + \gamma D$, the operator $e^{-\frac{1}{2}sD}\,\psi(C)$ is bounded on Z. We find that

$$h = e^{-\frac{1}{2}sD}(\psi(C)e^{-\frac{1}{2}sD}\,x) \in e^{-\frac{1}{2}sD}(Z)\ .$$

$\Rightarrow$) Let $s > 0$ be fixed. For each $\psi \in B_+(\mathbb{R})$ the self-adjoint operator $\psi(C)e^{-sD}$ maps $T_{Z,C}$ continuously into $S_{Z,D}$. Then Theorem (I.4.7),(ii) yields $\tau > 0$ such that the operator $e^{\tau C}\, e^{-sD}$ is bounded.
Say $e^{\tau C}\, e^{-sD} \leqq K_{s,\tau} I$, where $K_{s,\tau} \geqq 1$. Put

$$P_{nm} = \int_{n-1}^{n}\int_{m-1}^{m} d\,G_\lambda\, H_\mu\ , \qquad n,m \in \mathbb{N}\ .$$

It follows that

$$e^{(n-1)\tau}\, e^{-ms}\, P_{nm} \leqq e^{\tau C}\, e^{-sD}\, P_{nm} \leqq K_{s,\tau}\, P_{nm}\ .$$

So we derive the following relation for all $n,m \in \mathbb{N}$

$$\exp((n-1)\tau - ms - \log K_{s,\tau})P_{nm} \leqq P_{nm} .$$

Hence, $(n-1)\tau - ms - \log K_{s,\tau} \geqq 1$ implies $P_{nm} = 0$.
We obtain constants $a > 0$ and $b > 0$ such that

$$P_{nm} = 0 \quad \text{if} \quad 0 \leqq m < an-b , \qquad n,m \in \mathbb{N} \cup \{0\} .$$

The latter result leads to the relation $C \leqq \kappa I + \gamma D$ for certain constants $\kappa,\gamma > 0$. □

<u>Remark</u>. $\forall_{s>0}\ \exists_{t>0}$: $e^{tC}\ e^{-sD}$ is bounded $\Leftrightarrow$ $\exists_s\ \exists_t$: $e^{tC}\ e^{-sD}$ is bounded.

(III.2.9) <u>Lemma</u>

$ST_{Z;C,D} \triangleq T_{Z,C}$ iff there exists $s > 0$ such that for all $t > 0$ the operator $e^{sD}\ e^{-tC}$ is bounded on Z. Equivalently, iff for all $\varepsilon > 0$ there exists $\eta > 0$ such that $D \leqq \eta I + \varepsilon C$.

<u>Proof</u>. Since each operator $\psi(C)$, $\psi \in B_+(\mathbb{R})$, maps $ST_{Z;C,D}$ into itself, and since $T_{Z,C} = \bigcup_{\psi \in B_+(\mathbb{R})} \psi(C)\cdot Z$, the following equivalence is valid:

$$T_{Z,C} \triangleq ST_{Z;C,D} \Leftrightarrow Z \subset ST_{Z;C,D} .$$

$\Leftarrow$) If the operator $e^{sD}\ e^{-tC}$ is bounded for certain fixed $s > 0$ and for all $t > 0$, then we have

$$Z \subset e^{-sD}(T_{Z,C}) \subset ST_{Z;C,D} .$$

$\Rightarrow$) Since $Z \subset ST_{Z;C,D}$ and since each nonnegative self-adjoint operator $\theta(C,D)$, $\theta \in F_+(\mathbb{R}^2)$, maps $ST_{Z;C,D}$ into Z, it follows that $\theta(C,D)$ is a bounded self-adjoint operator on Z for all $\theta \in F_+(\mathbb{R}^2)$.
Let $\varepsilon > 0$. Define θ_ε by

$$\theta_\varepsilon(\lambda,\mu) = \begin{cases} m-\varepsilon n & \text{if } (\lambda,\mu) \in [n-1,n) \times [m-1,m) \text{ with } m > \varepsilon n \\ e^{-\lambda} & \text{elsewhere.} \end{cases}$$

We show that $\theta_\varepsilon \in F_+(\mathbb{R}^2)$.
So, let $s > 0$. Put $t = \min(1,\varepsilon s)$. From the observation

$$\sup_{\lambda\geqq 0,\mu\geqq 0} (\theta_\varepsilon(\lambda,\mu)e^{-s\mu+\varepsilon s\lambda}) \leqq e^{\varepsilon s} \sup_{\substack{n,m\in\mathbb{N}\\ m>\varepsilon n}} ((m-\varepsilon n)e^{-s(m-\varepsilon n)}) < \infty$$

it follows that $\theta_\varepsilon \in \mathbf{F}_+(\mathbb{R}^2)$. Hence $\theta_\varepsilon(C,D)$ is a bounded nonnegative operator on Z. So there exists $\kappa_\varepsilon > 0$ such that

$$\theta_\varepsilon(C,D) \leqq \kappa_\varepsilon I\,.$$

For $m > \varepsilon n$ it implies

(*) $\qquad (m-\varepsilon n)P_{nm} \leqq \kappa_\varepsilon\, P_{nm}$

with

$$P_{nm} = \int_{n-1}^{n}\int_{m-1}^{m} d\,G_\lambda\, H_\mu\,.$$

From (*) it follows that $P_{nm} = 0$ if $m > \varepsilon n + \kappa_\varepsilon$.

Summarized: $\forall_{\varepsilon>0}\ \exists_{\kappa_\varepsilon>0}$: $P_{nm} = 0$ if $m > \varepsilon n + \kappa_\varepsilon$.

The proof is completed by the following observation:

$$[\forall_{\varepsilon>0}\ \exists_{\kappa_\varepsilon>0}: P_{nm} = 0 \text{ if } m > \varepsilon n + \kappa_\varepsilon] \Leftrightarrow$$

$$\Leftrightarrow [\forall_{\varepsilon>0}\ \exists_{\eta>0}: D \leqq \varepsilon C + \eta I] \Leftrightarrow$$

$$\Leftrightarrow [\forall_{s>0}\ \forall_{t>0}: \text{the operator } e^{sD}\, e^{-tC} \text{ is bounded on } Z] \Leftrightarrow$$

$$\Leftrightarrow [\exists_{s>0}\ \forall_{t>0}: \text{the operator } e^{sD}\, e^{-tC} \text{ is bounded on } Z]\,.$$ □

From Lemmas (III.2.8) and (III.2.9) we obtain the following result.

(III.2.10) <u>Corollary</u>

Let P denote a projection which strongly commutes with C and D. Then

$$ST_{Z;C,D} \triangleq S_{P(Z),PDP} \oplus T_{(I-P)(Z),(I-P)C(I-P)}$$

iff

$$\forall_{s>0}\ \exists_{t>0}: Pe^{-sD}\, e^{tC}\, P \qquad \text{is a bounded operator on } Z$$

and

$$\exists_{s>0}\ \forall_{t>0}: (I-P)e^{sD}\, e^{-tC}(I-P) \quad \text{is a bounded operator on } Z.$$

Proof. From Lemmas (III.2.8) and (III.2.9) we obtain

$$ST_{Z;C,D} \triangleq S_{Z,D} \quad \text{iff} \quad \forall_{s>0} \exists_{t>0}: e^{tC} e^{-sD} \quad \text{is bounded on } Z$$

and

$$ST_{Z;C,D} \triangleq T_{Z,C} \quad \text{iff} \quad \exists_{s>0} \forall_{t>0}: e^{sD} e^{-tC} \quad \text{is bounded on } Z.$$

Further, for any projection P which strongly commutes with C and D

$$P(ST_{Z;C,D}) = ST_{P(Z);PCP,PDP}$$

and

$$(I-P)(ST_{Z;C,D}) = ST_{(I-P)(Z);(I-P)C(I-P),(I-P)D(I-P)} .$$

⇒) Observe that

$$ST_{P(Z);PCP,PDP} = S_{P(Z),PDP}$$

and

$$ST_{(I-P)(Z);(I-P)C(I-P),(I-P)D(I-P)} = T_{(I-P)(Z),(I-P)C(I-P)} .$$

Then the result follows from the previous lemmas.

⇐) Observe that

$$\exp(-tPCP) = P\exp(-tC)P , \quad \exp(-sPDP) = P\exp(-sD)P ,$$

$$\exp(-t(I-P)C(I-P)) = (I-P)\exp(-tC)(I-P) , \text{ etc.}$$

So by Lemmas (III.2.8) and (III.2.9) we have

$$ST_{P(Z);PCP,PDP} = S_{P(Z),PDP}$$

$$ST_{(I-P)(Z);(I-P)C(I-P);(I-P)D(I-P)} = T_{(I-P)(Z);(I-P)C(I-P)} . \qquad \square$$

(III.2.11) Lemma

Suppose there exists no orthogonal projection P which strongly commutes with C and D such that $ST_{Z;C,D}$ can be written as the direct sum

$$ST_{Z;C,D} = S_{P(Z),PDP} \oplus T_{(I-P)(Z),(I-P)C(I-P)} .$$

Let $a > 0$ and $b \geqq 0$. Then there exists $0 < c < a$ with the property that for each $d > 0$, natural numbers n, m can be found such that $cn+d < m < an-b$, and also $P_{nm} = {}_{n-1}\int^{n} {}_{m-1}\int^{m} dG_{\lambda} H_{\mu} \neq 0$.

Proof. Suppose the assertion were not true. It means

(*) $\forall_{0<c<a}\ \exists_{d>0}\ \forall_{n,m\in\mathbb{N}}: (cn+d < m < an-b) \wedge (P_{nm} = 0)$.

Define the projection

$$\Pi_{a,b} = \iint_{V_{a,b}} d\,G_\lambda H_\mu$$

where

$$V_{a,b} = \bigcup_{\substack{n,m\in\mathbb{N},\\ m\geqq an-b}} \{(\lambda,\mu) \in \mathbb{R}^+ \times \mathbb{R}^+ \mid \lambda \in [n-1,n) \wedge \mu \in [m-1,m)\} .$$

Then we have

$$\forall_{s>0}\ \exists_{t>0}: e^{tC}\, e^{-sD}\, \Pi_{a,b} \text{ is bounded on } Z$$

and thus it follows that

(+) $\Pi_{a,b}(ST_{Z;C,D}) = S_{\Pi_{a,b}(Z),\Pi_{a,b}C\Pi_{a,b}}$.

Let $s > 0$ be fixed and let $t > 0$ such that $t/s < a$. Then for $n,m \in \mathbb{N}$ we have

$$e^{sD}\, e^{-tC}(I - \Pi_{a,b})P_{nm} \leqq e^{sm}\, e^{-t(n-1)}(I - \Pi_{a,b})P_{nm} .$$

By assumption (*) there exists $d_{s,t} > 0$ such that $P_{nm} = 0$ if

$$m > \frac{t}{s}n + d_{s,t} \quad \underline{\text{and}} \quad m < an - b .$$

Therefore we derive for all $n,m \in \mathbb{N}$ with $m < an-b$

$$\exp(sm)\exp(-tn)\exp(t)\, P_{nm} \leqq \exp(t+sd_{s,t})\, P_{nm} .$$

Because of the choice of the projection $\Pi_{a,b}$ we get

(**) $\|e^{sD}\, e^{-tC}(I - \Pi_{a,b})\| \leqq \exp(t + sd_{s,t})$.

So we have proved that for all $t > 0$ with $t/s < a$ there exists $d_{s,t} > 0$ such that (**) holds. Hence (**) holds for all $t > 0$ with suitable $d_{s,t}$. By Lemma (III.2.9) and Corollary (III.2.10) we derive

(++) $(I - \Pi_{a,b})ST_{Z;C,D} = T_{(I-\Pi_{a,b})(Z),(I-\Pi_{a,b})C(I-\Pi_{a,b})}$.

Now (+) and (++) yield a contradiction with the assumption that no orthogonal projection P, as indicated in the lemma, exists. □

(III.2.12) Corollary

Suppose there exists no orthogonal projection P which strongly commutes with C and D such that $ST_{Z;C,D}$ can be written as the direct sum

$$ST_{Z;C,D} = S_{P(Z),PDP} \oplus T_{(I-P)(Z),(I-P)C(I-P)} .$$

Put $P_{nm} = \int_{n-1}^{n} \int_{m-1}^{m} d\,G_\lambda H_\mu$, $n,m \in \mathbb{N}$. Then there exist double sequences $(n_{k,\ell})_{k,\ell\in\mathbb{N}}$ and $(m_{k,\ell})_{k,\ell\in\mathbb{N}}$ of natural numbers with the following properties:

(i) $\forall_{k\in\mathbb{N}}\ \exists_{c_k,0<c_k<1/k}\ \forall_{\ell\in\mathbb{N}}:\ c_k < \dfrac{m_{k,\ell}}{n_{k,\ell}} < \dfrac{1}{k}$,

(ii) $\forall_{\ell\in\mathbb{N}}\ \forall_{k\in\mathbb{N}}:\ m_{k,\ell} > \ell$,

(iii) $\forall_{(n,m)\in\{(n_{k,\ell},m_{k,\ell})\,|\,k,\ell\in\mathbb{N}\}}:\ P_{nm} \neq 0$.

Proof. Let $k \in \mathbb{N}$ be fixed. In Lemma (III.2.11) we take $a = 1/k$ and $b = 0$. Then following Lemma (III.2.11) there exists c_k, $0 < c_k < 1/k$, such that for each $d = \ell \in \mathbb{N}$, we can find $n_{k,\ell}, m_{k,\ell} \in \mathbb{N}$ with the following properties:

$$c_k\, n_{k,\ell} + \ell < m_{k,\ell} < \frac{1}{k}\, n_{k,\ell}$$

and

$$P_{n_{k,\ell},m_{k,\ell}} \neq 0 .$$

It follows that

$$c_k < c_k + \frac{\ell}{n_{k,\ell}} < \frac{m_{k,\ell}}{n_{k,\ell}} < \frac{1}{k}$$

and also that

$$m_{k,\ell} > \ell + c_k\, n_{k,\ell} > \ell .$$ □

(III.2.13) Theorem

Suppose there exists no orthogonal projection P which strongly commutes with C and D such that $ST_{Z;C,D}$ can be written as the direct sum

$$ST_{Z;C,D} = S_{P(Z);PDP} \oplus T_{(I-P)(Z),(I-P)C(I-P)} .$$

Then we have

(i) There exists $x \in Z$ such that $x \in \mathcal{D}(\theta(C,D))$ for all $\theta \in F_+(\mathbb{R}^+)$ but <u>not</u> $\text{emb}(x) \in ST_{Z;C,D}$.

(ii) There exists a bounded set in $ST_{Z;C,D}$ for which there exists <u>no</u> $s > 0$ such that the set is contained and bounded in the Frechet space $e^{-sD}(T_{Z,C})$.

(iii) There exists a null sequence in $ST_{Z;C,D}$ for which there exists <u>no</u> $s > 0$ such that the sequence is a null sequence in the Frechet space $e^{-sD}(T_{Z,C})$.

<u>Proof</u>. We use the notations of Corollary (III.2.12).
In each nontrivial subspace $P_{n_{k,\ell},m_{k,\ell}}(Z)$ we choose $y_{k,\ell}$ with $\|y_{k,\ell}\| = 1$.
Let η denote the function of Lemma (III.2.7). Observe that the operator $\eta(C,D)$ is bounded on Z.
Let $s > 0$. Then take $k \in \mathbb{N}$ so large that $(\frac{1}{k})^2 < \frac{1}{2}s$. Put $t_k = \frac{c_k}{8}\, s$. Then we obtain for all $\ell \in \mathbb{N}$

$$\|e^{sD}\, e^{-t_k C}\, \eta(C,D) y_{k,\ell}\| \geq$$

$$\geq \min_{\substack{\lambda\in[n_{k,\ell}-1,n_{k,\ell}),\\ \mu\in[m_{k,\ell}-1,m_{k,\ell})}} (e^{s\mu}\, e^{-t_k\lambda}\, \eta(\lambda,\mu)) =$$

$$= \min_{\substack{\lambda\in[n_{k,\ell}-1,n_{k,\ell}),\\ \mu\in[m_{k,\ell}-1,m_{k,\ell})}} \left(\exp\left(\mu\left[-\left(\frac{1}{\frac{\lambda}{\mu}+1}\right)^2 + s - \frac{\lambda}{\mu}\, t_k\right]\right)\right) \geq$$

$$\geq \exp(m_{k,\ell} - 1)\left[-\left(\frac{n_{k,\ell} - 1}{m_{k,\ell}} + 1\right)^{-2} + s - \frac{n_{k,\ell}}{m_{k,\ell} - 1}\, t_k\right] \geq$$

$$\geq \exp(\ell - 1)\left[-\left(\frac{1}{k}\right)^2 + s - \frac{2}{c_k}\, t_k\right] \geq$$

$$\geq \exp \tfrac{1}{4}(\ell - 1)s .$$

(Observe that $\dfrac{n_{k,\ell}-1}{m_{k,\ell}} > k-1$ and $\dfrac{n_{k,\ell}}{m_{k,\ell}-1} < \dfrac{2}{c_k}$.)

So we achieved the following result:

$$(*)\qquad \forall_{s>0}\ \exists_{k\in\mathbb{N}}\ \exists_{t>0}\ \forall_{\ell\in\mathbb{N}}: \|e^{sD}\, e^{-tC}\, \eta(C,D) y_{k,\ell}\| \geq \exp \tfrac{1}{4}(\ell-1)s\ .$$

(i) Put

$$y = \sum_{k,\ell=1}^{\infty} \frac{1}{k\ell}\, y_{k,\ell}\ ,$$

and put $x = \eta(C,D)y$. Then $x,y \in Z$. All operators $\theta(C,D)\eta(C,D)$ are bounded. Because of the properties of the function η, we get $x \in \mathcal{D}(\theta(C,D))$ for all $\theta \in \mathbf{F}_{+}(\mathbb{R}^{+})$. However, from (*) we derive that for each $s > 0$ there can be found $k \in \mathbb{N}$ and $t > 0$ such that

$$\Big\| \sum_{\ell=L_1}^{L_2} e^{sD}\, e^{-tC}\, \eta(C,D)\, \frac{1}{k\ell}\, y_{k,\ell}\Big\| \geq \frac{1}{kL_2} \exp \tfrac{1}{4}(L_2-1)s$$

for all $L_2 > L_1$, $L_2, L_1 \in \mathbb{N}$.
Thus we find

$$\forall_{s>0}\ \exists_{t>0}: x \notin \mathcal{D}(e^{sD}\, e^{-tC})\ .$$

and hence $x \notin ST_{Z;C,D}$.

(ii) Let B_1 denote the set $\{w \in S_{Z,D} \mid \|w\| = 1\}$. Put $B = \eta(C,D)(B_1)$. Since $\eta(C,D)$ is a bounded operator which commutes with all operators e^{-sD}, $s > 0$, the operator $\eta(C,D)$ maps $S_{Z,D}$ into $S_{Z,D}$. Further, in a natural way the analyticity space $S_{Z,D}$ is embedded in the analyticity trajectory space $ST_{Z;C,D}$. So the set B is contained in $ST_{Z;C,D}$. Since $\theta(C,D)\eta(C,D)$ is a bounded operator for all $\theta \in \mathbf{F}_{+}(\mathbb{R}^2)$, the set B is bounded in $ST_{Z;C,D}$. If B were bounded in $e^{-sD}(T_{Z,C})$ for certain $s > 0$, then we should have

$$\forall_{t>0}\ \exists_{K_t>0}\ \forall_{w\in B}: \|e^{sD}\, e^{-tC}\, w\| \leq K_t\ .$$

Now observe that the double sequence $(y_{k,\ell})_{k,\ell\in\mathbb{N}}$ is contained in B_1. From (*) it follows that for each $s > 0$, we can find $k \in \mathbb{N}$ and $t > 0$ such that the sequence $(\eta(C,D)y_{k,\ell})_{\ell\in\mathbb{N}}$ satisfies

$$\forall_{K_t>0}\ \exists_{\ell\in\mathbb{N}}:\ \|e^{sD}\ e^{-tC}\ \eta(C,D)y_{k,\ell}\| \geq K_t\ .$$

This proves that there exists no $s > 0$ such that the set B is bounded in the Frechet space $e^{-sD}(\mathcal{T}_{Z,C})$.

(iii) Put

$$x_{k,\ell} \doteq \frac{1}{(k+\ell)^2}\,\eta(C,D)y_{k,\ell}\ , \qquad k,\ell \in \mathbb{N}\ .$$

In a standard way the double sequence $(x_{k,\ell})_{k,\ell\in\mathbb{N}}$ can be made into a sequence $(\tilde{x}_j)_{j\in\mathbb{N}}$. It is easy to check that each sequence $(\theta(C,D)x_j)_{j\in\mathbb{N}}$ is a null sequence in the Hilbert space Z. Also, $(\tilde{x}_j)_{j\in\mathbb{N}} \subset S\mathcal{T}_{Z;C,D}$ and hence $\tilde{x}_j \to 0$ in $S\mathcal{T}_{Z;C,D}$. However, for all $s > 0$ there exist $k \in \mathbb{N}$ and $t > 0$ such that the sequence $(e^{sD}\ e^{-tC}\ x_{k,\ell})_{\ell\in\mathbb{N}}$ is unbounded in Z. So there exists no $s > 0$ such that the sequence $(\tilde{x}_j)_{j\in\mathbb{N}}$ is a null sequence in the Frechet space $e^{-sD}(\mathcal{T}_{Z,C})$.
□

(III.2.14) <u>Theorem</u>

Under the assumptions of the previous theorem:

The space $S\mathcal{T}_{Z;C,D}$ is <u>not</u> (sequentially) complete.

<u>Proof</u>. Let x be as in Theorem (III.2.18),(i). Put

$$\Pi_n = \int_0^n \int_0^n d\,G_\lambda\,H_\mu\ , \qquad n \in \mathbb{N}\ .$$

Then for all $n \in \mathbb{N}$ we have $\mathrm{emb}(\Pi_n x) \in S\mathcal{T}_{Z;C,D}$. Further, since

$$\|\theta(C,D)(\Pi_n - \Pi_m)x\| \leq \|\theta(C,D)\eta(C,D)\|\,\|(\Pi_n - \Pi_m)y\|\ ,$$

the sequence $(\mathrm{emb}(\Pi_n x))_{n\in\mathbb{N}}$ is a Cauchy sequence in $S\mathcal{T}_{Z;C,D}$. The limit of $(\mathrm{emb}(\Pi_n x))_{n\in\mathbb{N}}$ is x, i.e.

$$\|\theta(C,D)(\Pi_n x - x)\| \to 0 \quad \text{as } n \to \infty\ .$$

However, $x \notin S\mathcal{T}_{Z;C,D}$. □

<u>Summarized</u>: $S\mathcal{T}_{Z;C,D}$ is complete iff there exists an orthogonal projection P which strongly commutes with C and D such that

$$ST_{Z;C,D} = S_{P(Z),PDP} \oplus T_{(I-P)(Z),(I-P)C(I-P)} .$$

Example. Let C and D be two nonnegative self-adjoint operators with common orthonormal basis of simultaneous eigenvectors $(v_{k,\ell})_{k,\ell\in\mathbb{N}}$. Hence, $Cv_{k,\ell} = \lambda_k v_{k,\ell}$ and $Dv_{k,\ell} = \mu_\ell v_{k,\ell}$. If $\lambda_k \uparrow \infty$ and $\mu_\ell \uparrow \infty$, then the space $ST_{Z;C,D}$ is not complete.

III.3. The trajectory-analyticity space $TS_{Z;C,D}$

From Definition (III.1.6) it follows that the space $TS_{Z;C,D}$ is the subspace of $TT_{Z;C,D}$ which consists of all mappings F for which

$$\forall_{s>0}: \lim_{t\downarrow 0} F(t,s) \in S_{Z,C} .$$

It means that the trajectory $s \mapsto F(0,s) \in T_{Z,D}$ is a mapping from $(0,\infty)$ into $S_{Z,C}$. Conversely, for any trajectory $G \in T_{Z,D}$ with the property that $G(s)$ belongs to $S_{Z,C}$ for all $s > 0$, the mapping $(t,s) \mapsto e^{-tC} G(s)$ belongs to $TS_{Z;C,D}$. Thus the space $TS_{Z;C,D}$ can be identified with the subspace of $T_{Z,D}$ which consists of all trajectories through $S_{Z,C}$.
We recall that the topology $\tau\sigma_{proj}$ for $TS_{Z;C,D}$ is generated by the seminorms

$$p_{\psi,s}(F) = \|\psi(C)F(0,s)\| , \quad F \in TS_{Z;C,D} ,$$

where $s > 0$ and $\psi \in \mathbf{B}_+(\mathbb{R})$.

Each bounded subset of a trajectory space is homeomorphic to a bounded subset of the related Hilbert space. A similar result is valid for trajectory-analyticity spaces.

(III.3.1) Lemma

The subset B of $TS_{Z;C,D}$ is bounded iff for each $s > 0$ there exists $t > 0$ such that the set $\{F(0,s) \mid F \in B\}$ is bounded in the Hilbert space $e^{-tC}(Z)$.

Proof. The following assertions are equivalent:

- B is bounded in $TS_{Z;C,D}$.
- Each seminorm $p_{\psi,s}$ is bounded on B.

- The set $\{F(0,s) \mid F \in B\}$ is bounded in $S_{Z,C}$ for each $s > 0$.

Then by Theorem (I.1.10) the result follows. □

Let $\theta \in F_+(\mathbb{R}^2)$ and let $x \in Z$. Then $\theta(C,D)e^{-sD}$, $s > 0$, is a bounded operator. Moreover, $\theta(C,D)e^{-sD}x$ belongs to $S_{Z,C}$ for all $s > 0$. So the trajectory $(t,s) \mapsto e^{-tC}\ \theta(C,D)e^{-sD}x$, $t \geq 0$, $s > 0$, belongs to $TS_{Z;C,D}$. This element of $TS_{Z;C,D}$ will be denoted by $\theta(C,D)x$.

(III.3.2) <u>Lemma</u>

The set $B \subset TS_{Z;C,D}$ is bounded iff there exists $\theta \in F_+(\mathbb{R}^2)$ and a bounded subset B_0 of Z, such that

$$B = \theta(C,D)B_0 = \{\theta(C,D)x \mid x \in B_0\}\ .$$

<u>Proof</u>

$\Leftarrow$) Let $s > 0$. Then there is $t > 0$ such that the operator $e^{tC}\ \theta(C,D)e^{-sD}$ is bounded on Z. So we obtain

$$\|e^{tC}\ \theta(C,D)e^{-sD}\ x\| \leq \|e^{tC}\ \theta(C,D)e^{-sD}\|\,\|x\|$$

for all $x \in B_0$. Since $B_0 \subset Z$ is bounded, B is a bounded subset of $TS_{Z;C,D}$.

$\Rightarrow$) Let $n,m \in \mathbb{N}$ and put $P_{nm} = {}_{n-1}\int^{n}\ {}_{m-1}\int^{m} d\,G_\lambda\ H_\mu$. For each $G \in TS_{Z;C,D}$ define $P_{nm}\ G := e^{sD}\ P_{nm}\ G(0,s)$. Then $P_{nm}\ G \in Z$. Further, put

$$r_{nm} = \sup\,\{\|P_{nm}\ G\| \mid G \in B\}\ .$$

Let $s > 0$. By assumption, there are $t > 0$ and $K_{s,t} > 0$ such that $\|G(-t,s)\| \leq K_{s,t}$ for all $G \in B$. It yields

$$\|P_{nm}\ G\| = \|e^{sD}\ e^{-tC}\ P_{nm}\ G(-t,s)\| \leq$$

$$\leq e^{ms}\ e^{-(n-1)t}\|G(-t,s)\| \leq e^{ms}\ e^{-(n-1)t}\ K_{s,t}\ .$$

Thus the following result is obtained:

$$(*) \qquad \forall_{s>0}\ \exists_{t>0}\ \exists_{K>0}\ \forall_{n,m\in\mathbb{N}}:\ nm\,r_{nm}\ e^{-ms}\ e^{nt} \leq K\ .$$

Define the function θ on $\mathbb{R}^+ \times \mathbb{R}^+$ by

$$\begin{cases} \theta(\lambda,\mu) = nm\, r_{nm} & \text{if } r_{nm} \neq 0,\ \lambda \in [n-1,n),\ \mu \in [m-1,m) \\ \theta(\lambda,\mu) = e^{-\lambda} & \text{elsewhere.} \end{cases}$$

We show that $\theta \in \mathbf{F}_+(\mathbb{R}^2)$. So let $s > 0$. Then by (*) there exists $0 < t < 1$ and $K > 0$ such that for all $\lambda \in [n-1,n)$ and all $\mu \in [m-1,m)$

$$\theta(\lambda,\mu)e^{t\lambda}\, e^{-s\mu} \leqq nm\, r_{nm}\, e^{nt}\, e^{-(m-1)s} \leqq e^{s} K$$

if $r_{nm} \neq 0$ and if $r_{nm} = 0$

$$\theta(\lambda,\mu)e^{\lambda t}\, e^{-\mu s} \leqq e^{-(n-1)}\, e^{nt} \leqq e^{t} .$$

For each $G \in B$ we define x_G by

$$x_G = \sum_{n,m,r_{nm}\neq 0} \left(\frac{1}{nm\, r_{nm}}\right)(P_{nm}\, G) .$$

Since

$$\sum_{n,m,r_{nm}\neq 0} \left(\frac{1}{nm\, r_{nm}}\, \|P_{nm}\, G\|\right)^2 \leqq \sum_{n,m=1}^{\infty} \frac{1}{n^2\, m^2} ,$$

it follows that $\|x_G\| \leqq \pi^2/6$ for all $G \in B$. Now put

$$B_0 = \{x_G \mid G \in B\} .$$

Then B_0 is bounded in Z, and $B = \theta(C,D)B_0$. □

(III.3.3) <u>Corollary</u>

Let $G \in TS_{Z;C,D}$. Then there exists $\theta \in \mathbf{F}_+(\mathbb{R}^2)$ and $x \in Z$ such that $G = \theta(C,D)x$.

The next step is to prove that B is homeomorphic to a bounded subset B_0 of Z, where an operator $\tilde{\theta}(C,D)$, $\tilde{\theta} \in \mathbf{F}_+(\mathbb{R}^2)$ extablishes the homeomorphism.

(III.3.4) <u>Theorem</u>

Let B be a bounded subset of $TS_{Z;C,D}$. Then there exists $\tilde{\theta} \in \mathbf{F}_+(\mathbb{R}^2)$ and a bounded subset B_0 of Z such that $\tilde{\theta}(C,D)$ is a homeomorphism from B_0 onto B. This means the following:

(i) $B = \tilde{\theta}(C,D)B_0$.

(ii) For any net $(F_\alpha) \subset B$ which converges to $F \in B$ there exists a net $(x_\alpha) \subset B_0$ with limit $x \in B_0$ such that $F_\alpha = \tilde{\theta}(C,D)x_\alpha$ and $F = \tilde{\theta}(C,D)x$.

(iii) For any sequence $(x_n)_{n\in\mathbb{N}} \subset B_0$ with limit $x \in B_0$ the sequence $(\tilde{\theta}(C,D)x_n)_{n\in\mathbb{N}}$ has the $\tau\sigma_{proj}$-limit $\tilde{\theta}(C,D)x$.

<u>Proof</u>. The set $B - B$ is bounded in $TS_{Z;C,D}$. So by the previous lemma there exists a bounded subset B_1 of Z and $\theta \in \mathbf{F}_+(\mathbb{R}^2)$ such that $B - B = \theta(C,D)B_1$. Define $\tilde{\theta} \in \mathbf{F}_+(\mathbb{R}^2)$ by

$$\tilde{\theta}(\lambda,\mu) = (1+\lambda^2)(1+\mu^2)\theta(\lambda,\mu) .$$

Then for all $F \in B$,

$$F = \theta(C,D)y_F = \tilde{\theta}(C,D)x_F$$

where

$$x_F = (I+C^2)^{-1}(I+D^2)^{-1}\, y_F ,$$

and $y_F \in B_2$ with $B_2 = \theta(C,D)^{-1}\, B$. Put

$$B_0 = \{x_F \mid F \in B\} .$$

Then $\tilde{\theta}(C,D)$ maps B_0 bijectively onto B.

(ii) Let $(F_\alpha) \subset B$ be a $\tau\sigma_{proj}$-convergent net with limit $F \in B$. Let $\varepsilon > 0$. Take a fixed constant $K > 0$ so large that $(1+K^2)^{-1} < \varepsilon$. It follows that

$$(*) \qquad \left(\int_0^\infty\int_K^\infty + \int_K^\infty\int_0^\infty\right) d(G_\lambda\, H_\mu(x_{F_\alpha}-x_F),(x_{F_\alpha}-x_F)) \leq$$

$$\leq (1+K^2)^{-2}\,\|y_{F_\alpha} - y_F\|^2 \leq \varepsilon^2\, C_1^2$$

where $C_1 = \sup\{\|y_G\| \mid G \in B_1\}$.

The remaining part of the integral can be estimated as follows:

$$(**) \qquad \int_0^K\int_0^K d(G_\lambda\, H_\mu(x_{F_\alpha}-x_F),x_{F_\alpha}-x_F) \leq$$

$$\leq e^{2Ks} \sup_{\lambda,\mu\in[0,K]} (\widetilde{\theta}(\lambda,\mu)^{-2}) \|F_\alpha(0,s) - F(0,s)\|^2 .$$

It follows that there exists α_0 such that for all $\alpha > \alpha_0$

$$\int_0^K \int_0^K d(G_\lambda H_\mu (x_{F_\alpha} - x_F), x_{F_\alpha} - x_F) < \varepsilon^2 .$$

By (*) and (**) we get for all $\alpha > \alpha_0$

$$\|x_{F_\alpha} - x_F\| < \varepsilon(1+C_1^2)^{\frac{1}{2}} .$$

(iii) Since for all $\psi \in \mathbf{B}_+(\mathbb{R})$, all $s > 0$ and all $\theta \in \mathbf{F}_+(\mathbb{R}^2)$ the operator $\psi(C)\theta(C,D)e^{-sD}$ is bounded, the proof is trivial. □

(III.3.5) <u>Corollary</u>

(i) Let $(H_n)_{n\in\mathbb{N}}$ be a sequence in $TS_{Z;C,D}$. Then the sequence $(H_n)_{n\in\mathbb{N}}$ is a Cauchy sequence iff there exists $\theta \in \mathbf{F}_+(\mathbb{R}^2)$ and a Cauchy sequence $(x_n)_{n\in\mathbb{N}} \subset Z$ such that $H_n = \theta(C,D)x_n$, $n \in \mathbb{N}$.

(ii) The space $TS_{Z;C,D}$ is sequentially complete.

(III.3.6) <u>Corollary</u>

(i) Let $K \subset TS_{Z;C,D}$. Then K is compact iff there exists a compact subset $K_0 \subset Z$, and $\theta \in \mathbf{F}_+(\mathbb{R}^2)$ such that $K = \theta(C,D)K_0$.

(ii) $K \subset TS_{Z;C,D}$ is compact iff K is sequentially compact.

(II.3.7) <u>Theorem</u>

The space $TS_{Z;C,D}$ is complete.

<u>Proof</u>. Let (F_α) be a Cauchy net in $TS_{Z;C,D}$. Then for each $s > 0$ and $t \geq 0$ the net $F_\alpha(t,s)$ is a Cauchy net in $S_{Z,C}$. Since $S_{Z,C}$ is complete, it yields an element $F(t,s) \in S_{Z,C}$ with $F_\alpha(t,s) \to F(t,s)$ in σ_{ind}-sense for each $s > 0$ and $t \geq 0$. Since $(e^{-tC} e^{-sD})_{t\geq 0, s\geq 0}$ is a two parameter semigroup of continuous mappings on $S_{Z,C}$, the mapping $(t,s) \mapsto F(t,s)$ is an element of $TS_{Z;C,D}$. Further, for all $\psi \in \mathbf{B}_+(\mathbb{R})$ and all $s > 0$

$$\|\psi(C)(F_\alpha(0,s) - F(0,s))\| \to 0 .$$ □

III.4. Pairing and duality of $ST_{Z;C,D}$ and $TS_{Z;C,D}$

In this section we introduce a pairing on $ST_{Z;C,D} \times TS_{Z;C,D}$. Through this pairing these spaces can be regarded as each other's dual.

(III.4.1) Definition

Let $h \in ST_{Z;C,D}$ and let $F \in TS_{Z;C,D}$. Then the complex number $\langle h,F\rangle$ is defined by

$$\langle h,F\rangle = (h(t,-s),F(-t,s))_Z .$$

Here first $s > 0$ must be taken so small that $t \mapsto h(t,-s)$ belongs to $T_{Z,C}$, and next $t > 0$ must be taken so small that $F(0,s) \in e^{-tC}(Z)$ where we put $F(-t,s) = e^{tC} F(0,s)$.

The definition of $\langle h,F\rangle$ does not depend on the choice of s and t because of the trajectory property as described in Section III.1.1 of both the elements in $ST_{Z;C,D}$ and in $TS_{Z;C,D}$.
We have the following representation theorem which extends Theorem (I.3.2).

(III.4.2) Theorem

(i) Let $F \in TS_{Z;C,D}$. Then the linear functional

$$h \mapsto \langle h,F\rangle$$

is continuous on $ST_{Z;C,D}$.

(ii) Let ℓ be a continuous linear functional on $ST_{Z;C,D}$. Then there exists $G \in TS_{Z;C,D}$ such that

$$\ell(h) = \langle h,G\rangle , \quad h \in ST_{Z;C,D} .$$

(iii) Let $h \in ST_{Z;C,D}$. Then the linear functional

$$F \mapsto \langle h,F\rangle$$

is continuous on $TS_{Z;C,D}$.

(iv) Let m be a continuous linear functional on $TS_{Z;C,D}$. Then there exists $g \in ST_{Z;C,D}$ such that

$$m(F) = \overline{\langle g,F\rangle} , \quad F \in TS_{Z;C,D} .$$

Proof

(i) By Corollary (III.3.3), F can be written as $F = \theta(C,D)x$ with $x \in Z$ and $\theta \in \mathbf{F}_+(\mathbb{R}^2)$. We obtain the estimation

$$|\langle h,F\rangle| = |(\theta(C,D)h,x)| \leq \|\theta(C,D)h\|\,\|x\| \,.$$

From Theorem (III.2.4) the wanted continuity follows.

(ii) Because of the definition of $\sigma\tau_{\text{ind}}$ for $ST_{Z;C,D}$, each linear functional $\ell \circ e^{-sD}$, $s > 0$, is continuous on $T_{Z,C}$. So there exists $G(s) \in S_{Z,C}$ with

$$(\ell \circ e^{-sD})(H) = \overline{\langle G(s),H\rangle} \,, \qquad H \in T_{Z,C} \,.$$

Since $(e^{-sD})_{s\geq 0}$ is a semigroup of continuous linear mappings on $S_{Z,C}$, we get

$$G(s+\sigma) = e^{-\sigma D}\, G(s) \,, \qquad s,\sigma > 0 \,.$$

Now define $G \in TS_{Z;C,D}$ by

$$G\colon (t,s) \to e^{-tC}\, G(s) \,, \qquad t \geq 0,\ s > 0 \,.$$

Then for each $h \in ST_{Z;C,D}$, $h \in e^{-sD}(T_{Z,C})$,

$$\ell(h) = (\ell \circ e^{-sD})(e^{sD}\, h) = (h(t,-s),G(-t,s))_Z \,.$$

(iii) Following Lemma (III.2.9), there are $x \in Z$, $s > 0$, and $\psi \in \mathbf{B}_+(\mathbb{R})$ such that $h = \psi(C)e^{-sD}x$. From the inequality

$$|\langle h,F\rangle| = |(x,\psi(C)F(0,s))| \leq \|x\|\,\|\psi(C)F(0,s)\|$$

the wanted continuity follows.

(iv) The topology $\tau\sigma_{\text{proj}}$ on $TS_{Z;C,D}$ is generated by the seminorms $p_{\psi,s}$, $\psi \in \mathbf{B}_+(\mathbb{R})$, $s > 0$. So the continuity of m on $TS_{Z;C,D}$ yields $\varphi \in \mathbf{B}_+(\mathbb{R})$ and $\sigma > 0$, such that

$$|m(F)| \leq p_{\varphi,\sigma}(F) = \|\varphi(C)F(0,\sigma)\|_Z \,, \qquad F \in TS_{Z;C,D} \,.$$

(Observe that the set of functions $\{(\lambda,\mu) \mapsto e^{-s\mu}\,\psi(\lambda) \mid \psi \in \mathbf{B}_+(\mathbb{R}),\ s > 0\}$ is directed.) The linear functional

$$m \circ \varphi(C)^{-1}\, e^{\sigma D}$$

is norm continuous on the dense linear subspace of Z given by

$$\{\varphi(C)F(0,\sigma) \mid F \in TS_{Z;C,D}\} .$$

So there exists $y \in Z$ such that

$$(m \circ \varphi(C)^{-1} e^{\sigma D})(\varphi(C)F(0,\sigma)) = (\varphi(C)F(0,\sigma),y)_Z .$$

Now introduce $g \in ST_{Z;C,D}$ by

$$g(t,s) = \varphi(C)e^{-tC} e^{-(s+\sigma)D} y , \quad t > 0,\ s \geqq -\sigma .$$

Then

$$m(F) = \overline{\langle g,F\rangle} .$$

□

(III.4.3) Definition

- The weak topology for $ST_{Z;C,D}$ is the topology generated by the seminorms $h \mapsto |\langle h,F\rangle|$ with $F \in TS_{Z;C,D}$.
- The weak topology for $TS_{Z;C,D}$ is the topology generated by the seminorms $F \mapsto |\langle h,F\rangle|$ with $h \in ST_{Z;C,D}$.

(III.4.4) Theorem (Banach-Steinhaus)

(i) Let $W \subset TS_{Z;C,D}$ be weakly bounded. Then W is strongly bounded.

(ii) Let $V \subset ST_{Z;C,D}$ be weakly bounded. Then V is strongly bounded.

Proof. We use the Banach-Steinhaus theorem for Hilbert spaces.

(i) Let $s > 0$ and let $\psi \in \mathbf{B}_+(\mathbb{R})$. Then for each $x \in Z$, there exists $K_{\psi,s,x} > 0$ such that for all $F \in W$

$$|\langle \psi(C)e^{-sD} x,F\rangle| = |(x,\psi(C)F(0,s))| \leqq K_{\psi,s,x} .$$

Hence there exists $K_{\psi,s} > 0$ such that

$$p_{\psi,s}(F) = \|\psi(C)F(0,s)\| \leqq K_{\psi,s} , \quad F \in W .$$

(ii) Let $\theta \in \mathbf{F}_+(\mathbb{R}^2)$. Then for each $x \in Z$, there exists $K_{\theta,x} > 0$ such that for all $h \in V$

$$|\langle h,\theta(C,D)x\rangle| = |(\theta(C,D)h,x)| \leqq K_{\theta,x} .$$

Hence there exists $K_\theta > 0$ such that

$$\|\theta(C,D)h\| \leq K_\theta , \quad h \in V . \qquad \square$$

In Section III.2 we have proved that Conjecture (III.2.5) is false. Due to this nonvalidity, only in the trivial case that $ST_{Z;C,D}$ is the direct sum of an analyticity space and a trajectory space, there is a correspondence between bounded subsets of $ST_{Z;C,D}$ and bounded subsets of Z. On the other hand, bounded subsets of $TS_{Z;C,D}$ are homeomorphic to bounded subsets of Z with respect to the topology $\tau\sigma_{proj}$ for $TS_{Z;C,D}$ and the norm topology for Z. Here we prove that bounded subsets of $TS_{Z;C,D}$ are homeomorphic to bounded subsets of Z if both spaces carry their weak topology. It is not likely that a similar result is valid for $ST_{Z;C,D}$.

(III.4.5) Theorem

Let B be a (weakly) bounded subset of $TS_{Z;C,D}$. Then there exists $\tilde{\theta} \in \mathbf{F}_+(\mathbb{R}^2)$ and a bounded subset B_0 of Z such that $\tilde{\theta}(C,D)$ is a homeomorphism from B_0 onto B, where B_0, B are endowed with the relative weak topology.

Proof. By Lemma (III.3.2), there exists $\theta \in \mathbf{F}_+(\mathbb{R}^2)$ and a bounded subset B_1 of Z such that $B = \theta(C,D)B_1$. Put

$$\tilde{\theta}(\lambda,\mu) = (1+\lambda^2)(1+\mu^2)\theta(\lambda,\mu) , \qquad \lambda,\mu \in \mathbb{R}^+ ,$$

and put

$$B_0 = (I+C^2)^{-1}(I+D^2)^{-1} B_1 .$$

Then $\tilde{\theta} \in \mathbf{F}_+(\mathbb{R}^2)$ and B_0 is a bounded subset of Z. We have $B = \tilde{\theta}(C,D)(B_0)$. Let (x_α) be a weakly convergent net in B_0 with limit $x \in B_0$. Then for all $y \in Z$, $(y,x_\alpha-x) \to 0$, and hence for all $h \in ST_{Z;C D}$

$$\langle h,\tilde{\theta}(C,D)(x_\alpha-x)\rangle = (\tilde{\theta}(C,D)h,x_\alpha-x) \to 0 .$$

So $\tilde{\theta}(C,D)$: $B_0 \to B$ is a weakly continuous bijection.
Now let (F_α) be a weakly convergent net in B with limit $F \in B$. Then there exist nets $(y_\alpha) \subset B_1$, $(x_\alpha) \subset B_0$ and vectors $y \in B_1$, $x \in B_0$ such that

$$F_\alpha = \theta(C,D)y_\alpha = \tilde{\theta}(C,D)x_\alpha \quad \text{and} \quad F = \theta(C,D)y = \tilde{\theta}(C,D)x .$$

We prove that (x_α) tends to x weakly.

Let $\varepsilon > 0$ and let $z \in Z$. Fix $A > 0$ so large that $(1+A^2)^{-1} < \varepsilon$. Then for all $z \in Z$

$$(*) \qquad \left| \left(\int_A^\infty \int_0^\infty + \int_0^\infty \int_A^\infty \right) d(G_\lambda H_\mu (x_\alpha - x), z) \right| \leq$$

$$\leq \left| \left(\int_A^\infty \int_0^\infty + \int_0^\infty \int_A^\infty \right) (1+\lambda^2)^{-1} (1+\mu^2)^{-1} \, d(G_\lambda H_\mu (y_\alpha - y), z) \right|$$

$$\leq \varepsilon (\|y_\alpha\| + \|y\|) \|z\| \leq 2K_{B_1} \|z\| \varepsilon ,$$

where

$$K_{B_1} = \sup \{ \|v\| \mid v \in B_1 \} .$$

Further, note that

$$z_{A,\tilde{\theta}} = \int_0^A \int_0^A \tilde{\theta}(\lambda,\mu)^{-1} \, g \, G_\lambda H_\mu z$$

belongs to $S_{Z,D}$, and hence $\mathrm{emb}(z_{A,\tilde{\theta}}) \in ST_{Z;C,D}$. So there exists α_0 such that for all $\alpha > \alpha_0$

$$|\langle \mathrm{emb}(z_{A,\tilde{\theta}}), F_\alpha - F \rangle| < \varepsilon$$

and hence

$$(**) \qquad \left| \int_0^A \int_0^A d(G_\lambda H_\mu (x_\alpha - x), z) \right| = |\langle \mathrm{emb}(z_{A,\tilde{\theta}}), F_\alpha - F \rangle| < \varepsilon .$$

Combining (*) and (**) yields $|(x_\alpha - x, z)| < \varepsilon(1 + 2K_{B_1} \|z\|)$ for all $\alpha > \alpha_0$. □

(III.4.6) <u>Corollary</u>

(i) Let $(F_n)_{n\in\mathbb{N}}$ be a sequence in $TS_{Z;C,D}$. Then $(F_n)_{n\in\mathbb{N}}$ is a weak Cauchy sequence in $TS_{Z;C,D}$ iff there exists a weak Cauchy sequence $(x_n)_{x\in\mathbb{N}}$ in Z and $\theta \in \mathbf{F}_+(\mathbb{R}^2)$ such that $F_n = \theta(C,D)x_n$.

(ii) The space $TS_{Z;C,D}$ is weakly sequentially complete.

(III.4.7) Corollary

Any bounded sequence in $\mathcal{T}S_{Z;C,D}$ contains a weakly convergent subsequence.

III.5. An inclusion diagram for compound spaces

As in the previous sections $(G_\lambda)_{\lambda\in\mathbb{R}}$ and $(H_\lambda)_{\lambda\in\mathbb{R}}$ denote the spectral resolutions of the identity of C and D. Further, by P we denote the orthogonal projection defined by

$$P = \iint_{\lambda\geq\mu} d\, G_\lambda\, H_\mu \ .$$

Note that P commutes strongly both with C and with D.

(III.5.1) Definition

The nonnegative self-adjoint operator $C \wedge D$ is defined by

$$C \wedge D = PCP + (I-P)D(I-P) \ ,$$

and the nonnegative self-adjoint operator $C \vee D$ by

$$C \vee D = (I-P)C(I-P) + PDP \ .$$

Remark. The operators $C \wedge D$ and $C \vee D$ satisfy the relations

$$C \wedge D = \iint_{\lambda,\mu\in\mathbb{R}} \max(\lambda,\mu)\, d\, G_\lambda\, H_\mu \ , \qquad C \vee D = \iint_{\lambda,\mu\in\mathbb{R}} \min(\lambda,\mu)\, d\, G_\lambda\, H_\mu \ .$$

We first prove the following lemma.

(III.5.2) Lemma

Let C_1 and C_2 denote the strongly commuting self-adjoint nonnegative operators in Z. Suppose $C_1 \leq C_2$. Then $\mathcal{T}_{Z,C_1}$ can be embedded in $\mathcal{T}_{Z,C_2}$. Notation: $\mathcal{T}_{Z,C_1} \hat{\subset} \mathcal{T}_{Z,C_2}$.

Proof. Since $C_1 \leq C_2$, the operator

$$e^{tC_1}\, e^{-tC_2} = e^{t(C_1-C_2)}$$

is bounded for each $t > 0$. Let $H \in \mathcal{T}_{Z,C_1}$. Then the trajectory

$$t \mapsto e^{tC_1} e^{-tC_2} H(t)$$

belongs to $\mathcal{T}_{Z,C_2}$.

(III.5.3) Theorem

(i) $\quad \mathcal{S}_{Z,C\wedge D} = \mathcal{S}_{Z,C+D} = \mathcal{S}_{Z,C} \cap \mathcal{S}_{Z,D}$

(ii) $\quad \mathcal{S}_{Z,C\vee D} = \mathcal{S}_{Z,C} + \mathcal{S}_{Z,D}$

(iii) $\quad \mathcal{T}_{Z,C\wedge D} \hat{=} \mathcal{T}_{Z,C} + \mathcal{T}_{Z,D}$

(iv) $\quad \mathcal{T}_{Z,C\vee D} \hat{=} \mathcal{T}_{Z,C} \cap \mathcal{T}_{Z,D}$

(in (iii) and (iv) "$\hat{=}$" means "is linearly homeomorphic to". In (ii) + denotes the usual sum in Z and in (iii) the usual sum in $\mathcal{T}_{Z,C+D}$. By Lemma (III.5.2) $\mathcal{T}_{Z,C}$ and $\mathcal{T}_{Z,D}$ are embedded in $\mathcal{T}_{Z,C+D}$.)

Proof. From the definition of the projection P it follows that the operators $Pe^{tD} e^{-tC} P$ and $(I-P)e^{tC} e^{-tD}(I-P)$ are bounded for each $t > 0$.

(i) Let $w \in \mathcal{S}_{Z,C\wedge D}$. Then there are $t > 0$ and $x \in Z$ such that

$$w = e^{-t(C\wedge D)}x = Pe^{-tC}Px + (I-P)e^{-tD}(I-P)x \ .$$

Put

$$\tilde{x} = Px + (I-P)e^{tC} e^{-tD}(I-P)x \ .$$

Then $w = e^{-tC}\tilde{x}$ and hence $w \in \mathcal{S}_{Z,C}$. In the same way we get $w \in \mathcal{S}_{Z,D}$. Hence

$$\mathcal{S}_{Z,C\wedge D} \subset \mathcal{S}_{Z,C} \cap \mathcal{S}_{Z,D} \ .$$

On the other hand, let $v \in \mathcal{S}_{Z,C} \cap \mathcal{S}_{Z,D}$. Then there exists $t > 0$ and $x,y \in Z$ such that $v = e^{-tC}x = e^{-tD}y$. So v can be written as

$$v = Pv + (I-P)v = Pe^{-tC}Px + (I-P)e^{-tD}(I-P)y$$

$$= e^{-t(C\wedge D)}(Px + (I-P)y) \in \mathcal{S}_{Z,C\wedge D} \ .$$

It follows that $\mathcal{S}_{Z,C} \cap \mathcal{S}_{Z,D} \subset \mathcal{S}_{Z,C\wedge D}$.

Since $C \wedge D \leqq C + D \leqq 2(C \wedge D)$, it is obvious that $S_{Z,C+D} = S_{Z,C\wedge D}$.

(ii) Let $w \in S_{Z,C\vee D}$. Then there are $x \in Z$ and $t > 0$ such that

$$w = e^{-t(C\vee D)}x = Pe^{-tD}Px + (I-P)e^{-tC}(I-P)x$$

$$= e^{-tD}Px + e^{-tC}(I-P)x \; .$$

So $w \in S_{Z,C} + S_{Z,D}$, and $S_{Z,C\vee D} \subset S_{Z,C} + S_{Z,D}$.
On the other hand, let $t > 0$ and let $x,y \in Z$. Put $v = e^{-tC}x + e^{-tD}y$.
Then we can write

$$v = e^{-t(C\wedge D)}[e^{t(C\wedge D)}e^{-tC}x + e^{t(C\wedge D)}e^{-tD}y]$$

and hence $v \in S_{Z,C\wedge D}$. It follows that $S_{Z,C} + S_{Z,D} \subset S_{Z,C\vee D}$.

(iii) Again, observe that $C \wedge D \leqq C + D \leqq 2(C \wedge D)$. So by Lemma (III.5.2) we get

$$T_{Z,C\wedge D} \hat{\subset} T_{Z,C+D} \hat{\subset} T_{Z,2(C\wedge D)} \; .$$

Since $T_{Z,C\wedge D} \hat{=} T_{Z,2(C\wedge D)}$, it follows that $T_{Z,C\wedge D} \hat{=} T_{Z,C+D}$. Since $C \leqq C \wedge D$ and $D \leqq C \wedge D$, it follows from Lemma (III.5.2) that $T_{Z,C} \hat{\subset} T_{Z,C\wedge D}$ and $T_{Z,D} \hat{\subset} T_{Z,C\wedge D}$. We obtain the inclusion $T_{Z,C} + T_{Z,D} \hat{\subset} T_{Z,C\wedge D}$.
Conversely, let $G \in T_{Z,C\wedge D}$. Then for all $t > 0$, we define

$$PG\colon t \mapsto PG(t) \quad \text{and} \quad (I-P)G\colon t \mapsto (I-P)G(t) \; .$$

We have for all $t,\tau > 0$

$$(PG)(t+\tau) = Pe^{-\tau(C\wedge D)}PG(t) = e^{-\tau C}(PG)(t)$$

and, similarly,

$$(I-P)G(t+\tau) = e^{-\tau D}(I-P)G(t) \; .$$

Hence $PG \in T_{Z,C}$ and $(I-P)G \in T_{Z,D}$, and therefore consequently, $G \in T_{Z,C} + T_{Z,D}$.
It follows that $T_{Z,C\wedge D} \subset T_{Z,C} + T_{Z,D}$.

(iv) Let $H \in T_{Z,C} \cap T_{Z,D}$, where $T_{Z,C}, T_{Z,D}$ is regarded as a subspace of $T_{Z,C+D}$. It follows that the trajectory

$$t \mapsto e^{tC}H(t)$$

belongs to $T_{Z,D}$, and the trajectory

$$t \mapsto e^{tD}\, H(t)$$

belongs to $T_{Z,C}$. Now put

$$\hat{H}: t \mapsto e^{t(C\wedge D)}\, H(t)\ .$$

Then $\hat{H} \in T_{Z,C\vee D}$, and hence $T_{Z,C} \cap T_{Z,D} \hat{\subset} T_{Z,C\vee D}$. Since $C \vee D \leqq C$ and $C \vee D \leqq D$, we get

$$T_{Z,C\vee D} \hat{\subset} T_{Z,C} \quad \text{and} \quad T_{Z,C\vee D} \hat{\subset} T_{Z,D}\ . \qquad \square$$

In Section III.1 we have introduced the spaces $ST_{Z;C,D}$ and $TS_{Z;C,D}$ as linear subspaces of $TT_{Z;C,D}$. Since $TT_{Z;C,D} \hat{=} TT_{Z;D,C}$, also the spaces $ST_{Z;D,C}$ and $TS_{Z;D,C}$ are contained in $TT_{Z;C,D}$. So it makes sense to consider intersections and algebraic sums of the spaces $ST_{Z;C,D}$ and $ST_{Z;D,C}$, and of the spaces $TS_{Z;C,D}$ and $TS_{Z;D,C}$. It turns out that the compound spaces $ST_{Z;C\wedge D,C\vee D}$, $ST_{Z;C\vee D,C\wedge D}$, $TS_{Z;C\wedge D,C\vee D}$ and $TS_{Z;C\vee D,C\wedge D}$, which are well-defined because $C \vee D$ and $C \wedge D$ commute, describe these sums and intersections.
The result of the next theorem will eventually lead to a kernel theorem for the extendible mappings in Section III.6.

(III.5.4) Theorem

(i) $ST_{Z;C,D} \cap ST_{Z;D,C} \hat{=} ST_{Z;C\vee D,C\wedge D}$

(ii) $ST_{Z;C,D} + ST_{Z;D,C} \hat{=} ST_{Z;C\wedge D,C\vee D}$

(iii) $TS_{Z;C,D} \cap TS_{Z;D,C} \hat{=} TS_{Z;C\wedge D,C\vee D}$

(iv) $TS_{Z;C,D} + TS_{Z;D,C} \hat{=} TS_{Z;C\vee D,C\wedge D}$

Proof

(i) Let $k \in ST_{Z;C,D} \cap ST_{Z;D,C}$. Then $k \in TT_{Z;C,D}$ and

$$\exists_{\tau>0}: s \mapsto k(-\tau,s) \in T_{Z,D}$$

and

$$\exists_{\sigma>0}: t \mapsto k(t,-\sigma) \in T_{Z,C}\ .$$

Define $\hat{k}$ by

$$\hat{k}: (t,s) \mapsto Pk(t,s) + (I-P)k(s,t) \ , \qquad t > 0,\ s \geq -s_0$$

with s_0 chosen sufficiently small. We have

$$\hat{k}(t+\tau,s+\sigma) = [Pe^{-\tau C}\ e^{-\sigma D}\ P + (I-P)e^{-\sigma C}\ e^{-\tau D}(I-P)]\hat{k}(t,s)$$

$$= e^{-\tau(C\wedge D)}\ e^{-\sigma(C\vee D)}\ \hat{k}(t,s) \ ,$$

for all admissible t, τ, s and σ. Furthermore,

$$t \mapsto \hat{k}(t,-s_0) \in T_{Z,C\vee D} \ .$$

It follows that $\hat{k} \in ST_{Z;C\wedge D,C\vee D}$ and hence

$$ST_{Z;C,D} \cap ST_{Z;D,C} \mathrel{\hat{\subset}} ST_{Z;C\vee D,C\wedge D} \ .$$

Conversely, let $h \in ST_{Z;C\vee D,C\wedge D}$. Then we define

$$\hat{h}: (t,s) \mapsto Ph(t,s) + (I-P)h(s,t) \ .$$

From the definition of the projection P it follows that $\hat{h} \in ST_{Z;C,D}$. Similarly,

$$(t,s) \mapsto (I-P)h(t,s) + Ph(s,t)$$

belongs to $ST_{Z;D,C}$. Hence $ST_{Z;C\vee D,C\wedge D} \mathrel{\hat{\subset}} ST_{Z;C,D} \cap ST_{Z;D,C}$.

(ii) Let $h_1 \in ST_{Z;C,D}$ and $h_2 \in ST_{Z;D,C}$. Then there exists $s_0 > 0$ such that

$$t \mapsto h_1(t,-s_0) \in T_{Z,C} \quad \text{and} \quad t \to h_2(t,-s_0) \in T_{Z,D} \ .$$

Now define

$$s_{h_1,h_2}(t,s) = Ph_1(t,s) + Pe^{-t(C\wedge D)}\ e^{tD}\ Pe^{-s(C\vee D)}\ e^{-s_0 C}\ Ph_2(t,-s_0)$$

$$+ (I-P)e^{-t(C\wedge D)}\ e^{tC}(I-P)e^{-s(C\vee D)}\ e^{-s_0 D}\ Ph_1(t,-s_0) + (I-P)h_2(t,s) \ .$$

Then straightforward computation shows that $s_{h_1,h_2} \in ST_{Z;C\wedge D,C\vee D}$. Thus it can be seen that $ST_{Z;C,D} + ST_{Z;D,C} \mathrel{\hat{\subset}} ST_{Z;C\wedge D,C\vee D}$.
Conversely, let $h \in ST_{Z;C\wedge D,C\vee D}$. Then

$$(t,s) \mapsto Ph(t,s) \in ST_{Z;C,D}$$

and

$$(t,s) \mapsto (I-P)h(t,s) \in ST_{Z;D,C} .$$

It follows that each element of $ST_{Z;C\wedge D,C\vee D}$ can be written as the sum of an element of $ST_{Z;C,D}$ and an element of $ST_{Z;D,C}$.

(iii) Let $G \in TS_{Z;C,D} \cap TS_{Z;D,C}$. Then for all $s > 0$ and $t > 0$

$$G(0,s) \in S_{Z,C} \quad \text{and} \quad G(t,0) \in S_{Z,D} .$$

Define $\hat{G}$ by

$$\hat{G}(t,s) = Pe^{-tC} PG(0,s) + (I-P)e^{-tD}(I-P)G(s,0)$$

where $t \geq 0$ and $s > 0$. One can easily check that $\hat{G}$ belongs to $TT_{Z;C\wedge D,C\vee D}$. Moreover, for all $s > 0$

$$\hat{G}(0,s) = PG(0,s) + (I-P)G(s,0) \in S_{Z,C\wedge D}$$

because

$$PG(0,s) = Pe^{-\tau C} PG(-\tau,s) \quad \text{and} \quad (I-P)G(s,0) = (I-P)e^{-\tau D}(I-P)G(s,-\tau)$$

with $\tau > 0$ dependent on s. Thus we obtain the relation

$$TS_{Z;C,D} \cap TS_{Z;D,C} \subset TS_{Z;C\wedge D,C\vee D}$$

Let, conversely, $H \in TS_{Z;C\wedge D,C\vee D}$. Then the mapping

$$(t,s) \mapsto PH(t,s) + (I-P)H(s,t)$$

belongs to $TS_{Z;C,D}$. Similarly, the mapping

$$(t,s) \mapsto (I-P)H(t,s) + PH(s,t)$$

belongs to $TS_{Z;D,C}$. Thus, the space $TS_{Z;C\wedge D,C\vee D}$ can be embedded both in $TS_{Z;C,D}$ and in $TS_{Z;D,C}$.

(iv) Let $K_1 \in TS_{Z;D,C}$ and let $K_2 \in TS_{Z;C,D}$. Then $\tilde{K}_1$ defined by $\tilde{K}_1(s) = K_1(s,0)$ belongs to $T_{Z,D}$, and $\tilde{K}_2$ defined by $\tilde{K}_2(s) = K_2(s,0)$ belongs to $T_{Z,C}$. So by Theorem (III.5.3) we get $\tilde{K}_1 + \tilde{K}_2 \in T_{Z,C\wedge D}$.
For all $s > 0$ we have

$$\tilde{K}_1(s) \in S_{Z,C} \quad \text{and} \quad \tilde{K}_2(s) \in S_{Z,D} ,$$

and hence $\tilde{K}_1(s) + \tilde{K}_2(s) \in S_{Z,C\vee D}$, and also $(\tilde{K}_1 + \tilde{K}_2)(s) \in S_{Z,C\vee D}$. So if we put

$$s_{K_1,K_2}(t,s) = e^{-t(C\vee D)}(\tilde{K}_1 + \tilde{K}_2)(s) \ , \qquad t \geq 0,\ s > 0,$$

then

$$s_{K_1,K_2} \in TS_{Z;C\vee D,C\wedge D}$$

Thus we have derived the inclusion $TS_{Z;C,D} + TS_{Z;D\ C} \hat{\subset} TS_{Z;C\vee D,C\wedge D}$.
In order to derive the reverse inclusion, let $G \in TS_{Z;C\vee D,C\wedge D}$. Then G_1 defined by $G_1(t,s) = PG(t,s)$, $t \geq 0$, $s > 0$, belongs to $TS_{Z;C,D}$. Similarly, G_2 defined by $G_2(t,s) = (I-P)G(t,s)$, $t \geq 0$, $s > 0$, belongs to $TS_{Z;D,C}$. Since $G = G_1 + G_2$, we have

$$TS_{Z;C\vee D,C\wedge D} \hat{\subset} TS_{Z;C,D} + TS_{Z;D,C} \ . \qquad \square$$

(III.5.5) Lemma

(i) $ST_{Z;D,C} \hat{\subset} TS_{Z;C,D}$.

(ii) If $C \geq D$, then $ST_{Z;D,C} \hat{=} S_{Z,C}$ and $TS_{Z;D,C} \hat{=} T_{Z,C}$.

Proof

(i) Let $h \in ST_{Z;D,C}$. Define the mapping $\hat{h} \in TT_{Z;C,D}$ by

$$\hat{h}(t,s) = h(s,t) \ , \qquad t,s > 0 \ .$$

There exists $t_0 > 0$ such that the mapping $s \mapsto \hat{h}(-t_0,s)$ belongs to $T_{Z,D}$. It follows that $h \in TS_{Z;C,D}$.

(ii) Let $C \geq D$. From Lemma (III.2.29) we obtain that $ST_{Z;D,C} = S_{Z,C}$. Moreover, it is clear that $TS_{Z;D,C} \hat{\subset} T_{Z,C}$.
Let $H \in T_{Z,C}$. Define $\hat{H}$ by

$$\hat{H}(t,s) = e^{-tD}\, H(s) \ , \qquad t \geq 0,\ s > 0 \ .$$

Then $\hat{H}(0,s) = H(s)$. Let $\psi \in B_+(\mathbb{R})$. Since $\psi(D)e^{-\frac{1}{2}sC}$ is bounded for all $s > 0$ it follows that $H(s) \in \mathcal{D}(\psi(D))$. Hence $\hat{H} \in TS_{Z;D,C}$, and $T_{Z,C} \hat{\subset} TS_{Z;D,C}$. $\square$

(III.5.6) Inclusion diagram

$$
\begin{array}{ccccccc}
SS_{Z;C\vee D,C\vee D} & \hat{\subset} & ST_{Z;C\wedge D,C\vee D} & \hat{\subset} & TS_{Z;C\vee D,C\wedge D} & \hat{=} & TT_{Z;C\wedge D,C\wedge D} \\
\Vert & & \cup\!\rangle & & \cup\!\rangle & & \cup\!\rangle \\
SS_{Z;C\vee D,C\vee D} & \hat{\subset} & ST_{Z;D,C\vee D} & \hat{\subset} & TS_{Z;C\vee D,D} & \hat{=} & TT_{Z;D,D} \\
\cup\!\rangle & & \cup\!\rangle & & \cup\!\rangle & & \Vert \\
SS_{Z;C,C} & \hat{\subset} & ST_{Z;D,C} & \hat{\subset} & TS_{Z;C,D} & \hat{\subset} & TT_{Z;D,D} \\
\Vert & & \cup\!\rangle & & \cup\!\rangle & & \cup\!\rangle \\
SS_{Z;C,C} & \hat{=} & ST_{Z;C\vee D,C} & \hat{\subset} & TS_{Z;C,C\vee D} & \hat{\subset} & TT_{Z;C\vee D,C\vee D} \\
\cup\!\rangle & & \cup\!\rangle & & \cup\!\rangle & & \Vert \\
SS_{Z;C\wedge D,C\wedge D} & \hat{=} & ST_{Z;C\vee D,C\wedge D} & \hat{\subset} & TS_{Z;C\wedge D,C\vee D} & \hat{\subset} & TT_{Z;C\vee D,C\vee D} \\
\cap\!\rangle & & \cap\!\rangle & & \cup\!\rangle & & \Vert \\
SS_{Z;D,D} & \hat{=} & ST_{Z;C\vee D,D} & \hat{\subset} & TS_{Z;D,C\vee D} & \hat{\subset} & TT_{Z;C\vee D,C\vee D} \\
\Vert & & \cap\!\rangle & & \cup\!\rangle & & \cap\!\rangle \\
SS_{Z;D,D} & \hat{\subset} & ST_{Z;C,D} & \hat{\subset} & TS_{Z;D,C} & \hat{\subset} & TT_{Z;C,C} \\
\cap\!\rangle & & \cap\!\rangle & & \cup\!\rangle & & \Vert \\
SS_{Z;C\vee D,C\vee D} & \hat{\subset} & ST_{Z;C,C\vee D} & \hat{\subset} & TS_{Z;C\vee D,C} & \hat{=} & TT_{Z;C,C} \\
\Vert & & \cap\!\rangle & & \cup\!\rangle & & \cap\!\rangle \\
SS_{Z;C\vee D,C\vee D} & \hat{\subset} & ST_{Z;C\wedge D,C\vee D} & \hat{\subset} & TS_{Z;C\vee D,C\wedge D} & \hat{=} & TT_{Z;C\wedge D,C\wedge D}
\end{array}
$$

Remark. $SS_{Z;C,C} \hat{=} S_{Z,C}$, $TT_{Z;C,C} \hat{=} T_{Z,C}$, etc.

We note that in the diagram given on the preceding page all embeddings are continuous and have dense ranges. We omit the proof of this assertion, although the statement is interesting in itself. The proof can be found in [ETh], Ch. III.

III.6. Topological tensor products and kernel theorems

In this section X and Y denote separable Hilbert spaces, and A and B denote self-adjoint operators in X and Y, respectively. The fourth section of Chapter I contains detailed characterizations of five types of continuous linear mappings represented by

$$L(S_{X,A},S_{Y,B}),\ L(S_{X,A},T_{Y,B}),\ L(T_{X,A},S_{Y,B}),\ L(T_{X,A},T_{Y,B})$$

and

$$E(S_{X,A},S_{Y,B}),$$

the space of extendible mappings. In this section we prove a so-called kernel theorem for each of these types. A kernel theorem says that all continuous linear mappings of one type arise from the "kernels" out of a specific topological tensor product. Therefore, five topological tensor products are introduced.

The Hilbert space $X \otimes Y$ which consists of all Hilbert-Schmidt operators from X into Y, can be regarded as a sesquilinear topological tensor product of X and Y. We note that if and only if X or Y is finite dimensional, the space of bounded linear operators $B(X,Y)$ equals $X \otimes Y$. So only for finite dimensional X or Y the kernel theorem holds for the continuous linear operators from X into Y.

Let $(e_k)_{k\in\mathbb{N}}$ be an orthonormal basis in X. For any $W \in X \otimes Y$ we have

$$\sum_{k=1}^{\infty} \|We_k\|^2 < \infty .$$

The quantity $\sum_{k=1}^{\infty} \|We_k\|^2$ does not depend on the choice of the basis $(e_k)_{k\in\mathbb{N}}$. The space $X \otimes Y$ is a Hilbert space with the inner product

$$(W_1,W_2)_{X\otimes Y} = \sum_{k=1}^{\infty} (W_1e_k,W_2e_k)_Y , \qquad W_1,W_2 \in X \otimes Y .$$

The associated norm is the so-called Hilbert-Schmidt norm

$$\|W\|^2_{X\otimes Y} = \sum_{k=1}^{\infty} \|We_k\|^2_Y , \qquad W \in X \otimes Y .$$

Let $x \in X$ and let $y \in Y$. Then the rank-one operator $x \otimes y$ acts on X through $(x \otimes y)z = (z,x)_X y$. The algebraic tensor product, denoted by $X \otimes_a Y$, consists of all finite linear combinations of this kind of rank-one operators, i.e.

$$W \in X \otimes_a Y \Leftrightarrow \exists_{(x_k)\in X} \exists_{(y_k)\in Y} : W = \sum_{k=1}^{K} x_k \otimes y_k .$$

The completion of $X \otimes_a Y$ with respect to the norm $\|\cdot\|_{X\otimes Y}$ can be identified with the Hilbert space $X \otimes Y$.

For later reference we mention the following properties, which are taken from [RS].

(III.6.1) Properties

(i) $\forall_{x_1,x_2\in X} \forall_{y_1,y_2\in Y} : (x_1 \otimes y_1,x_2 \otimes y_2)_{X\otimes Y} = (x_1,x_2)_X(y_2,y_1)_Y .$

(ii) $\forall_{\lambda\in\mathbb{C}} \forall_{x\in X} \forall_{y\in Y} : \lambda(x \otimes y) = (\bar{\lambda}x) \otimes y = x \otimes (\lambda y) .$

Hence the mapping $X \times Y \to X \otimes Y$ defined by $[x;y] \mapsto x \otimes y$ is anti-linear in x and linear in y.

(iii) $\forall_{W\in X\otimes Y} \forall_{x\in X} \forall_{y\in Y} : (W,x\otimes y)_{X\otimes Y} = (Wx,y)_Y .$

Let J_1 and J_2 denote bounded operators in $\mathbf{B}(X)$ and $\mathbf{B}(Y)$, respectively. Then $J_1 \otimes J_2$ denotes the bounded linear operator on $X \otimes Y$ defined by

$$J_1 \otimes J_2(W) = J_2 W J_1^* .$$

(iv) We have $\|J_1 \otimes J_2\| = \|J_1\|\|J_2\|$.

(v) J_1 and J_2 injective $\Rightarrow J_1 \otimes J_2$ injective.

Let $x \in \mathcal{D}(A)$. Then for all $y \in \mathcal{Y}$ we define

$$A \otimes I(x \otimes y) = Ax \otimes y \ .$$

With the aid of linear extension the operator $A \otimes I$ is well defined on the algebraic tensor product $\mathcal{D}(A) \otimes_a I$. From [Wei], Ch. VIII, it follows that $A \otimes I$ with domain $\mathcal{D}(A) \otimes_a I$ is nonnegative and essentially self-adjoint in $\mathcal{X} \otimes \mathcal{Y}$. We denote the unique self-adjoint extension by $A \otimes I$, also. It is not hard to see that the spectral resolution of the identity $I \otimes I$ corresponding to $A \otimes I$ is given by $(E_\lambda \otimes I)_{\lambda \in \mathbb{R}}$ where $(E_\lambda)_{\lambda \in \mathbb{R}}$ denotes the resolution of the identity corresponding to A.
Similarly, the nonnegative self-adjoint operator $I \otimes B$ can be introduced. Let $(F_\mu)_{\mu \in \mathbb{R}}$ denote the spectral resolution of B. Then $(I \otimes F_\mu)_{\mu \in \mathbb{R}}$ is the one for $I \otimes B$. We have

$$(E_\lambda \otimes I)(I \otimes F_\mu) = E_\lambda \otimes F_\mu \ , \qquad \lambda, \mu \in \mathbb{R} \ .$$

In the inclusion diagram (III.5.7) we replace $\mathcal{Z}$ by $\mathcal{X} \otimes \mathcal{Y}$, the operator C by $A \otimes I$ and the operator D by $I \otimes B$. Thus we obtain all required spaces which describe each of the five mentioned topological tensor products.
First we direct our attention to the spaces $\mathcal{SS}_{\mathcal{X}\otimes\mathcal{Y};A\otimes I, I\otimes B}$ and $\mathcal{TT}_{\mathcal{X}\otimes\mathcal{Y};A\otimes I, I\otimes B}$. Note that they can be identified with the spaces $\mathcal{S}_{\mathcal{X}\otimes\mathcal{Y},A\otimes I+I\otimes B}$ and $\mathcal{T}_{\mathcal{X}\otimes\mathcal{Y},A\otimes I+I\otimes B}$, respectively.

(III.6.2) <u>Definition</u>

The sequilinear mapping $\otimes : \mathcal{S}_{\mathcal{X},A} \times \mathcal{S}_{\mathcal{Y},B} \to \mathcal{SS}_{\mathcal{X}\otimes\mathcal{Y};A\otimes I, I\otimes B}$ is defined by

$$[w,v] \mapsto w \otimes v \ .$$

The definition is consistent because for each $w \in \mathcal{S}_{\mathcal{X},A}$ and each $v \in \mathcal{S}_{\mathcal{Y},B}$ there exist $x \in \mathcal{X}$, $y \in \mathcal{Y}$ and $\sigma, \tau > 0$ such that $w = e^{-\tau A}x$ and $v = e^{-\sigma B}y$. Then

$$w \otimes v = (e^{-\tau A}x) \otimes (e^{-\sigma B}y) = (e^{-\tau A} \otimes I)(I \otimes e^{-\sigma B})(x \otimes y) \ .$$

Hence $w \otimes v \in \mathcal{SS}_{\mathcal{X}\otimes\mathcal{Y};A\otimes I, I\otimes B}$.

(III.6.3) <u>Theorem</u>

$\mathcal{SS}_{\mathcal{X}\otimes\mathcal{Y};A\otimes I, I\otimes B}$ is a complete topological tensor product of $\mathcal{S}_{\mathcal{X},A}$ and $\mathcal{S}_{\mathcal{Y},B}$. By this we mean

(i) $SS_{X\otimes Y;A\otimes I,I\otimes B}$ is complete.

(ii) The mapping $\otimes : S_{X,A} \times S_{Y,B} \to SS_{X\otimes Y;A\otimes I,I\otimes B}$ is continuous.

(iii) The algebraic tensor product $S_{X,A} \otimes_a S_{Y,B}$ is dense in $SS_{X\otimes Y;A\otimes I,I\otimes B}$.

Proof

(i) The space $S_{X\otimes Y,A\otimes I+I\otimes B} = SS_{X\otimes Y;A\otimes I,I\otimes B}$ is complete.

(ii) It is enough to check the continuity of $\otimes$ at $[0,0]$. In the appendix to this chapter it will be proved that for each $\psi \in B_+(\mathbb{R})$ there exists $\chi \in B_+(\mathbb{R})$ such that $\chi(\lambda)\chi(\mu) \geqq \psi(\lambda+\mu)$.
It follows that the topology for $SS_{X\otimes Y;A\otimes I,I\otimes B}$ is generated by the seminorms

$$w \mapsto \|(\chi(A) \otimes \chi(B))w\|_{X\otimes Y}\ , \qquad \chi \in B_+(\mathbb{R})\ .$$

So let $\chi \in B_+(\mathbb{R})$. Then for all $w \in S_{X,A}$ and all $v \in S_{Y,B}$

$$\|\chi(A) \otimes \chi(B)(w \otimes v)\|_{X\otimes Y} = \|\chi(A)w\|_X \|\chi(B)v\|_Y\ ,$$

from which the continuity of $\otimes$ on $S_{X,A} \times S_{Y,B}$ follows.

(iii) For each $t,s > 0$, $e^{-tA}(X) \otimes_a e^{-sB}(Y)$ is dense in $e^{-tA}(X) \otimes e^{-sB}(Y)$.
From this the wanted result follows. □

(III.6.4) Definition

The canonical sesquilinear mapping $\otimes : T_{X,A} \times T_{Y,B} \to TT_{X\otimes Y;A\otimes I,I\otimes B}$
$[F,G] \to F \otimes G$ is defined by

$$(F \otimes G)(t,s) = F(t) \otimes G(s)\ .$$

The definition is consistent because for all $t,\tau > 0$ and $s,\sigma > 0$

$$(F \otimes G)(t+\tau,s+\sigma) = e^{-\tau A}\, F(t) \otimes e^{-\sigma B}\, G(s)$$

$$= (e^{-\tau A} \otimes I)(I \otimes e^{-\sigma B})(F \otimes G)(t,s)\ .$$

(III.6.5) Theorem

$TT_{X\otimes Y;A\otimes I,I\otimes B} \mathrel{\hat=} T_{X\otimes Y,A\otimes I+I\otimes B}$ is a complete topological tensor product of $T_{X,A}$ and $T_{Y,B}$. By this we mean

(i) $TT_{X\otimes Y;A\otimes I,I\otimes B}$ is complete.

(ii) The mapping $\otimes : T_{X,A} \times T_{Y,B} \to TT_{X\otimes Y;A\otimes I,I\otimes B}$ is continuous.

(iii) The algebraic tensor product $T_{X,A} \otimes_a T_{Y,B}$ is dense in $TT_{X\otimes Y;A\otimes I,I\otimes B}$.

<u>Proof</u>

(i) The completeness follows from the observation that the spaces $TT_{X\otimes Y;A\otimes I,I\otimes B}$ and $T_{X\otimes Y,A\otimes I+I\otimes B}$ are homeomorphic.

(ii) Let $t,s > 0$. Then for all $F \in T_{X,A}$ and all $G \in T_{Y,B}$ we have

$$\|(F \otimes G)(t,s)\|_{X\otimes Y} = \|F(t)\|_X \|G(s)\|_Y .$$

From this the continuity of $\otimes$ at $[0;0]$ follows.

(iii) $X \otimes_a Y$ is dense in $X \otimes Y$, and $X \otimes Y$ is densely embedded in $TT_{X\otimes Y;A\otimes I,I\otimes B}$. □

Next we show that the elements of the topological tensor products $SS_{X\otimes Y;A\otimes I,I\otimes B}$ and $TT_{X\otimes Y;A\otimes I,I\otimes B}$ can be interpreted as linear mappings in $L(T_{X,A},S_{Y,B})$ and in $L(S_{X,A},T_{Y,B})$, respectively. We give necessary and sufficient conditions on the semigroups $(e^{-tA})_{t>0}$ and $(e^{-sB})_{s>0}$ which ensure the validity of the kernel theorems which are current in distribution theory.

<u>Case</u> a: The space $L(T_{X,A},S_{Y,B})$

Let $\hat{\theta} \in SS_{X\otimes Y;A\otimes I,I\otimes B}$. Then we define the action of $\hat{\theta}$ as a linear operator from $T_{X,A}$ into $S_{Y,B}$ in the following way.
Let $F \in T_{X,A}$. We define θF by

(a) $$\theta F = \hat{\theta}(-\tau,0)F(\tau) .$$

For $\tau > 0$ sufficiently small this definition makes sense and does not depend on τ. Moreover, there exists $\sigma > 0$ such that

$$\hat{\theta}(-\tau,0)F(\tau) = e^{-\sigma B}(\hat{\theta}(-\tau,-\sigma)F(\tau))$$

and hence $\theta F \in S_{Y,B}$.

(III.6.6) Theorem

(i) For each $\hat{\theta} \in SS_{X\otimes Y;A\otimes I,I\otimes B} = S_{X\otimes Y,A\otimes I+I\otimes B}$ the linear operator θ from $T_{X,A}$ into $S_{Y,B}$ is continuous.

(ii) For each $\hat{\theta} \in SS_{X\otimes Y;A\otimes I,I\otimes B}$, $F \in T_{X,A}$, $G \in T_{Y,B}$,

$$\langle \theta F,G\rangle_{Y} = \langle \hat{\theta},F\otimes G\rangle_{X\otimes Y} .$$

(iii) If for each $t > 0$ at least one of the operators e^{-tA}, e^{-tB} is HS, then $SS_{X\otimes Y;A\otimes I,I\otimes B}$ comprises all elements of $\mathbf{L}(T_{X,A},S_{Y,B})$.

(iv) $SS_{X\otimes Y;A\otimes I,I\otimes A}$ comprises all elements of $\mathbf{L}(T_{X,A},S_{X,A})$ iff for each $t > 0$ the operator e^{-tA} is HS.

Proof

(i) The operator θ defined in (a) satisfies condition (iii) of Theorem (I.4.7). Namely, since $\hat{\theta} \in X \otimes Y$ we have $\theta_r = \hat{\theta}$. Since $\hat{\theta} \in SS_{X\otimes Y;A\otimes I,I\otimes B}$ we have for $t > 0$ sufficiently small, $\hat{\theta}(-t,-t) \in X \otimes Y$, and hence $e^{tB}\, \theta_r\, e^{tA}$ is bounded on $e^{-tA}(X) \subset X$.

(ii) For $\tau,\sigma > 0$ sufficiently small, we derive

$$\langle \theta F,G\rangle_{Y} = (\hat{\theta}(-\tau,-\sigma)F(\tau),G(\sigma))_{Y} =$$

$$= (\hat{\theta}(-\tau,-\sigma),F(\tau) \otimes G(\sigma))_{X\otimes Y} = \langle \hat{\theta},F\otimes G\rangle_{X\otimes Y} .$$

(iii) Let $\Gamma \in \mathbf{L}(T_{X,A},S_{Y,B})$. By Theorem (I.4.7), condition (iii), there exists $\tau > 0$ such that the operator $e^{\tau B}\, \Gamma_r\, e^{\tau A}$ is bounded. Then put

$$\hat{\Gamma}: (t,s) \mapsto e^{-(s+\tau)B}(e^{\tau B}\, \Gamma_r\, e^{\tau A})e^{-(t+\tau)A}$$

where $t \geqq -\tau$ and $s \geqq -\tau$. Clearly, $\hat{\Gamma} \in SS_{X\otimes Y;A\otimes I,I\otimes B}$. By Definition (a), $\hat{\Gamma}$ represents the operator Γ.

(iv) The if-part is a special case of (iii).
For the only-if part consider the operator $\Gamma = e^{-tA}: T_{X,A} \to S_{X,A}$ for any $t > 0$. If $\Gamma \in SS_{X\otimes X;A\otimes I,I\otimes A}$ it has to be HS. □

Case (b) The space $\mathbf{L}(S_{X,A}, T_{Y,B})$

Let $\hat{K} \in TT_{X\otimes Y;A\otimes I,I\otimes B}$. Then we define the action of $\hat{K}$ as a linear operator from $S_{X,A}$ into $T_{Y,B}$ in the following way:

(b) $\quad (Kw)(s) = \hat{K}(\tau,s)e^{\tau A}\, w\,, \qquad s > 0\,.$

For each $w \in S_{X,A}$ and each $s > 0$ this definition makes sense if $\tau > 0$ is taken so small that $w \in e^{\tau A}(X)$. Moreover, $(Kw)(s)$ does not depend on τ.

(III.6.7) Theorem

(i) For each $\hat{K} \in TT_{X\otimes Y;A\otimes I,I\otimes B} \triangleq T_{X\otimes Y,A\otimes I+I\otimes B}$ the linear operator $K\colon S_{X,A} \to T_{Y,B}$ defined by (b) is continuous.

(ii) For each $\hat{K} \in TT_{X\otimes Y;A\otimes I,I\otimes B}$, $w \in S_{X,A}$, $v \in S_{Y,B}$,

$$\langle w,Kv\rangle_Y = \langle w\otimes v,\hat{K}\rangle_{X\otimes Y}\,.$$

(iii) If for each $t > 0$ at least one of the operators e^{-tA}, e^{-tB} is HS, then $TT_{X\otimes Y;A\otimes I,I\otimes B}$ comprises all elements of $\mathbf{L}(S_{X,A}, T_{Y,B})$.

(iv) $TT_{X\otimes Y;A\otimes I,I\otimes A}$ comprises all elements of $\mathbf{L}(S_{X,A}, T_{X,A})$ iff for each $t > 0$ the operator e^{-tA} is HS.

Proof

(i) We use Condition (ii) of Theorem (I.4.5).
Let $t > 0$ and $s > 0$. Then $e^{-sB}\, K\, e^{-tA} = \hat{K}(t,s) \in X \otimes Y$.

(ii) Take $\tau > 0$ and $\sigma > 0$ so small that $w \in e^{-\tau A}(X)$ and $v \in e^{-\sigma B}(Y)$. Then we have

$$\langle v,Kw\rangle_Y = (e^{\sigma B}\, v, \hat{K}(\tau,\sigma)e^{\tau A}\, w)_Y$$
$$= ((e^{\tau A} \otimes e^{\sigma B})(w \otimes v), \hat{K}(\tau,\sigma))_{X\otimes Y} = \langle w\otimes v,\hat{K}\rangle_{X\otimes Y}\,.$$

(iii) Let $L \in \mathbf{L}(S_{X,A}, T_{Y,B})$. According to Theorem (I.4.5), condition (ii), the operator $e^{-sB}\, L\, e^{-tA}$ is bounded for all $t,s > 0$. Define the mapping $\hat{L}$ on $(0,\infty) \times (0,\infty)$ by

$$\hat{L}\colon (t,s) \mapsto e^{-sB}\, L\, e^{-tA}\,.$$

By assumption $\hat{L}(t,s) \in X \otimes Y$, and hence $\hat{L} \in TT_{X\otimes Y;A\otimes I,I\otimes B}$. Further, $\hat{L}$ represents the operator L as described in (b).

(iv) The if-part is a special case of (iii).
For the only-if part consider the mapping emb: $S_{X,A} \hookrightarrow T_{X,A}$. It follows that $(t,s) \mapsto e^{-tA}\, e^{-sA}$ belongs to $TT_{X\otimes Y;A\otimes I,I\otimes A}$. Hence e^{-tA} is HS for all $t > 0$. □

The Theorems (III.6.6) and (III.6.7) present the kernel theorems which are current in the current distribution theories. Here three more kernel theorems will be introduced in which the so-called compound spaces play a crucial role.

(III.6.8) <u>Definition</u>

The canonical sesquilinear mapping $\otimes : T_{X,A} \times S_{Y,B} \to ST_{X\otimes Y;A\otimes I,I\otimes B}$ $[F,v] \to F \otimes v$, is defined by

$$(F \otimes v)(t,s) = F(t) \otimes (e^{-sB} v)\ , \qquad t > 0,\ s \geq -s_0$$

where $s_0 > 0$ must be chosen so small that $v \in e^{-s_0 B}(Y)$. The definition is consistent because $F \otimes v \in TT_{X\otimes Y;A\otimes I,I\otimes B}$ and $t \mapsto (F \otimes v)(t,-s_0) \in T_{X\otimes Y,A\otimes I}$.

(III.6.9) <u>Corollary</u>

The sesquilinear mapping $\otimes$ defined in (III.6.8) maps $T_{X,A} \times S_{Y,B}$ into $TS_{X\otimes Y;I\otimes B,A\otimes I}$.

<u>Proof</u>. Follows from Lemma (III.5.5.i) where it has been shown that

$$ST_{X\otimes Y;A\otimes I,I\otimes B} \mathrel{\hat{\subset}} TS_{X\otimes Y;I\otimes B,A\otimes I}\ .$$

□

(III.6.10) <u>Theorem</u>

$TS_{X\otimes Y;I\otimes B,A\otimes I}$ is a complete topological tensor product of $T_{X,A}$ and $S_{Y,B}$.

(i) $TS_{X\otimes Y;I\otimes B.A\otimes I}$ is complete.

(ii) The mapping $\otimes : T_{X,A} \times S_{Y,B} \to TS_{X\otimes Y;I\otimes B,A\otimes I}$ is continuous.

(iii) The algebraic tensor product $T_{X,A} \otimes_a S_{Y,B}$ is dense in $TS_{X\otimes Y;I\otimes B,A\otimes I}$.

Proof

(i) Completeness follows from Theorem (III.3.7).

(ii) For all $t > 0$ and all $\psi \in B_+(\mathbb{R})$ we have

$$\|(I \otimes \psi(B))(F \otimes w)(0,s)\|_{X\otimes Y} = \|F(s)\|_X \|\psi(B)w\|_Y$$

with $F \in T_{X,A}$, $w \in S_{Y,B}$. Thus continuity of $\otimes$ follows.

(iii) $S_{X\otimes Y, I\otimes B}$ is dense in $TS_{X\otimes Y; I\otimes B, A\otimes I}$. The result follows since $X \otimes_a S_{Y,B}$ is dense in $S_{X\otimes Y, I\otimes B}$. □

(III.6.11) Definition

The canonical sesquilinear mapping $\otimes : S_{X,A} \times T_{Y,B} \to ST_{X\otimes Y; I\otimes B, A\otimes I}$ $[w,G] \to w \otimes G$ is defined by

$$(w \otimes G)(t,s) = (e^{-sA}\, w) \otimes G(t) \ , \qquad t > 0,\ s \geq -s_0 \ .$$

Here $s_0 > 0$ must be chose so small that $w \in e^{-s_0 A}(X)$. This definition is consistent because $w \otimes G \in TT_{X\otimes Y; I\otimes B, A\otimes I}$ and $t \mapsto (w \otimes G)(t,-s_0) \in T_{X\otimes Y, I\otimes B}$.

(III.6.12) Corollary

The sesquilinear mapping $\otimes$ of (III.6.11) maps $S_{X,A} \times T_{Y,B}$ into $TS_{X\otimes Y; A\otimes I, I\otimes B}$.

Proof. Cf. Lemma (III.5.5). □

We define the mapping inv from $TT_{Y\otimes X; B\otimes I, I\otimes A}$ onto $TT_{X\otimes Y; I\otimes B, A\otimes I}$ by

$$(\text{inv}(F))(t,s) = F(t,s)^* \ , \qquad s > 0,\ t > 0 \ .$$

It can be checked that inv restricted to $TS_{Y\otimes X; B\otimes I, I\otimes A}$ is an antilinear homeomorphism from $TS_{Y\otimes X; B\otimes I, I\otimes A}$ onto $TS_{X\otimes Y; I\otimes B, A\otimes I}$, and, also, that inv restricted to $ST_{Y\otimes X; B\otimes I, I\otimes A}$ is an antilinear homeomorphism from $ST_{Y\otimes X; B\otimes I, I\otimes A}$ onto $ST_{X\otimes Y; I\otimes B, A\otimes I}$. So, as a consequence of Theorem III.6.10 we have

(III.6.13) Theorem

$TS_{X\otimes Y; A\otimes I, I\otimes B}$ is a complete topological tensor product of $S_{X,A}$ and $T_{Y,B}$.

(i) $TS_{X\otimes Y;A\otimes I,I\otimes B}$ is complete.

(ii) The mapping $\otimes : S_{X,A} \times T_{Y,B} \to TS_{X\otimes Y;A\otimes I,I\otimes B}$.

(iii) The algebraic tensor product $S_{X,A} \otimes_a T_{Y,B}$ is dense in $TS_{X\otimes Y;A\otimes I,I\otimes B}$. □

Next we show that the elements of $TS_{X\otimes Y;I\otimes B,A\otimes I}$ and $TS_{X\otimes Y;A\otimes I,I\otimes B}$ represent linear mappings in $\mathbf{L}(S_{X,A},S_{Y,B})$ and $\mathbf{L}(T_{X,A},T_{Y,B})$, respectively. We give necessary and sufficient conditions on the semigroups $(e^{-tA})_{t>0}$ and $(e^{-sB})_{s>0}$ which ensure the validity of a kernel theorem for $\mathbf{L}(S_{X,A},S_{Y,B})$ and $\mathbf{L}(T_{X,A},T_{Y,B})$.

<u>Case</u> (c) The space $\mathbf{L}(S_{X,A},S_{Y,B})$.

Let $\hat{P} \in TS_{X\otimes Y;I\otimes B,A\otimes I}$. Then we define the action of $\hat{P}$ as a linear operator from $S_{X,A}$ into $S_{Y,B}$ in the following way:

(c) $$Pw = \hat{P}(0,\sigma)e^{\sigma A}\, w\ .$$

For any $w \in S_{X,A}$ this definition makes sense if $\sigma > 0$ is taken so small that $w \in e^{-\sigma A}(X)$. Further, since $\hat{P}(0,\sigma) \in \mathcal{D}(I \otimes \psi(B))$ for all $\psi \in \mathbf{B}_+(\mathbb{R})$, we obtain $\hat{P}(0,\sigma)e^{\sigma A}\, w \in S_{Y,B}$. It is clear that Definition (c) does not depend on $\sigma > 0$.

(III.6.14) <u>Theorem</u>

(i) For each $\hat{P} \in TS_{X\otimes Y;I\otimes B,A\otimes I}$ the linear operator P from $S_{X,A}$ into $S_{Y,B}$ defined by (c), is continuous.

(ii) Let $\hat{P} \in TS_{X\otimes Y;I\otimes B,A\otimes I}$, let $w \in S_{X,A}$ and let $G \in T_{Y,B}$. Then

$$\langle Pw,G\rangle_Y = \langle\!\langle w\otimes G,\hat{P}\rangle\!\rangle_{X\otimes Y}\ .$$

(iii) If for each $t > 0$, $s > 0$, at least one of the operators e^{-tA}, e^{-sB} is HS, then $TS_{X\otimes Y;I\otimes B,A\otimes I}$ comprises all elements of $\mathbf{L}(S_{X,A},S_{Y,B})$.

(iv) The space $TS_{X\otimes X;I\otimes A,A\otimes I}$ comprises all elements of $\mathbf{L}(S_{X,A})$ iff for each $t > 0$ the operator e^{-tA} is HS.

Proof

(i) We use condition (iii) of Theorem (I.4.2).

Let $s > 0$. Then for all $\psi \in B_+(\mathbb{R})$, $\hat{P}(0,s) \in \mathcal{D}(I \otimes \psi(B))$. So the operator $\psi(B)\hat{P}(0,s)$ belongs to $X \otimes Y$, and therefore $\psi(B)Pe^{-sA}$ is bounded.

(ii) Note first that $w \otimes G \in ST_{X\otimes Y;\, I\otimes B,\, A\otimes I}$. We have

$$\begin{aligned}
\langle Pw,G\rangle_Y &= (\hat{P}(-\tau,\sigma)e^{\sigma A}\, w, G(\tau))_Y \\
&= \overline{(e^{\sigma A}\, w \otimes G(\tau), \hat{P}(-\tau,\sigma))}_{X\otimes Y} \\
&= \overline{((w \otimes G)(\tau,-\sigma), \hat{P}(-\tau,\sigma))}_{X\otimes Y} \\
&= \overline{\langle w \otimes G, \hat{P}\rangle}_{X\otimes Y} \,.
\end{aligned}$$

Here $\tau > 0$ must be chosen so small that $\hat{P}(0,\sigma) \in I \otimes e^{\tau B}(X \otimes Y)$.

(iii) Let $Q: S_{X,A} \to S_{Y,B}$ be continuous. According to Theorem (I.4.2), condition (iii), for all $s > 0$ there exists $t_s > 0$ such that the linear operator $e^{t_s B}\, Q\, e^{-sA}: X \to Y$ is bounded. Now define $\hat{Q}$ by

$$\hat{Q}(t,s) = e^{-tB}\, Q\, e^{-sA}\,, \qquad s > 0,\ t \geqq 0\,.$$

Then $\hat{Q} \in TT_{X\otimes Y;\, I\otimes B,\, A\otimes I}$. Moreover, for all $s > 0$

$$\hat{Q}(0,s) \in \mathcal{D}(I \otimes \psi(B))$$

because

$$\psi(B)\hat{Q}(0,s) = \psi(B)e^{-t_s B}\, \hat{Q}(-t_s,s)\,.$$

So it follows that $\hat{Q} \in TS_{X\otimes Y;\, I\otimes B,\, A\otimes I}$, and $\hat{Q}$ represents the linear mapping $Q \in L(S_{X,A}, S_{Y,B})$.

(iv) The if-part is a special case of (iii).

For the only-if part consider the identity mapping I. It follows that for all $s > 0$ and $t > 0$ the operator $e^{-tA}\, I\, e^{-sB}$ is HS. □

<u>Case</u> (d) The space $\mathbf{L}(T_{X,A},T_{Y,B})$

Let $\hat{\Phi} \in TS_{X\otimes Y;A\otimes I,I\otimes B}$. Then we define the action of $\hat{\Phi}$ as a linear mapping from $T_{X,A}$ into $T_{Y,B}$ in the following way:

$$\text{(d)} \qquad (\Phi F): s \mapsto \hat{\Phi}(-t_s,s)F(t_s) \ , \qquad s > 0 \ ,$$

where $t_s > 0$ has to be taken so small that $\hat{\Phi}(0,s) \in (e^{-t_sA} \otimes I)(X \otimes Y)$. It is clear that definition (d) does not depend on the choice of t_s.

Another way of introducing ΦF is the following. Since $F \in T_{X,A}$, there exists $x \in X$ and $\psi \in \mathbf{B}_+(\mathbb{R})$ such that $F = \psi(A)x$. Now put

$$(\Phi F)(s) = [\hat{\Phi}(0,s)\psi(A)]x \ , \qquad s > 0 \ .$$

It is not hard to check that this definition does not depend on the choice of x and ψ. It is equivalent to definition (d).

(III.6.15) <u>Theorem</u>

(i) For each $\hat{\Phi} \in TS_{X\otimes Y;A\otimes I,I\otimes B}$ the linear operator Φ from $T_{X,A}$ into $T_{Y,B}$ as defined in (d), is continuous.

(ii) Let $\hat{\Phi} \in TS_{X\otimes Y;A\otimes I,I\otimes B}$, let $F \in T_{X,A}$ and let $v \in S_{Y,B}$. Then

$$\langle v,\Phi F\rangle_Y = \langle\langle F\otimes v,\hat{\Phi}\rangle\rangle_{X\otimes Y} \ .$$

(iii) If for each s,t > 0 at least one of the operators e^{-tA}, e^{-sB} is HS, then $TS_{X\otimes Y;A\otimes I,I\otimes B}$ comprises all elements of $\mathbf{L}(T_{X,A},T_{Y,B})$.

(iv) The space $TS_{X\otimes X;A\otimes I,I\otimes A}$ comprises all elements of $\mathbf{L}(T_{X,A})$ iff the operator e^{-tA} is HS for all t > 0.

<u>Proof</u>

(i) We use condition (ii) of Theorem (I.4.8).
Let s > 0. Then $\hat{\Phi}(0,s) \in S_{X\otimes Y,A\otimes I}$. According to Case (a) $\hat{\Phi}(o,s)$ is a continuous mapping from $T_{X,A}$ into Y.

(ii) First observe that $F \otimes v \in ST_{X\otimes Y;A\otimes I,I\otimes B}$. We have

$$\langle v,\Phi F\rangle_{Y} = (e^{sB} v,\hat{\Phi}(-t_s,s)F(t_s))_{Y} =$$

$$= ((F \otimes v)(t_s,-s),\hat{\Phi}(-t_s,s))_{X\otimes Y} = \langle F\otimes v,\hat{\Phi}\rangle_{X\otimes Y} .$$

(iii) Let Ψ: $T_{X,A} \to T_{Y,B}$ be a continuous linear mapping. According to Theorem (I.4.8), condition (iii), for all s > 0 there exists $t_s > 0$ such that the operator $e^{-sB}\Psi e^{t_sA} : X \supset e^{-t_sA}(X) \to Y$ is bounded and hence HS. We define $\hat{\Psi}$ by

$$\hat{\Psi}(t,s) = e^{-sB}\Psi e^{-tA} , \qquad s > 0,\ t \geqq 0 .$$

Then $\hat{\Psi} \in TT_{X\otimes Y;A\otimes I,I\otimes B}$. Moreover, for all s > 0

$$\hat{\Psi}(0,s) \in \mathcal{D}(\psi(A) \otimes I)$$

because

$$\hat{\Psi}(0,s)\psi(A) \subset \Psi(-t_s,s)\psi(A)e^{-t_sA} .$$

It follows that $\hat{\Psi} \in TS_{X\otimes Y;A\otimes I,I\otimes B}$, and $\hat{\Psi}$ represents Ψ.

(iv) For the only-if part, consider the identity I. It follows that, for all s > 0 and t > 0, $e^{-sA}\,I\,e^{-tA}$ is HS. □

In Section (III.5) for a pair of strongly commuting nonnegative self-adjoint operators C and D we have defined the operators $C \wedge D$:

$$C \wedge D = \iint_{\mathbb{R}^2} \max(\lambda,\mu)\,d\,G_\lambda H_\mu$$

and the operator $C \vee D$:

$$C \vee D = \iint_{\mathbb{R}^2} \min(\lambda,\mu)\,d\,G_\lambda H_\mu .$$

We have shown that

$$TS_{Z;C,D} \cap TS_{Z;D,C} \cong TS_{Z;C\wedge D,C\vee D}$$

and

$$ST_{Z;C,D} + ST_{Z;D,C} \cong ST_{Z;C\wedge D,C\vee D} .$$

Here we apply these results to the spaces $TS_{X\otimes Y;I\otimes B,A\otimes I}$, $TS_{X\otimes Y;A\otimes I,I\otimes B}$, $ST_{X\otimes Y;I\otimes B,A\otimes I}$ and $ST_{X\otimes Y;A\otimes I,I\otimes B}$. We introduce the notation:

$$A \textcircled{\scriptsize$\wedge$} B = (A \otimes I) \wedge (I \otimes B) \;, \qquad A \textcircled{\scriptsize$\vee$} B = (A \otimes I) \vee (I \otimes B) \;.$$

(III.6.16) <u>Definition</u>

The mapping ext: $S_{X,A} \otimes_a S_{Y,B} \to TS_{X\otimes Y; A \textcircled{\scriptsize$\wedge$} B, A \textcircled{\scriptsize$\vee$} B}$ is defined by

$$\text{ext}(w \otimes v)(t,s) \mapsto e^{-t(A \textcircled{\scriptsize$\wedge$} B)}\, e^{-s(A \textcircled{\scriptsize$\vee$} B)}(w \otimes v)$$

where $s > 0$ and $t \geq -\tau$ with $\tau > 0$ so small that $w \in e^{-\tau A}(X)$ and $v \in e^{-\tau B}(Y)$.

(III.6.17) <u>Theorem</u>

The space $TS_{X\otimes Y; A \textcircled{\scriptsize$\wedge$} B, A \textcircled{\scriptsize$\vee$} B}$ is a complete topological tensor product of the spaces $S_{X,A}$ and $S_{Y,B}$.

(i) $TS_{X\otimes Y; A \textcircled{\scriptsize$\wedge$} B, A \textcircled{\scriptsize$\vee$} B}$ is complete.

(ii) The mapping ext $\circ \otimes : S_{X,A} \times S_{Y,B} \to TS_{X\otimes Y; A \textcircled{\scriptsize$\wedge$} B, A \textcircled{\scriptsize$\vee$} B}$ is continuous.

(iii) $\text{ext}(S_{X,A} \otimes_a S_{Y,B})$ is dense in $TS_{X\otimes Y; A \textcircled{\scriptsize$\wedge$} B, A \textcircled{\scriptsize$\vee$} B}$.

<u>Proof</u>

(i) All spaces of this kind are complete.

(ii) It is sufficient to check continuity at $[0;0]$. Let $\psi \in B_+(\mathbb{R})$ and let $s > 0$. Then we have the estimation

$$\|\psi(A \textcircled{\scriptsize$\wedge$} B) e^{-t(A \textcircled{\scriptsize$\vee$} B)}(w \otimes v)\|_{X\otimes Y} \leq$$

$$\leq \|\psi(A) w\|_X \|v\|_Y + \|w\|_X \|\psi(B) v\|_Y$$

for all $w \in S_{X,A}$ and $v \in S_{Y,B}$. Thus continuity of ext $\circ \otimes$ follows.

(iii) Follows from the inclusion diagram (III.5.7). □

<u>Remark</u>. For all $w \in S_{X,A}$, $v \in S_{Y,B}$, $F \in T_{X,A}$ and $G \in T_{Y,B}$

$$w \otimes G + F \otimes v \in ST_{X\otimes Y; A \textcircled{\scriptsize$\wedge$} B, A \textcircled{\scriptsize$\vee$} B} \;.$$

<u>Case</u> (e) The space $\mathbf{E}(S_{X,A}, S_{Y,B})$

Let $\hat{E} \in TS_{X\otimes Y; A \,Ⓐ\, B, A \,Ⓥ\, B}$. Then for each $w \in S_{X,A}$ we define

(e1) $$Ew = ([e^{\sigma(A \,Ⓥ\, B)}(e^{-\sigma A} \otimes I)]\, \hat{E}(0,\sigma))(e^{\sigma A} w)$$

where $\sigma > 0$ has to be taken so small that $w \in e^{-\sigma A}(X)$.

Definition (e1) does not depend on the choice of σ. Further, $Ew \in S_{Y,B}$ because

$$\hat{E}(0,\sigma) \in \mathcal{D}(\psi(A \,Ⓐ\, B)) \subset \mathcal{D}(I \otimes \psi(B))$$

for all $\psi \in \mathbf{B}_+(\mathbb{R})$. Observe that the operator $e^{\sigma(A \,Ⓥ\, B)}(e^{-\sigma A} \otimes I)$ is bounded on $X \otimes Y$ for each $\sigma > 0$.

Let $F \in T_{X,A}$. We define the extension $\bar{E}$ of E by

(e2) $$(\bar{E}F): s \mapsto ([e^{s(A \,Ⓥ\, B)}(I \otimes e^{-sB})][(e^{t_s A} \otimes I)e^{-t_s(A \,Ⓐ\, B)}]\, \hat{E}(-t_s, s))F(t_s)$$

where, for each $s > 0$, $t_s > 0$ must be taken so small that

$$\hat{E}(0,s) \in e^{-t_s(A \,Ⓐ\, B)}(X \otimes Y) .$$

Observe that for all $s > 0$ the operators $e^{s(A \,Ⓐ\, B)}(I \otimes e^{-sB})$ and $(e^{t_s A} \otimes I)e^{-t_s(A \,Ⓐ\, B)}$ are bounded on $X \otimes Y$. It follows that $\bar{E}F \in T_{Y,B}$.

We omit the proof of the final theorem of this section. This proof is an immediate consequence of Theorems (III.6.14) and (III.6.15).

(III.6.18) <u>Theorem</u>

(i) For each $\hat{E} \in TS_{X\otimes Y; A \,Ⓐ\, B, A \,Ⓥ\, B}$ the linear operator E defined in (e1) is an extendible mapping with extension $\bar{E}$ defined in (e2).

(ii) For each $\hat{E} \in TS_{X\otimes Y; A \,Ⓐ\, B, A \,Ⓥ\, B}$, $w \in S_{X,A}$, $v \in S_{Y,B}$, $F \in T_{X,A}$ and $G \in T_{Y,B}$

$$\langle\langle w \otimes G + F \otimes v, \hat{E}\rangle\rangle_{X\otimes Y} = \langle Ew, G\rangle_Y + \langle v, \bar{E}F\rangle_Y .$$

(iii) If for each $s,t > 0$ at least one of the operators e^{-tA} or e^{-sB} is Hilbert-Schmidt, then $TS_{X\otimes Y; A \,Ⓐ\, B, A \,Ⓥ\, B}$ comprises all elements of $\mathbf{E}(S_{X,A}, S_{Y,B})$.

(iv) $TS_{X\otimes X;A \text{Ⓐ} A,A \text{Ⓥ} A}$ comprises all elements of $\mathbf{E}(S_{X,A})$ iff the operator e^{-tA} is HS for all $t > 0$.

Appendix

Let A, B denote nonnegative self-adjoint operators in the separable Hilbert spaces X and Y, respectively. The seminorms which generate the topology σ_{ind} for the space $S_{X\otimes Y,A\otimes I+I\otimes B}$ are given by

$$p_\psi(\theta) = \|\psi(A \otimes I + I \otimes B)\theta\|_{X\otimes Y}\ , \qquad \theta \in S_{X\otimes Y,A\otimes I+I\otimes B}$$

where $\psi \in \mathbf{B}_+(\mathbb{R})$. We have

$$\psi(A\otimes I + I\otimes B) = \iint_{\mathbb{R}^2} \psi(\lambda+\mu)d\, E_\lambda \otimes F_\mu$$

where $(E_\lambda)_{\lambda\in\mathbb{R}}$, $(F_\mu)_{\mu\in\mathbb{R}}$ denote the spectral resolution corresponding to A and B, respectively.

In this appendix we want to show that also the set of seminorms

$$q_\varphi(\theta) = \|(\varphi(A) \otimes \varphi(B))\theta\|_{X\otimes Y}\ , \qquad \theta \in S_{X\otimes Y,A\otimes I+I\otimes B}\ , \quad \varphi \in \mathbf{B}_+(\mathbb{R})\ ,$$

induce the topology σ_{ind}.

Without damaging the generality we can assume that the functions in $\mathbf{B}_+(\mathbb{R})$ are monotone increasing. Then it follows that

$$\psi(\lambda)\psi(\mu) \leqq \psi(\lambda+\mu)^2\ , \qquad \lambda,\mu \in \mathbb{R}^+\ ,$$

for all $\psi \in \mathbf{B}_+(\mathbb{R})$. It follows that for each $\psi \in \mathbf{B}_+(\mathbb{R})$,

$$(*) \qquad \|(\psi(A) \otimes \psi(B))\theta\|_{X\otimes Y} \leqq \|\psi(A \otimes I + I \otimes B)^2\, \theta\|_{X\otimes Y}$$

for all $\theta \in S_{X\otimes Y,A\otimes I+I\otimes B}$.

Next, let $\varphi \in \mathbf{B}_+(\mathbb{R})$. We may as well assume that φ is increasing and larger than 1. Define the function χ by

$$\chi(\lambda) = \varphi(2n) \quad \text{if } \lambda \in [n-1,n)\ , \qquad n \in \mathbb{N}\ .$$

Then for all $\lambda,\mu \in \mathbb{R}^+$ with $\lambda \in [n-1,n)$ and $\mu \in [m-1,m)$

$$\chi(\lambda)\chi(\mu) \geqq \varphi(2n)\varphi(2m) \geqq \varphi(n+m) \geqq \varphi(\lambda+\mu) \ .$$

Moreover, $\chi \in \mathbf{B}_+(\mathbb{R})$, because for each $t > 0$

$$\sup_{\lambda\geqq 0} \chi(\lambda)e^{-t\lambda} \leqq \sup_{k\in\mathbb{N}} \varphi(2k)e^{-t(k-1)} =$$

$$= e^{t} \sup_{\lambda\geqq 0} (\varphi(\lambda)e^{-\frac{1}{2}t\lambda}) < \infty \ .$$

Finally, for all $\theta \in S_{X\otimes Y, A\otimes I+I\otimes B}$,

$$(**) \qquad \|\varphi(A \otimes I + I \otimes B)\theta\|_{X\otimes Y} \leqq \|(\chi(A) \otimes \chi(B))\theta\|_{X\otimes Y} \ .$$

From (*) and (**), the wanted result can be obtained.

CHAPTER IV

Algebras of continuous linear mappings on analyticity spaces and trajectory spaces

Introduction

In the mathematical description of quantum mechanics the observables are represented by self-adjoint operators which are unbounded in general. So it does not seem natural to consider only algebras of bounded operators in such descriptions. A lot of investigation has been done in the field of algebras of unbounded operators and their abstractions. In the papers [Ju], [Po], [Al] and [Di] one looks for generalizations of C*-algebras, which are more in line with the use of operator algebras in theoretical physics. In each generalization genuine problems are: the definition of spectrum, the definition of positivity of elements, the existence of sufficient projections, the description of state space and characterization of ideals.
In this chapter we investigate the unbounded operator algebras $\mathsf{L}(S_{X,A})$, $\mathsf{L}(T_{X,A})$ and $\mathsf{E}(S_{X,A})$. Endowed with their natural topologies, they are locally convex algebras, i.e. locally convex topological vector spaces in which multiplication is separately continuous. The algebra $\mathsf{E}(S_{X,A})$ is an involutive algebra, it seems the most interesting of the three. We remark that our operator algebras seem to have little affinity with the algebras introduced in the above mentioned papers.
Although there are great dissimilarities between our theory and Hilbert space theory, we discuss the algebras $\mathsf{L}(S_{X,A})$, $\mathsf{L}(T_{X,A})$ and $\mathsf{E}(S_{X,A})$ along the lines of bounded operator theory for Hilbert spaces. So in each of these algebras we define five topologies which correspond to the usual five topologies in the Banach algebra $\mathsf{B}(X)$. Also, we discuss the relationship between the algebraic and the toplogical structure of our algebras. The first three sections cover this type of considerations.

In Section IV.4 we present a general method to obtain ideals in $\mathbf{L}(S_{X,A})$, $\mathbf{L}(T_{X,A})$ and $\mathbf{E}(S_{X,A})$ from ideals in $\mathbf{B}(X)$. Thus it can be seen that $TS_{X\otimes X;I\otimes A,A\otimes I}$ is an ideal in $\mathbf{L}(S_{X,A})$, which plays the same role as the ideal $X \otimes X$ in the algebra $\mathbf{B}(X)$. Similar considerations are valid for the spaces $TS_{X\otimes X;A\otimes I,I\otimes A}$ and $TS_{X\otimes X;A \wedge A,A \vee A}$. Moreover, these considerations lead to a description of the preduals $\mathbf{L}_*(S_{X,A})$, $\mathbf{L}_*(T_{X,A})$ and $\mathbf{E}_*(S_{X,A})$. To this end, note that the ideal in $\mathbf{B}(X)$ established by the trace class operators is a representation of $\mathbf{B}_*(X)$.

If $S_{X,A}$ is a nuclear space, then in a natural way $S_{X,A}$ can be identified with a sequence space, which is dense in the Hilbert space of square summable sequences. Hence to each continuous linear mapping of $S_{X,A}$ there corresponds an infinite matrix. Section IV.5 is devoted to a detailed characterization of the matrices corresponding to the elements of $\mathbf{L}(S_{X,A})$. We thus obtain an identification between $\mathbf{L}(S_{X,A})$ and a class of well specified matrices. It enables us to construct a large variety of elements in $\mathbf{L}(S_{X,A})$. We mention here the class of unbounded weighted shifts. These operators are the simplest nontrivial operators in $\mathbf{L}(S_{X,A})$. Also, the matrix calculus has proved to be very important in the classification problem: Look for conditions on the pair (Y,B) such that $S_{X,A} = S_{Y,B}$ for given A and X. Results in this direction can be found in [EG 2-3].

In Section IV.6 we employ the matrix calculus in the construction of nuclear analyticity spaces on which a prescribed set of linear operators acts continuously. For a countable number of bounded linear operators and for a finite number of commuting self-adjoint operators such nuclear analyticity spaces indeed exist.

Finally, in Section IV.7, we indicate a possibility to interprete parts of classical quantum statistical mechanics by means of the mathematical apparatus developed for the algebra $\mathbf{E}(S_{X,A})$ and its product $\mathbf{E}_*(S_{X,A})$. Here we consider only nuclear $S_{X,A}$. We characterize the positive functionals, the normal states, and we show that the pure states are precisely the extreme points in the set of normal states. We consider it advantageous that the $*$-algebra $\mathbf{E}(S_{X,A})$ contains unbounded self-adjoint operators. Also it is an advantage that $\mathbf{E}(S_{X,A})$ behaves very much like a C*-algebra, topologically seen. However, it is a disadvantage that there is no abstact characterization of the algebras of type $\mathbf{E}(S_{X,A})$. So the choice of the nuclear analyticity space $S_{X,A}$ is not arbitrary. It presupposes a relationship between the analyticity space and the quantum mechanical system under consideration. This relationship is not yet very clear.

IV.1. The algebra $\mathbf{L}(S_{X,A})$

The algebra $\mathbf{L}(S_{X,A})$ consists of all continuous linear mappings from $S_{X,A}$ into $S_{X,A}$. In Section I.4 the following characterizations have been given.

(IV.1.1) Characterization

Let L be a densely defined linear operator in X with $\mathcal{D}(L) \supset S_{X,A}$.

(i) $L \in \mathbf{L}(S_{X,A})$ iff $\forall_{t>0}\ \forall_{\psi \in \mathbf{B}_+(\mathbb{R})}$: the operator $\psi(A)\,L\,e^{-tA}$ is a densely defined bounded linear operator in X.

(ii) $L \in \mathbf{L}(S_{X,A})$ iff $\forall_{t>0}\ \exists_{\tau>0}$: the operator $e^{\tau A}\,L\,e^{-tA}$ is a densely defined bounded linear operator in X.

We define the uniform topology $\lambda\sigma_u$ for $\mathbf{L}(S_{X,A})$ as follows.

(IV.1.2) Definition

The uniform topology $\lambda\sigma_u$ for $\mathbf{L}(S_{X,A})$ is the locally convex topology generated by the seminorms $s_{\psi,t}$, $\psi \in \mathbf{B}_+(\mathbb{R})$, $t > 0$,

$$s_{\psi,t}(L) = \|(\psi(A)\,L\,e^{-tA}\| , \quad L \in \mathbf{L}(S_{X,A}) .$$

Here $\|\cdot\|$ denotes the norm of the Banach algebra $\mathbf{B}(X)$.

(IV.1.3) Lemma

If $S_{X,A}$ is a nuclear space, then $(S_{X,A}) \mathrel{\hat{=}} TS_{X\otimes X;\, I\otimes A,\, A\otimes I}$ and the topology $\tau\sigma_{proj}$ for $TS_{X\otimes X;\, I\otimes A,\, A\otimes I}$ is equivalent to the topology $\lambda\sigma_u$ for $\mathbf{L}(S_{X,A})$.

Proof. Let $S_{X,A}$ be a nuclear space. It means that e^{-tA} is a Hilbert-Schmidt operator for all $t > 0$. So by Theorem (III.6.14) the space $TS_{X\otimes X;\, I\otimes A,\, A\otimes I}$ comprises the whole $\mathbf{L}(S_{X,A})$.

The topology $\tau\sigma_{proj}$ is generated by the seminorms

$$p_{\psi,s}(P) = \|\psi(A)P(0,s)\|_{X\otimes X}\ , \qquad s > 0,\ \psi \in B_+(\mathbb{R})\ .$$

Let $L \in \mathbf{L}(S_{X,A})$ with corresponding representant $\hat{L} \in TS_{X\otimes X;I\otimes A,A\otimes I}$. Then for all $t > 0$ and all $\psi \in B_+(\mathbb{R})$ the following relation is valid:

$$s_{\psi,t}(L) \leq p_{\psi,t}(\hat{L}) \leq \|e^{-\frac{1}{2}tA}\|_{X\otimes X}\ s_{\psi,\frac{1}{2}t}(L)\ .$$

This proves the assertion of the lemma. □

The algebra $\mathbf{L}(S_{X,A})$ with topology $\lambda\sigma_u$ is a locally convex topological vector space. In the following theorem we list other properties.

(IV.1.4) Lemma

(i) The space $\mathbf{L}(S_{X,A})$ is $\lambda\sigma_u$-complete.

(ii) Multiplication is separately $\lambda\sigma_u$-continuous.

Proof

(i) Let (L_α) be a Cauchy net in $\mathbf{L}(S_{X,A})$. Let $t > 0$. Then $(L_\alpha\ e^{-tA})$ is a Cauchy net in $\mathbf{B}(X)$. So there exists $L(t)$ in $\mathbf{B}(X)$ such that $L(t)$ is the limit of the net $(L_\alpha\ e^{-tA})$. We define the operator L on $S_{X,A}$ by

$$Lf := L(t)(e^{tA}\ f)\ , \qquad f \in S_{X,A}\ ,$$

where $t > 0$ must be chose so small that $f \in e^{-tA}(X)$. The definition of L does not depend on the choice of t, because

$$Lf = [\lim_\alpha L_\alpha\ e^{-tA}]\ e^{tA}\ f = X - \lim_\alpha\ (L_\alpha\ f)\ .$$

Moreover, each of the operators $\psi(A)$, $\psi \in B_+(\mathbb{R})$ is closed in X. It follows that $Lf \in \mathcal{D}(\psi(A))$, and $\psi(A)Lf = X - \lim_\alpha\ (\psi(A)L_\alpha\ f)$. Hence

$$Lf \in \bigcap_{\psi\in B_+(\mathbb{R})} \mathcal{D}(\psi(A)) = S_{X,A}\ .$$

By condition (iv) of Theorem (I.4.2) we obtain $L \in \mathbf{L}(S_{X,A})$.

(ii) Let $M \in \mathbf{L}(S_{X,A})$ be fixed. The continuity of the linear mapping

$$L \mapsto LM\ , \qquad L \in \mathbf{L}(S_{X,A})$$

follows from the inequality

$$\|\psi(A)LM\, e^{-tA}\| \leq \|\psi(A)L\, e^{-\tau A}\| \|e^{\tau A}\, M e^{-tA}\|$$

where $\tau > 0$ must be taken sufficiently small.

Next we show the continuity of the mapping $L \mapsto ML$.

Let $\varepsilon > 0$, let $t > 0$ and let $\psi \in \mathbf{B}_+(\mathbb{R})$. There exists an open neighbourhood Ω in $S_{X,A}$ such that

$$\|\psi(A)\, M\, v\| < \frac{\varepsilon}{2}$$

as soon as $v \in \Omega$. The existence of Ω follows from the continuity of M.

Let (L_α) be a net in $\mathbf{L}(S_{X,A})$ which converges to zero in the topology $\lambda\sigma_u$.

By Theorem (I.1.4) there exists $\varphi \in \mathbf{B}_+(\mathbb{R})$ such that

$$\{w \in S_{X,A} \mid \|\varphi(A)u\| < 1\} \subset \Omega\ .$$

Further, there exists α_1 such that $\|\varphi(A)L_\alpha\, e^{-tA}\| < 1$ for all $\alpha > \alpha_1$. It follows that

$$\|\psi(A)ML_\alpha\, e^{-tA}\, x\| < \frac{\varepsilon}{2}$$

for all $x \in X$ with $\|x\| = 1$. So we derive

$$\|\psi(A)ML\, e^{-tA}\| \leq \frac{\varepsilon}{2} < \varepsilon$$

for all $\alpha > \alpha_1$. □

(IV.1.5) Theorem

$\mathbf{L}(S_{X,A})$ with topology $\lambda\sigma_u$ is a complete locally convex algebra. □

Since the elements of $\mathbf{L}(S_{X,A})$ are linear mappings on $S_{X,A}$, in a natural way the elements of $S_{X,A}$ endow $\mathbf{L}(S_{X,A})$ with the so-called strong topology, i.e. the topology of pointwise convergence.

(IV.1.6) Definition

The strong topology $\lambda\sigma_s$ for $\mathbf{L}(S_{X,A})$ is the locally convex topology generated by the seminorms $s_{w,\psi}$,

$$s_{w,\psi}(L) = \|\psi(A)\,L\,w\| , \qquad L \in \mathbf{L}(S_{X,A}) ,$$

where we take $w \in S_{X,A}$ and $\psi \in \mathbf{B}_+(\mathbb{R})$.

We note that the net (P_α) in $\mathbf{L}(S_{X,A})$ is $\lambda\sigma_s$-convergent with limit P if and only if the net $(P_\alpha w)$ is σ_{ind}-convergent with limit Pw for all $w \in S_{X,A}$. The topology $\lambda\sigma_s$ is the coarsest topology for which the linear mappings

$$L \mapsto Lw , \qquad L \in \mathbf{L}(S_{X,A}) , \quad w \in S_{X,A} ,$$

are continuous with respect to the topology σ_{ind}.

(IV.1.7) <u>Lemma</u>

The topology $\lambda\sigma_u$ is equivalent to the topology of uniform convergence on bounded subsets of $S_{X,A}$.

<u>Proof</u>. Let B be a bounded subset of $S_{X,A}$. Let (P_α) be a $\lambda\sigma_u$-convergent net in $\mathbf{L}(S_{X,A})$ with limit P. Then there is $t > 0$ such that the set $e^{tA}(B)$ is bounded in X. We have for all $w \in B$, all $\psi \in \mathbf{B}_+(\mathbb{R})$ and all α,

$$\|\psi(A)(P_\alpha - P)w\| \leq \|\psi(A)(P_\alpha - P)e^{-tA}\| \sup_{w \in B} (\|e^{tA} w\|) .$$

On the other hand, let $\varepsilon > 0$ and $t > 0$. Suppose the net $(P_\alpha) \subset \mathbf{L}(S_{X,A})$ converges on the bounded subset $\{e^{-tA} x \mid \|x\| = 1\} \subset S_{X,A}$ to the limit P. Then for each $\psi \in \mathbf{B}_+(\mathbb{R})$ there exists α_1 such that

$$\|\psi(A)(P_\alpha - P)e^{-tA} x\| < \frac{\varepsilon}{2} , \qquad \alpha > \alpha_1 ,$$

for all $x \in X$ with $\|x\| = 1$. It follows that

$$\|\psi(A)(P_\alpha - P)e^{-tA}\| \leqq \frac{\varepsilon}{2} < \varepsilon , \qquad \alpha > \alpha_1 . \qquad \square$$

(IV.1.8) <u>Definition</u>

The weak topology $\lambda\sigma_w$ is the locally convex topology for $\mathbf{L}(S_{X,A})$ generated by the seminorms $s_{w,F}$,

$$s_{w,F}(L) = |\langle Lw, F\rangle| , \qquad L \in \mathbf{L}(S_{X,A}) ,$$

where we take $w \in S_{X,A}$ and $F \in T_{X,A}$.

The net (L_α) in $\mathbf{L}(S_{X,A})$ converges weakly to $L \in \mathbf{L}(S_{X,A})$ if and only if $\langle (L_\alpha - L)w, F\rangle \to 0$ for all $w \in S_{X,A}$ and all $F \in T_{X,A}$. The weak topology $\lambda\sigma_w$ is the coarsest topology for which all linear functionals

$$L \mapsto \langle Lw, F\rangle \,, \qquad L \in \mathbf{L}(S_{X,A})$$

are continuous. We remark that $\lambda\sigma_s$ is the topology of uniform weak convergence on bounded subsets of $T_{X,A}$.

(IV.1.9) Theorem (Principle of uniform boundedness)

Let B be a subset of $\mathbf{L}(S_{X,A})$. The following statements are equivalent:

(i) B is $\lambda\sigma_u$-bounded.

(ii) B is $\lambda\sigma_s$-bounded.

(iii) B is $\lambda\sigma_w$-bounded.

Proof. It is clear that (i) $\Rightarrow$ (ii) $\Rightarrow$ (iii).
The equivalence (ii) $\Leftrightarrow$ (iii) follows from the Banach-Steinhaus theorem for $S_{X,A}$, which is stated in Theorem (I.3.4).
In order to prove the implication (ii) $\Rightarrow$ (i), let $t > 0$ and let $\psi \in \mathbf{B}_+(\mathbb{R})$. Then for all $x \in X$ the set

$$\{\psi(A)\, L\, e^{-tA}\, x \mid L \in B\}$$

is bounded in X. The Banach-Steinhaus theorem for $\mathcal{B}(X)$ yields a constant $\alpha_{t,\psi} > 0$ such that

$$\|\psi(A)\, L\, e^{-tA}\| < \alpha_{t,\psi} \,.$$

for all $L \in B$. □

(IV.1.10) Corollary

Let $(P_n)_{n\in\mathbb{N}}$ be a sequence in $\mathbf{L}(S_{X,A})$.

(i) Suppose for each $w \in S_{X,A}$ the sequence $(P_n w)_{n\in\mathbb{N}}$ is σ_{ind}-convergent. Then the linear mapping $P\colon w \mapsto \lim_{n\to\infty} P_n w$ belongs to $\mathbf{L}(S_{X,A})$.

(ii) In (i) we can replace the σ_{ind}-convergence by weak convergence.

Proof

(i) The sequence $(P_n)_{n\in\mathbb{N}}$ is strongly bounded in $\mathbf{L}(S_{X,A})$ and hence it is uniformly bounded by Theorem (IV.1.9). It follows that for each $t > 0$ there exists $\alpha_t > 0$ such that $\|P_n\, e^{-tA}\| \leq \alpha_t$ for all $n \in \mathbb{N}$. It is obvious that P is a linear mapping from $S_{X,A}$ into itself. Further we have for all $t > 0$ and all $x \in X$ with $\|x\| = 1$

$$\|P\, e^{-tA}\, x\| \leq \|(P-P_n)e^{-tA}\, x\| + \alpha_t \leq \alpha_t + 1$$

for $n \in \mathbb{N}$ (dependent on x) sufficiently large. By condition (iii) of Theorem (I.4.2) it follows that $P \in \mathbf{L}(S_{X,A})$.

(ii) The proof of (ii) runs similarly to the proof of (i). □

(IV.1.11) Corollary

$\mathbf{L}(S_{X,A})$ is sequentially $\lambda\sigma_s$- and $\lambda\sigma_w$-complete.

Proof. The proof can be derived from the previous corollary by using the σ_{ind}-completeness, respectively weak sequential completeness. of $S_{X,A}$. □

We did not yet characterize the bounded subsets of $\mathbf{L}(S_{X,A})$.

(IV.1.12) Theorem

Let $B \subset \mathbf{L}(S_{X,A})$. Then the following three statements are equivalent:

(i) B is bounded.

(ii) $\forall_{t>0}\ \forall_{\psi\in \mathbf{B}_+(\mathbb{R})}$: $\{\psi(A)\, L\, e^{-tA} \mid L \in B\}$ is bounded in $\mathbf{B}(X)$.

(iii) $\forall_{t>0}\ \exists_{\tau>0}$: $\{e^{\tau A}\, L\, e^{-tA} \mid L \in B\}$ is bounded in $\mathbf{B}(X)$.

Proof. The equivalence (i) ⇔ (ii) is just the translation of the meaning of boundedness in terms of the seminorms $s_{\psi,t}$ for $\lambda\sigma_u$.
Also, from (i) ⇔ (ii) the implication (iii) ⇒ (i) is clear. So we only prove the implication (i) ⇒ (iii).
Let $t > 0$. Then the set $\{L\, e^{-tA}\, x \mid x \in X,\ \|x\| = 1 \text{ and } L \in B\}$ is bounded in $S_{X,A}$. From the characterization of bounded subsets of $S_{X,A}$ given in Theorem (I.1.10), we obtain $\tau > 0$ such that the set

$$\{e^{\tau A} L e^{-tA} x \mid \|x\| = 1,\ L \in B\}$$

is bounded in X. □

Sequential convergence in $L(S_{X,A})$ can be characterized, similarly.

(IV.1.13) <u>Theorem</u>

Let $(L_n)_{n\in\mathbb{N}} \subset L(S_{X,A})$. The following three statements are equivalent.

(i) $(L_n)_{n\in\mathbb{N}}$ is a null sequence with respect to the topology $\lambda\sigma_u$.

(ii) $\forall_{t>0}\ \forall_{\psi\in B_+(\mathbb{R})}:\ \lim\limits_{n\to\infty} \|\psi(A)L_n\, e^{-tA}\| \to 0$.

(iii) $\forall_{t>0}\ \exists_{\tau>0}:\ \lim\limits_{n\to\infty} \|e^{\tau A}\, L_n\, e^{-tA}\| \to 0$.

<u>Proof</u>. We only show the implication (i) ⇒ (iii).

Let $t > 0$. By the previous theorem there exists $t' > 0$ and a constant $K_{t,t'} > 0$ such that for all $n \in \mathbb{N}$,

$$\|e^{t'A}\, L_n\, e^{-tA}\| \leqq K_{t,t'}\ .$$

Let $\varepsilon > 0$ and fix $0 < \tau < t'$. Take $R > 0$ so large that

$$\exp\,(-\,R(t'-\tau)\ < \tfrac{1}{2}\,\varepsilon(K_{t,t'})^{-1}$$

and also $n \in \mathbb{N}$ so large that

$$\|e^{\tau A}\, P_R\, L_n\, e^{-tA}\| < \tfrac{1}{2}\varepsilon$$

where $P_R = \chi_{[0,R)}(A)$. (Observe that the function $\lambda \mapsto e^{\tau\lambda}\ \chi_{(0,R]}(\lambda)$ is bounded.) We obtain

$$\begin{aligned}\|e^{\tau A}\, L_n\, e^{-tA}\| &\leqq \|e^{\tau A}\, P_R\, L_n\, e^{-tA}\| + \|e^{\tau A}(I-P_R)L_n\, e^{-tA}\| \\ &\leqq \tfrac{1}{2}\varepsilon + \|e^{-(t'-\tau)A}(I-P_R)\|\,\|e^{t'A}\, L_n\, e^{-tA}\| \\ &\leqq \tfrac{1}{2}\varepsilon + \tfrac{1}{2}\varepsilon(K_{t,t'})^{-1}\, K_{t,t'} = \varepsilon\ .\end{aligned}$$

□

(IV.1.14) Corollary

Multiplication is jointly sequentially continuous in $\mathbf{L}(S_{X,A})$ with respect to the topology $\lambda\sigma_u$.

Proof. Let $(L_n)_{n\in\mathbb{N}}$ and $(K_n)_{n\in\mathbb{N}}$ be convergent sequences in $\mathbf{L}(S_{X,A})$ with respective limits L and K. Let $t > 0$ and let $\psi \in \mathbf{B}_+(\mathbb{R})$. Then there exists $\tau > 0$ and $C_{t,\tau} > 0$ such that

$$\|e^{\tau A} K_n e^{-tA}\| < C_{t,\tau} \quad \text{for all } n \in \mathbb{N}$$

and

$$\lim_{n\to\infty} \|e^{\tau A}(K_n - K)e^{-tA}\| = 0 .$$

So the estimation

$$\|\psi(A)(L_n K_n - LK)e^{-tA}\| \leq \|\psi(A)(L_n - L)e^{-\tau A}\| \|e^{\tau A} K_n e^{-tA}\|$$

$$+ \|\psi(A) L e^{-\tau A}\| \|e^{\tau A}(K_n - K)e^{-tA}\|$$

yields the wanted result. □

The topologies $\lambda\sigma_u$, $\lambda\sigma_s$ and $\lambda\sigma_w$ for $\mathbf{L}(S_{X,A})$ correspond to the uniform, the strong and the weak topology for $\mathbf{B}(X)$. For $\mathbf{B}(X)$ there is also defined the ultra-strong topology and the ultra-weak topology. These topologies are induced on $\mathbf{B}(X)$ by its pre-dual $\mathbf{B}_*(X) \mathrel{\hat{=}} \mathbf{T}(X)$. Here $\mathbf{T}(X)$ denotes the Banach algebra of the trace class operators. We recall that $\Lambda \in \mathbf{T}(X)$ if for a certain orthonormal basis $(e_k)_{k\in\mathbb{N}}$ for X the series $\sum_{k=1}^{\infty} (|\Lambda| e_k, e_k)$ converges, where $|\Lambda| = (\Lambda^*\Lambda)^{\frac{1}{2}}$. Cf. [Wei], Ch. VII or [Scha], Ch. III.
Similarly we define the ultra-strong topology $\lambda\sigma_{us}$ and the ultra-weak topology $\lambda\sigma_{uw}$ for $\mathbf{L}(S_{X,A})$. We show its correspondence with the pre-dual $\mathbf{L}_*(S_{X,A})$ of $\mathbf{L}(S_{X,A})$.

(IV.1.15) Definition

The ultra-strong topology $\lambda\sigma_{us}$ for $\mathbf{L}(S_{X,A})$ is the locally convex topology generated by the seminorms $s_{(w_\ell),\psi}$,

$$s_{(w_\ell),\psi}(L) = \left(\sum_{\ell=1}^{\infty} \|\psi(A) L w_\ell\|^2\right)^{\frac{1}{2}}$$

where $\psi \in \mathbf{B}_+(\mathbb{R})$ and where the sequence $(w_\ell)_{\ell\in\mathbb{N}} \subset S_{X,A}$ satisfies

$$\sum_{\ell=1}^{\infty} \|e^{tA} w_\ell\|^2$$

for some $t > 0$.

We note that $s_{(w_\ell),\psi}$ is a well-defined seminorm on $\mathbf{L}(S_{X,A})$ because for all $L \in \mathbf{L}(S_{X,A})$ we have

$$s_{(w_\ell),\psi}(L) \leqq \|\psi(A)\, L\, e^{-tA}\| \left(\sum_{\ell=1}^{\infty} \|e^{tA} w_\ell\|^2\right)^{\frac{1}{2}} .$$

The topology $\lambda\sigma_{us}$ is the coarsest topology for which the mappings

$$L \mapsto (Lw_1, Lw_2, \ldots) \ , \qquad L \in \mathbf{L}(S_{X,A})$$

from $\mathbf{L}(S_{X,A})$ into $S_{d(X),d(A)}$ are continuous. Here,

$$d(X) = \bigoplus_{\ell=1}^{\infty} X \ ,$$

and $d(A)$ denotes the nonnegative self-adjoint operator in $d(X)$ defined by

$$\mathcal{D}(d(A)) = \left\{(x_\ell) \in \bigoplus_{\ell=1}^{\infty} \mathcal{D}(A) \;\middle|\; \sum_{\ell=1}^{\infty} \|Ax_\ell\|^2 < \infty\right\} ,$$

$$dA(x_\ell) = (Ax_\ell) \ , \qquad (x_\ell) \in \mathcal{D}(d(A)) \ .$$

(IV.1.16) Definition

The ultra-weak topology $\lambda\sigma_{uw}$ for $\mathbf{L}(S_{X,A})$ is the locally convex topology generated by the seminorms $s_{(w_\ell),(F_\ell)}$

$$s_{(w_\ell),(F_\ell)}(L) = \Big|\sum_{\ell=1}^{\infty} \langle Lw_\ell, F_\ell\rangle\Big| \ ,$$

where $(w_\ell) \in S_{d(X),d(A)}$, i.e. $\exists_{t>0}: \sum_{\ell=1}^{\infty} \|e^{tA} w_\ell\|^2 < \infty$, and

where $(F_\ell) \in T_{d(X),d(A)}$, i.e. $\forall_{t>0}: \sum_{\ell=1}^{\infty} \|F_\ell(t)\|^2 < \infty$.

The ultra-weak topology $\lambda\sigma_{uw}$ for $\mathbf{L}(S_{X,A})$ is the coarsest topology for which the linear forms

$$L \mapsto \sum_{\ell=1}^{\infty} \langle Lw_\ell, F_\ell \rangle \ , \qquad L \in \mathbf{L}(S_{X,A})$$

are continuous for all $(w_\ell) \in S_{d(X),d(A)}$ and $(F_\ell) \in T_{d(X),d(A)}$.

The trace class operators on X establish a Banach algebra with the trace norm $\|\cdot\|_1$ given by

$$\|\Lambda\|_1 = \sum_{k=1}^{\infty} (|\Lambda| e_k, e_k) \ .$$

Here $(e_k)_{k\in\mathbb{N}}$ denotes an orthonormal basis for X. The trace norm does not depend on the choice of the orthonormal basis. For each operator $B \in \mathbf{B}(X)$ the operators $B\Lambda$ and ΛB belong to $\mathbf{T}(X)$. The Banach algebra $\mathbf{T}(X)$ establishes a representation of the pre-dual of $\mathbf{B}(X)$ by means of the pairing

$$(\Lambda, B)_1 = \sum_{k=1}^{\infty} (B^* \Lambda e_k, e_k) \ .$$

It means that each continuous linear functional on $\mathbf{T}(X)$ is of the form

$$\Lambda \mapsto (\Lambda, B)_1 \ , \qquad \Lambda \in \mathbf{T}(X)$$

for a suitably chosen $B \in \mathbf{B}(X)$. For more details we refer to the monographs [Scha], [Tr] and [Wei].
The following characterization of $\mathbf{T}(X)$ is well-known (cf. [Wei], Ch. VII).

(IV.1.17) <u>Lemma</u>

Let $\Lambda \in \mathbf{T}(X)$. Then there exist sequences $(x_\ell),(y_\ell) \in d(X)$ such that

$$\Lambda x = \sum_{\ell=1}^{\infty} (x, x_\ell) y_\ell \ , \qquad x \in X \ .$$

In a similar way we proceed with repect to the algebra $\mathbf{L}(S_{X,A})$, and we introduce the space $\mathbf{T}(T_{X,A})$. To this end we recall that $Q \in \mathbf{L}(T_{X,A})$ if for all $t > 0$ there exists $\tau > 0$ such that the operator $e^{-tA}\, Q\, e^{\tau A}$ extends to an everywhere defined bounded operator on X. Avoiding too many notations we denote this extension by $e^{-tA}\, Q\, e^{\tau A}$, also.

(IV.1.18) Definition

The space $\mathbf{T}(T_{X,A})$ is the subspace of $\mathbf{L}(T_{X,A})$ which consists of all elements $\theta \in \mathbf{L}(T_{X,A})$ with the property

$$\exists_{t>0}\ \forall_{s>0}:\ e^{-sA}\,\theta\, e^{tA} \in \mathbf{T}(X)\ .$$

The pairing of $\mathbf{T}(T_{X,A})$ and $\mathbf{L}(S_{X,A})$ is as follows:

(IV.1.19) $$\langle\!\langle \theta, L\rangle\!\rangle_1 = (e^{-\tau A}\,\theta\, e^{tA}, e^{\tau A}\,\theta\, e^{-tA})_1\ .$$

Here first $t > 0$ must be taken so small that $e^{-sA}\,\theta\, e^{tA} \in \mathbf{T}(X)$ for all $s > 0$, and next $\tau > 0$ so small that $e^{\tau A}\, L\, e^{-tA} \in \mathbf{B}(X)$. By standard arguments it can be proved that the definition of $\langle\!\langle \theta, L\rangle\!\rangle_1$ does not depend on the choice of t and τ. From Lemma (IV.1.17) we obtain the following characterization.

(IV.1.20) Lemma

Let $\theta \in \mathbf{T}(T_{X,A})$. Then there exists a sequence $(w_\ell) \in S_{d(X),d(A)}$ and a sequence $(F_\ell) \in T_{d(X),d(A)}$ such that

$$\theta G = \sum_{\ell=1}^{\infty} \overline{\langle w_\ell, G\rangle}\, F_\ell\ , \qquad G \in T_{X,A}\ .$$

(Heuristically, $\theta = \sum_{\ell=1}^{\infty} (w_\ell \otimes F_\ell)$.)

Proof. First we take $t > 0$ fixed such that $e^{-sA}\,\theta\, e^{tA} \in \mathbf{T}(X)$ for all $s > 0$. Let $s > 0$. Then by Lemma (IV.1.17) there exist sequences $(x_\ell) \in d(X)$ and $(F_\ell(s)) \in d(X)$ such that

$$(e^{-sA}\,\theta\, e^{tA})x = \sum_{\ell=1}^{\infty} (x, x_\ell) F_\ell(s)\ .$$

It follows easily that $s \mapsto F_\ell(s)$ is a trajectory in $T_{X,A}$ for each $\ell \in \mathbb{N}$, and hence $(F_\ell) \in T_{d(X),d(A)}$. Put $w_\ell = e^{-tA} x_\ell$, $\ell \in \mathbb{N}$. Then $(w_\ell) \in S_{d(X),d(A)}$. The final step

$$\theta G = \sum_{\ell=1}^{\infty} \overline{\langle w_\ell, G\rangle}\, F_\ell\ , \qquad G \in T_{X,A}$$

is left as an easy exercise. □

Now let $\theta = \sum_{\ell=1}^{\infty} (w_\ell \otimes F_\ell)$ as indicated in the previous lemma. Then for all $L \in \mathbf{L}(S_{X,A})$

$$\langle\!\langle \theta, L \rangle\!\rangle_1 = \sum_{\ell=1}^{\infty} \langle L w_\ell, F_\ell \rangle .$$

So we obtain the following result.

(IV.1.21) Theorem

(i) The ultra-weak topology $\lambda\sigma_{uw}$ for $\mathbf{L}(S_{X,A})$ is the coarsest topology for which all linear functionals $L \mapsto \langle\!\langle \theta, L \rangle\!\rangle_1$, $L \in \mathbf{L}(S_{X,A})$, are continuous. So we have $\lambda\sigma_{uw} = \sigma(\mathbf{L}(S_{X,A}), \mathbf{T}(T_{X,A}))$.

(ii) Let $\theta \in \mathbf{T}(T_{X,A})$. Then the linear functional $L \mapsto \langle\!\langle \theta, L \rangle\!\rangle_1$ is continuous with respect to the topologies $\lambda\sigma_u$ and $\lambda\sigma_{us}$.

Proof

(i) The proof of (i) follows from the definition of the topology $\lambda\sigma_{uw}$ and the remarks just before Lemma (IV.1.20).

(ii) There exist $(w_\ell) \in S_{d(X),d(A)}$ and $(F_\ell) \in T_{d(X),d(A)}$ such that $\theta = \sum_{\ell=1}^{\infty} w_\ell \otimes F_\ell$. Following the characterization of the elements of a trajectory space and an analyticity space, there exist $\psi \in \mathbf{B}_+(\mathbb{R})$ and $t > 0$ such that $(F_\ell) = \psi(d(A))(y_\ell)$, i.e. for all $\ell \in \mathbb{N}$

$$F_\ell = \psi(A) y_\ell \quad \text{with} \quad (y_\ell) \in d(X)$$

and

$$(w_\ell) \in \exp(-t\, d(A))(d(X)) .$$

So we derive

$$\exists_{t>0}\, \exists_{\psi \in \mathbf{B}_+(\mathbb{R})} : \psi(A)^{-1}\, \theta\, e^{tA} \in \mathbf{T}(X) ,$$

i.e.

$$\psi(A)^{-1}\, \theta\, e^{tA} = \sum_{\ell=1}^{\infty} (e^{tA} w_\ell) \otimes y_\ell .$$

So we get

$$|\langle\!\langle \theta, L \rangle\!\rangle_1| = |(\psi(A)^{-1}\, \theta\, e^{tA}, \psi(A)\, L\, e^{-tA})_1| \leq \| \psi(A)^{-1}\, \theta\, e^{tA} \|_1\, s_{t,\psi}(L) .$$

Therefore, the linear functional is continuous in the uniform topology of $L(S_{X,A})$. Also, we have the estimation

$$|\langle\theta,L\rangle_1| = |\sum_{\ell=1}^{\infty} \langle Lw_\ell, F_\ell\rangle| = |\sum_{\ell=1}^{\infty} (\psi(A)Lw_\ell, y_\ell)|$$

$$\leq \left(\sum_{\ell=1}^{\infty} \|\psi(A)Lw_\ell\|^2\right)^{\frac{1}{2}} \left(\sum_{\ell=1}^{\infty} \|y_\ell\|^2\right)^{\frac{1}{2}}$$

$$= \left(\sum_{\ell=1}^{\infty} \|y_\ell\|^2\right)^{\frac{1}{2}} s_{(w_\ell),\psi}(L) .$$

From this estimation the ultra-strong continuity of the linear functional $L \mapsto \langle\theta,L\rangle_1$ follows. □

The topologies which we have introduced in this section fit into the following inclusion scheme

$$\begin{array}{ccccc} \lambda\sigma_u & \supset & \lambda\sigma_{us} & \supset & \lambda\sigma_{uw} \\ & & \cup & & \cup \\ & & \lambda\sigma_s & & \lambda\sigma_w \end{array}$$

Here $\supset$ means 'finer than'.

(IV.1.22) Corollary

Let $B \subset L(S_{X,A})$. Then the following assertions are equivalent.

(i) B is $\lambda\sigma_u$-bounded.

(ii) B is $\lambda\sigma_{us}$-bounded.

(iii) B is $\lambda\sigma_{uw}$-bounded.

Proof. It is clear that (i) ⇒ (ii) ⇒ (iii). It is also clear that (iii) implies the weak boundedness of the set B. By Theorem (IV.1.9) it follows that B is uniformly bounded, i.e. (iii) ⇒ (i). □

At the end of this section we devote some attention to the nuclear case. If $S_{X,A}$ is nuclear, then the operators e^{-tA}, $t > 0$, are Hilber-Schmidt. As we have seen in Chapter III, the space $TS_{X\otimes X;I\otimes A,A\otimes I}$ comprises the whole algebra $L(S_{X,A})$.

Let $\theta \in \mathbf{T}(T_{X,A})$. We define the mapping $\hat{\theta}$ on $(0,\infty) \times (0,\infty)$ by

$$\hat{\theta}(t,s) = e^{-tA}\,\theta\, e^{-sA} .$$

Because of the definition of $\mathbf{T}(T_{X,A})$ the mapping $\hat{\theta}$ can be extended to $(0,\infty) \times [-\sigma,\infty)$ for certain $\sigma > 0$, in such a way that $\hat{\theta}(t,s) \in X \otimes X$ for all $t > 0$ and all $s \geq -\sigma$. It follows that $\hat{\theta}$ belongs to $ST_{X\otimes X;\, I\otimes A,\, A\otimes I}$. Conversely, let $\hat{\xi} \in ST_{X\otimes X;\, I\otimes A,\, A\otimes I}$. Then there exists $\delta > 0$ such that $\hat{\xi}(t,-\delta) \in X \otimes X$ for all $t > 0$. Since $e^{-tA} \in X \otimes X$ for all $t > 0$ and hence $e^{-tA} \in \mathbf{T}(X)$ for all $t > 0$, it follows that

$$\hat{\xi}(t,-\delta) \in \mathbf{T}(X) , \qquad t > 0 .$$

If we define the operator ξ on $T_{X,A}$ by

$$\xi G\colon\ t \mapsto \hat{\xi}(t,-\delta)G(\delta) , \qquad t > 0,\ G \in T_{X,A} ,$$

it follows that $\xi \in \mathbf{T}(T_{X,A})$.
We obtain the identification $\mathbf{T}(T_{X,A}) \,\hat{=}\, ST_{X\otimes X;\, I\otimes A,\, A\otimes I}$. So in the nuclear case the pairing between $\mathbf{L}(S_{X,A})$ and $\mathbf{T}(T_{X,A})$ appears to be a representation of the usual pairing of the dual system $(TS_{X\otimes X;\, I\otimes A,\, A\otimes I}, ST_{X\otimes X;\, I\otimes A,\, A\otimes I})$.

(IV.1.23) <u>Theorem</u>

Let $S_{X,A}$ be a nuclear space. Then the topology $\lambda\sigma_{uw}$ for $\mathbf{L}(S_{X,A})$ is equivalent to the topology $\tau\sigma_w$ for $TS_{X\otimes X;\, I\otimes A,\, A\otimes I}$, i.e. the weak topology on $TS_{X\otimes X;\, I\otimes A,\, A\otimes I}$ induced by its duality with $ST_{X\otimes X;\, I\otimes A,\, A\otimes I}$.
Moreover, the topology $\lambda\sigma_{us}$ is equivalent to the topology $\tau\sigma_{proj}$, and $\tau\sigma_{proj}$ is equivalent to $\lambda\sigma_u$. Consequently, $\lambda\sigma_u$ and $\lambda\sigma_{us}$ are equivalent topologies whenever $S_{X,A}$ is a nuclear space.

<u>Proof</u>. The first assertion follows immediately from the observation that $ST_{X\otimes X;\, I\otimes A,\, A\otimes I}$ comprises the whole $\mathbf{T}(T_{X,A})$.
We prove the second assertion.
Since $S_{X,A}$ is nuclear, the operator A has discrete spectrum $0 \leq \lambda_1 \leq \lambda_2 \leq \ldots$ with corresponding eigenvectors $(v_j)_{j\in\mathbb{N}}$ which establish an orthonormal basis for X. We obtain for all $t > 0$ and all $\psi \in \mathbf{B}_+(\mathbb{R})$

$$\|\psi(A)\,L\,e^{-tA}\|_{X\otimes X} = \left(\sum_{\ell=1}^{\infty} \|\psi(A)\,L\,e^{-tA}\,v_\ell\|^2\right)^{\frac{1}{2}}$$

$$= s_{(e^{-tA}\,v_\ell),\psi}(L)\ , \qquad L \in \mathbf{L}(S_{X,A})\ .$$

Hence $\tau\sigma_{proj}$ is coarser than $\lambda\sigma_{uw}$.

Conversely, let $(w_\ell) \in e^{-td(A)}(d(X))$ and let $\psi \in \mathbf{B}_+(\mathbb{R})$. Then we have the estimation

$$s_{(w_\ell),\psi}(L) = \left(\sum_{\ell=1}^{\infty} \|\psi(A)Lw_\ell\|^2\right)^{\frac{1}{2}}$$

$$= \left(\sum_{\ell=1}^{\infty}\sum_{j=1}^{\infty} |(e^{tA}\,w_\ell, (\psi(A)\,L\,e^{-tA})^*\,v_j)|^2\right)^{\frac{1}{2}}$$

$$\leq \left(\sum_{\ell=1}^{\infty} \|e^{tA}\,w_\ell\|^2\right)^{\frac{1}{2}} \|\psi(A)\,L\,e^{-tA}\|_{X\otimes X}\ .$$

Hence the topology $\lambda\sigma_{us}$ is coarser than the topology $\tau\sigma_{proj}$. The remaining part of the assertion follows from Lemma (IV.1.3). □

IV.2. The algebra $\mathbf{L}(T_{X,A})$

The algebra $\mathbf{L}(T_{X,A})$ consists of all continuous linear mappings from $T_{X,A}$ into $T_{X,A}$. From Section I.4 the following characterizations can be obtained.

(IV.2.1) Characterization

Let L be a linear operator in X.

(i) L extends to a member of $\mathbf{L}(T_{X,A})$ iff $\forall_{t>0}\ \forall_{\psi\in\mathbf{B}_+(\mathbb{R})}$: the operator $e^{-tA}\,L\,\psi(A)$ is a densely defined bounded operator on X.

(ii) L extends to a member of $\mathbf{L}(T_{X,A})$ iff $\forall_{t>0}\ \exists_{\tau>0}$: the operator $e^{-tA}\,L\,e^{\tau A}$ is a densely defined bounded operator on X.

We observe that for each $L \in \mathbf{L}(T_{X,A})$ the dual mapping L' belongs to $\mathbf{L}(S_{X,A})$. The mapping L' can be regarded as a densely defined linear operator in X. So its adjoint $(L')^*$ is also well defined. For all $x \in \mathcal{D}((L')^*)$ and all $w \in S_{X,A}$ we have

$$\langle w, Lx\rangle = (L'w, x) = (w, (L')^* x) .$$

Hence $(L')^* x = Lx$ for $x \in \mathcal{D}((L')^*)$. It follows that each mapping $L \in \mathbf{L}(\mathcal{T}_{X,A})$ can be seen as an extension of a well-defined linear operator in X viz. $(L')^*$. Moreover, we observe that $(L')^{**}|_{\mathcal{S}_{X,A}} = L'$.

(IV.2.2) Definition

The uniform topology $\lambda\tau_u$ for $\mathbf{L}(\mathcal{T}_{X,A})$ is generated by the seminorms

$$t_{\psi,t}(L) = \|e^{-tA} L\psi(A)\| , \qquad L \in \mathbf{L}(\mathcal{T}_{X,A}) ,$$

where we take $\psi \in \mathbf{B}_+(\mathbb{R})$ and $t > 0$. Here $\|\cdot\|$ denotes the norm of $\mathbf{B}(X)$.

The mapping $L \mapsto L'$ is an anti-linear bijection from $\mathbf{L}(\mathcal{T}_{X,A})$ onto $\mathbf{L}(\mathcal{S}_{X,A})$. For $L_1, L_2 \in \mathbf{L}(\mathcal{T}_{X,A})$ we have $(L_1 L_2)' = L_2' L_1'$. The self-adjoint operators e^{-tA} and $\psi(A)$, $t > 0$, $\psi \in \mathbf{B}_+(\mathbb{R})$, can be naturally extended to linear operators of $\mathbf{L}(\mathcal{T}_{X,A})$, which will be denoted by e^{-tA} and $\psi(A)$, also. So we obtain

$$(e^{-tA} L\psi(A))^* = \psi(A) L' e^{-tA} .$$

It follows that for all $t > 0$ and all $\psi \in \mathbf{B}_+(\mathbb{R})$

$$t_{\psi,t}(L) = \|e^{-tA} L\psi(A)\| = \|\psi(A) L' e^{-tA}\| = s_{\psi,t}(L') .$$

Hence the mapping $L \mapsto L'$ is an anti-linear homeomorphism from $\mathbf{L}(\mathcal{T}_{X,A})$ onto $\mathbf{L}(\mathcal{S}_{X,A})$ with respect to the topologies $\lambda\tau_u$ and $\lambda\sigma_u$. Due to this homeomorphism we obtain from Lemma (IV.1.4)

(IV.2.3) Lemma

(i) The space $\mathbf{L}(\mathcal{T}_{X,A})$ is $\lambda\tau_u$-complete.

(ii) Multiplication is separately $\lambda\tau_u$-continuous.

(IV.2.4) Theorem

$\mathbf{L}(\mathcal{T}_{X,A})$ with topology $\lambda\tau_u$ is a complete locally convex algebra.

Suppose for a moment that $\mathcal{T}_{X,A}$ is a nuclear space. By the kernel theorem for $\mathbf{L}(\mathcal{T}_{X,A})$ we have

$$\mathbf{L}(T_{X,A}) \mathrel{\hat{=}} TS_{X\otimes X,A\otimes I,I\otimes A} .$$

As usual the sign ^ indicates that the above spaces can be identified. $TS_{X\otimes X;A\otimes I,I\otimes A}$ is a representation of $\mathbf{L}(T_{X,A})$. So they are not equal.

On the space $TT_{X\otimes X;I\otimes A,A\otimes I}$ we introduce the mapping ^{c} by

(IV.2.5) $Q^c(t,s) = Q(t,s)^*$, $t > 0,\ s > 0$.

Then $Q^c \in TT_{X\otimes X;A\otimes I,I\otimes A}$. With the identifications of Section III.2 we have

$$TT_{X\otimes X;I\otimes A,A\otimes I} \mathrel{\hat{=}} T_{X\otimes X,I\otimes A+A\otimes I} \mathrel{\hat{=}} TT_{X\otimes X;A\otimes I,I\otimes A} .$$

So the mapping ^{c} is a bijection from $TT_{X\otimes X;A\otimes I,I\otimes A}$ onto itself and from $TT_{X\otimes X;I\otimes A,A\otimes I}$ onto itself. Further, we can restrict the mapping ^{c} to $TS_{X\otimes X;A\otimes I,I\otimes A}$. It follows that $^c: Q \mapsto Q^c$ is an anti-linear bijection from $TS_{X\otimes X;A\otimes I,I\otimes A}$ onto $TS_{X\otimes X;I\otimes A,A\otimes I}$. Similarly it follows that $Q \mapsto Q^c$ is an anti-linear bijection from $TS_{X\otimes X;I\otimes A,A\otimes I}$ onto $TS_{X\otimes X;A\otimes I,I\otimes A}$. Moreover, because of the assumed nuclearity of $T_{X,A}$ we have

$$\mathbf{L}(S_{X,A}) \mathrel{\hat{=}} TS_{X\otimes X;I\otimes A,A\otimes I} .$$

Let $K \in \mathbf{L}(T_{X,A})$ with representant $\hat{K} \in TS_{X\otimes X;A\otimes I,I\otimes A}$. Then $K' \in \mathbf{L}(S_{X,A})$ is represented by $\hat{K}^c$. Similarly, for $L \in \mathbf{L}(S_{X,A})$ with representant $\hat{L} \in TS_{X\otimes X;I\otimes A,A\otimes I}$ the dual mapping $L' \in \mathbf{L}(T_{X,A})$ is represented by $\hat{L}^c$. Finally we observe that for all $Q \in TS_{X\otimes X;A\otimes I,I\otimes A}$

$$\|(\psi(A) \otimes I)Q(0,s)\|_{X\otimes X} = \|(I \otimes \psi(A))Q(0,s)^*\|_{X\otimes X}$$

where $s > 0$ and $\psi \in \mathbf{B}_+(\mathbb{R})$. So the mapping ^{c} is an anti-linear homeomorphism between $TS_{X\otimes X;I\otimes A,A\otimes I}$ and $TS_{X\otimes X;A\otimes I,I\otimes A}$. From Lemma (IV.1.3) we obtain

(IV.2.6) Lemma

Let $T_{X,A}$ be a nuclear space. Then the topology $\tau\sigma_{proj}$ for $TS_{X\otimes X;A\otimes I,I\otimes A}$ is equivalent to the topology $\lambda\tau_u$ for $\mathbf{L}(T_{X,A})$.

The elements of $\mathbf{L}(T_{X,A})$ are linear mappings on $T_{X,A}$. Therefore it is natural to introduce the strong topology for $\mathbf{L}(T_{X,A})$.

(IV.2.7) Definition

The strong topology $\lambda\tau_s$ for $\mathbf{L}(T_{X,A})$ is the locally convex topology generated by the seminorms $t_{F,t}$

$$t_{F,t}(L) = \|(LF)(t)\|_X , \quad L \in \mathbf{L}(T_{X,A})$$

where we take $F \in T_{X,A}$ and $t > 0$.

We note that the net (K_α) in $\mathbf{L}(T_{X,A})$ is $\lambda\tau_s$-convergent with limit K if and only if the net $(K_\alpha F)$ in $T_{X,A}$ is τ_{proj}-convergent with limit KF for all $F \in T_{X,A}$. The topology $\lambda\tau_s$ is the coarsest topology such that for all $F \in T_{X,A}$ the linear mapping from $\mathbf{L}(T_{X,A})$ in $T_{X,A}$ given by

$$L \mapsto LF , \quad L \in \mathbf{L}(T_{X,A}) ,$$

is continuous with respect to the topology τ_{proj} for $T_{X,A}$.

(IV.2.8) Lemma

The topology $\lambda\tau_u$ is equivalent to the topology of uniform convergence on bounded subsets of $T_{X,A}$.

Proof. Let (K_α) be a $\lambda\tau_u$-convergent net in $\mathbf{L}(T_{X,A})$ with limit K, and let B be a bounded subset of $T_{X,A}$. Following Section I.2 there exists $\psi \in \mathbf{B}_+(\mathbb{R})$ and a bounded set $B_0 \subset X$ such that $B = \psi(A)B_0$. For all $x \in B_0$ we derive

$$\|((K_\alpha - K)\psi(A)x)(t)\| \leqq \|e^{-tA}(K_\alpha - K)\psi(A)\| \|x\| .$$

Hence the net (K_α) converges uniformly on B.
Conversely, let $\psi \in \mathbf{B}_+(\mathbb{R})$ and let $\varepsilon > 0$. Suppose the net (L_α) converges in $\lambda\tau_s$-sense to L, uniformly on the bounded subset $\{\psi(A)x \mid \|x\| = 1\} \subset T_{X,A}$. Then for each $t > 0$ there exists α_t such that for all $\alpha > \alpha_t$ and all $x \in X$ with $\|x\| = 1$

$$\|((K_\alpha - K)\psi(A)x)(t)\| < \frac{\varepsilon}{2} .$$

It follows that for all $\alpha > \alpha_t$

$$\|e^{-tA}(K_\alpha - K)\psi(A)\| \leqq \frac{\varepsilon}{2} < \varepsilon . \qquad \square$$

Remark. Of course the operators $e^{-tA} L\psi(A)$ are not defined on the whole of X. They can be extended to bounded linear operators on X. Since no confusion will arise, we denote their extensions by $e^{-tA} L\psi(A)$, also.

(IV.2.9) Definition

The weak topology $\lambda\tau_w$ is the locally convex topology for $\mathbf{L}(T_{X,A})$ generated by the seminorms $t_{F,w}$,

$$t_{F,w}(L) = |\langle w,LF\rangle| \ , \qquad L \in \mathbf{L}(T_{X,A}) \ ,$$

where we take $w \in S_{X,A}$ and $F \in T_{X,A}$.

The net (K_α) in $\mathbf{L}(T_{X,A})$ converges weakly to $K \in \mathbf{L}(T_{X,A})$ if and only if $\langle w,(K_\alpha-K)F\rangle \to 0$ for all $w \in S_{X,A}$ and all $F \in T_{X,A}$. The weak topology $\lambda\tau_w$ is the coarsest topology for which all linear mappings

$$K \mapsto \overline{\langle w,KF\rangle} \ , \qquad K \in \mathbf{L}(T_{X,A})$$

are continuous.

The mapping $L \mapsto L'$ from $\mathbf{L}(T_{X,A})$ onto $\mathbf{L}(S_{X,A})$ is, in general, not continuous if both spaces carry their strong topology $\lambda\tau_s$ and $\lambda\sigma_s$. Since

$$t_{F,w}(L) = |\langle w,LF\rangle| = |\langle L'w,F\rangle| = s_{w,F}(L') \ ,$$

the mapping $L \mapsto L'$ is an anti-linear homeomorphism from $\mathbf{L}(T_{X,A})$ onto $\mathbf{L}(S_{X,A})$ if both spaces carry their weak topology $\lambda\tau_w$ and $\lambda\sigma_w$. From Theorem (IV.1.9) we obtain

(IV.2.10) Theorem (Principle of uniform boundedness)

Let B be a subset of $\mathbf{L}(T_{X,A})$. The following statements are equivalent:

(i) B is $\lambda\tau_u$-bounded.

(ii) B is $\lambda\tau_s$-bounded.

(iii) B is $\lambda\tau_w$-bounded.

Similar to Corollary (IV.1.10) we have the following result.

(IV.2.11) Corollary

Let $(K_n)_{n\in\mathbb{N}}$ be a sequence in $\mathbf{L}(T_{X,A})$.

(i) Suppose for each $G \in T_{X,A}$ the sequence $(K_n G)_{n\in\mathbb{N}}$ converges in $T_{X,A}$ with respect to the topology τ_{proj}. Then the linear mapping $K: G \mapsto \lim_{n\to\infty} K_n G$ belongs to $\mathbf{L}(T_{X,A})$.

(ii) In (i) we can replace the convergence with respect to τ_{proj} by weak convergence in $T_{X,A}$.

Proof.

(i) Following Theorem (IV.2.10) the sequence $(K_n)_{n\in\mathbb{N}}$ is bounded. This means that for each $t > 0$ there exists $\beta_t > 0$ such that $\|e^{-tA} K_n\| \leq \beta_t$ for all $n \in \mathbb{N}$. It is clear that K is a linear mapping from $T_{X,A}$ into $T_{X,A}$. We have to show that K is continuous. For all $x \in X$ with $\|x\| = 1$ we have

$$\|e^{-tA} Kx\| \leq \|e^{-tA}(K-K_n)x\| + \beta_t \leq \beta_t + 1$$

if $n \in \mathbb{N}$ is taken sufficiently large. By condition (iv) of Theorem (I.4.8) it follows that K belongs to $\mathbf{L}(T_{X,A})$.

(ii) The proof of (ii) runs similarly to the proof of (i), and is omitted. □

(IV.2.12) Corollary

$\mathbf{L}(T_{X,A})$ is sequentially $\lambda\tau_s$- and $\lambda\tau_w$-complete.

As we have shown already, the transposition mapping $K \mapsto K'$ from $\mathbf{L}(T_{X,A})$ onto $\mathbf{L}(S_{X,A})$ is an anti-linear homeomorphism if both algebras carray either their uniform or their weak topologies. From the previous section we obtain the following characterization of bounded subsets and null sequences of $\mathbf{L}(T_{X,A})$.

(IV.2.13) Theorem

Let $B \subset \mathbf{L}(T_{X,A})$. The following statements are equivalent:

(i) B is bounded.

(ii) $\forall_{t>0}\ \forall_{\psi\in \mathbf{B}_+(\mathbb{R})}$: the set $\{e^{-tA} L\psi(A) \mid L \in B\}$ is bounded in $\mathbf{B}(X)$.

(iii) $\forall_{t>0}\ \exists_{\tau>0}$: the set $\{e^{-tA} L e^{\tau A} \mid L \in B\}$ is bounded in $\mathbf{B}(X)$.

Proof. Cf. Theorem (IV.1.12). □

(IV.2.14) Theorem

Let $(K_n)_{n\in\mathbb{N}} \subset \mathbf{L}(\mathcal{T}_{X,A})$. The following statements are equivalent.

(i) $(K_n)_{n\in\mathbb{N}}$ is a null sequence with respect to $\lambda\tau_u$.

(ii) $\forall_{t>0}\ \forall_{\psi\in\mathbf{B}_+(\mathbb{R})}$: $(e^{-tA} K_n\, \psi(A))_{n\in\mathbb{N}}$ is a null sequence in $\mathbf{B}(X)$ with respect to the norm topology.

(iii) $\forall_{t>0}\ \exists_{\tau>0}$: $(e^{-tA} K_n\, e^{\tau A})_{n\in\mathbb{N}}$ is a null sequence in $\mathbf{B}(X)$ with respect to the norm topology.

Proof. Cf. Theorem (IV.1.13). □

(IV.2.15) Corollary

Multiplication is jointly sequentially continuous in $\mathbf{L}(\mathcal{T}_{X,A})$ with respect to the norm topology $\lambda\tau_u$.

Proof. Cf. Corollary (IV.1.14). □

Next, we introduce the substitutes for the ultra-strong topology $\lambda\sigma_{us}$ and the ultra-weak topology $\lambda\sigma_{uw}$. These topologies will be denoted by $\lambda\tau_{us}$ and $\lambda\tau_{uw}$.

(IV.2.16) Definition

The ultra-strong topology $\lambda\tau_{us}$ for $\mathbf{L}(\mathcal{T}_{X,A})$ is the locally convex topology generated by the seminorms $t_{(F_\ell),t}$

$$t_{(F_\ell),t}(K) = \left(\sum_{\ell=1}^{\infty} \|(KF_\ell)(t)\|^2\right)^{\frac{1}{2}}$$

where $t > 0$ and where the sequence $(F_\ell)_{\ell\in\mathbb{N}} \in \mathcal{T}_{d(X),d(A)}$. (For the definition of $d(X)$ and $d(A)$ see Definition (IV.1.15).)

Because for all $K \in \mathbf{L}(\mathcal{T}_{X,A})$ we have

$$t_{(F_\ell),t}(K) \leq \|e^{-tA} K\psi(A)\| \left(\sum_{\ell=1}^{\infty} \|x_\ell\|^2\right)^{\frac{1}{2}}$$

where $F_\ell = \psi(A)x_\ell$, $\ell \in \mathbb{N}$, with $(x_\ell) \in d(X)$, $\lambda\tau_{us}$ is coarser than $\lambda\tau_u$. The topology $\lambda\tau_{us}$ is the coarsest topology for which the mappings

$$K \mapsto (KF_1, KF_2, \ldots) \ , \qquad K \in \mathbf{L}(T_{X,A})$$

from $\mathbf{L}(T_{X,A})$ into $T_{d(X),d(A)}$ are continuous for any fixed $F \in T_{d(X),d(A)}$.

(IV.2.17) Definition

The ultra-weak topology $\lambda\tau_{uw}$ for $\mathbf{L}(T_{X,A})$ is the locally convex topology generated by the seminorms $t_{(F_\ell),(w_\ell)}$

$$t_{(F_\ell),(w_\ell)}(K) = \Big| \sum_{\ell=1}^{\infty} \langle w_\ell, KF_\ell\rangle \Big| \ , \qquad K \in \mathbf{L}(T_{X,A}) \ .$$

Here $(F_\ell) \in T_{d(X),d(A)}$ and $(w_\ell) \in S_{d(X),d(A)}$.

The topology $\lambda\tau_{uw}$ is the coarsest topology for which the linear forms

$$K \mapsto \sum_{\ell=1}^{\infty} \overline{\langle w_\ell, KF_\ell\rangle}$$

are continuous for each $(w_\ell) \in S_{d(X),d(A)}$ and $(F_\ell) \in T_{d(X),d(A)}$.

Let $(w_\ell) \in S_{d(X),d(A)}$ and $(F_\ell) \in T_{d(X),d(A)}$. Then for $t > 0$ sufficiently small, we have

$$t_{(F_\ell),(w_\ell)}(K) \leq \Big(\sum_{\ell=1}^{\infty} \| e^{tA} \, w_\ell \|^2 \Big)^{\frac{1}{2}} \Big(\sum_{\ell=1}^{\infty} \| (KF_\ell)(t) \|^2 \Big)^{\frac{1}{2}} \ .$$

So the ultra-weak topology is coarser than the ultra-strong topology. Also, it is clear that the weak topology $\lambda\tau_w$ is coarser than $\lambda\tau_{uw}$. These considerations lead to the following inclusion scheme:

(IV.2.18)
$$\begin{array}{ccccc} \lambda\tau_u & \supset & \lambda\tau_{us} & \supset & \lambda\tau_{uw} \\ & & \cup & & \cup \\ & & \lambda\tau_s & \supset & \lambda\tau_w \end{array}$$

In order to establish a candidate for the pre-dual of the algebra $\mathbf{L}(S_{X,A})$ we have introduced the space $\mathbf{T}(T_{X,A})$ in the previous section. Similarly, we introduce the space $\mathbf{T}(S_{X,A})$.

(IV.2.19) Definition

The space $\mathbf{T}(S_{X,A})$ is the subspace of $\mathbf{L}(S_{X,A})$ which consists of all elements $\Omega \in \mathbf{L}(S_{X,A})$ with the property

$$\exists_{t>0}\ \forall_{s>0}:\ e^{tA}\,\Omega\, e^{-sA} \in \mathbf{T}(X)\ .$$

As in the previous section, $\mathbf{T}(X)$ denotes the Banach algebra of the trace class operators on X.

Let $\theta \in \mathbf{T}(T_{X,A})$. Following Definition (IV.1.18) there exists $t > 0$ such that for all $s > 0$

$$e^{-sA}\,\theta\, e^{tA} \in \mathbf{T}(X)\ .$$

Since $\mathbf{T}(T_{X,A}) \subset \mathbf{L}(T_{X,A})$, the dual mapping θ' belongs to $\mathbf{L}(S_{X,A})$. We compute θ' as follows:

$$\langle w,\theta G\rangle = (e^{sA}\, w, (e^{-sA}\,\theta\, e^{tA})G(t))_X =$$

$$= ((e^{-sA}\,\theta\, e^{tA})^*\, e^{sA}\, w, G(t))_X\ .$$

Hence

$$\theta'|_{e^{-sA}(X)} = e^{-tA}(e^{-sA}\,\theta\, e^{tA})^*\, e^{sA}$$

where any sufficiently small $s > 0$ can be taken. The definition of θ' is independent of s and t. We have

$$\exists_{t>0}\ \forall_{s>0}:\ e^{tA}\,\theta'\, e^{-sA} = (e^{-sA}\,\theta\, e^{tA})^* \in \mathbf{T}(X)\ .$$

So the transposition mapping is an anti-linear homeomorphism from $\mathbf{T}(T_{X,A})$ onto $\mathbf{T}(S_{X,A})$.
From Lemma (IV.1.20) we obtain the following characterization.

(IV.2.20) Lemma

Let $\Omega \in \mathbf{T}(S_{X,A})$. Then there exists a sequence $(F_\ell) \in T_{d(X),d(A)}$ and a sequence $(w_\ell) \in S_{d(X),d(A)}$ such that

$$\Omega w = \sum_{\ell=1}^{\infty} (w,F_\ell)w_\ell\ , \qquad w \in S_{X,A}\ .$$

Heuristically, $\Omega = \sum_{\ell=1}^{\infty} (F_\ell \otimes w_\ell)$.)

<u>Proof</u>. Since $\Omega' \in \mathbf{T}(T_{X,A})$, there exists a sequence $(w_\ell) \in S_{d(X),d(A)}$ and a sequence $(F_\ell) \in T_{d(X),d(A)}$ such that

$$\Omega' F = \sum_{\ell=1}^{\infty} \overline{\langle w_\ell, F\rangle}\, F_\ell , \qquad F \in T_{X,A} .$$

It follows that

$$\Omega w = \sum_{\ell=1}^{\infty} \langle w, F_\ell\rangle\, w_\ell . \qquad \square$$

Let $\Omega \in \mathbf{T}(S_{X,A})$, $\Omega = \sum_{\ell=1}^{\infty} F_\ell \otimes w_\ell$. Then for all $K \in \mathbf{L}(T_{X,A})$

$$\langle\!\langle \Omega, K\rangle\!\rangle_1 := \sum_{\ell=1}^{\infty} \langle w_\ell, KF_\ell\rangle .$$

Thus we obtain a pairing between $\mathbf{T}(S_{X,A})$ and $\mathbf{L}(T_{X,A})$.

(IV.2.21) <u>Theorem</u>

The ultra-weak topology $\lambda\tau_{uw}$ for $\mathbf{L}(T_{X,A})$ is the coarsest topology for which the linear functionals $K \mapsto \overline{\langle\!\langle \Omega, K\rangle\!\rangle_1}$ are continuous for each $\Omega \in \mathbf{T}(S_{X,A})$. So we have $\lambda\tau_{uw} = \sigma(\mathbf{L}(T_{X,A}), \mathbf{T}(S_{X,A}))$.

From the inclusion scheme (IV.2.18) it follows that the linear functionals are also τ_u- and τ_{us}-continuous. We have

(IV.2.22) <u>Theorem</u>

Let $B \subset \mathbf{L}(T_{X,A})$. The following statements are equivalent:

(i) B is $\lambda\tau_u$-bounded.

(ii) B is $\lambda\tau_{us}$-bounded.

(iii) B is $\lambda\tau_{uw}$-bounded.

(IV.2.23) Corollary

The space $\mathbf{L}(T_{X,A})$ is sequentially complete with each of the topologies $\lambda\tau_{us}$ and $\lambda\tau_{uw}$.

The Banach algebra $\mathbf{T}(X)$ is a two-sided ideal in $\mathbf{B}(X)$. Here we present analogons for $\mathbf{L}(S_{X,A})$, $\mathbf{T}(T_{X,A})$ and $\mathbf{L}(T_{X,A})$, $\mathbf{T}(S_{X,A})$. In Section IV.4 some general considerations on ideals can be found.

(IV.2.24) Theorem

(i) $\mathbf{T}(S_{X,A})$ is a two-sided ideal in $\mathbf{L}(S_{X,A})$.

(ii) $\mathbf{T}(T_{X,A})$ is a two-sided ideal in $\mathbf{L}(T_{X,A})$.

Proof. We only prove assertion (i). Then the proof of assertion (ii) follows from the properties of the transposition $' : \mathbf{L}(S_{X,A}) \to \mathbf{L}(T_{X,A})$.
Let $L_1, L_2 \in \mathbf{L}(S_{X,A})$, and let $\theta \in \mathbf{T}(S_{X,A})$. Then there exists $s > 0$ such that for all $\tau > 0$, $e^{sA}\,\theta\, e^{-\tau A} \in \mathbf{T}(X)$. Further, there exists $s_1 > 0$ such that $e^{s_1 A}\, L_1\, e^{-sA} \in \mathbf{B}(X)$, and for all $t > 0$ there exists $\tau > 0$ such that $e^{\tau A}\, L_2\, e^{-tA} \in \mathbf{B}(X)$. It follows that for all $t > 0$ the operator

$$e^{s_1 A}(L_1\,\theta\, L_2)e^{-tA} = (e^{s_1 A}\, L_1\, e^{-sA})(e^{sA}\,\theta\, e^{-\tau A})(e^{\tau A}\, L_2\, e^{-tA})$$

belongs to $\mathbf{T}(X)$. □

(IV.2.25) Theorem

Let $\theta \in \mathbf{T}(T_{X,A})$ and let $L \in \mathbf{L}(S_{X,A})$. Then we have

(i) $\langle\theta, L\rangle_1 = \overline{\langle\theta', L'\rangle}_1$.

(ii) $\langle\theta, L_1\, L\, L_2\rangle_1 = \langle L_1'\,\theta\, L_2', L\rangle$ for each fixed pair $L_1, L_2 \in \mathbf{L}(S_{X,A})$.

Proof

(i)
$$\langle\theta, L\rangle_1 = (e^{-tA}\,\theta\, e^{sA}, e^{tA}\, L\, e^{-sA})_1$$

$$= \overline{((e^{-tA}\,\theta\, e^{sA})^*, (e^{tA}\, L\, e^{-sA})^*)}_1$$

$$= \langle\theta', L'\rangle_1 .$$

(ii) Let $L_1, L_2 \in \mathbf{L}(S_{X,A})$ be fixed. Then we have

$$\langle\!\langle \theta, L_1 L \rangle\!\rangle_1 = (e^{-tA}\,\theta\, e^{sA}, e^{tA}\, L_1\, e^{-t_1 A}\, e^{t_1 A}\, L\, e^{-sA})_1$$

$$= ((e^{tA}\, L_1\, e^{-t_1 A})^{*}\, e^{-tA}\,\theta\, e^{sA}, e^{t_1 A}\, L\, e^{-sA})_1$$

$$= \langle\!\langle L_1'\,\theta, L \rangle\!\rangle_1 \,.$$

Here $t_1 > 0$ must be taken so small that $e^{t_1 A}\, L\, e^{-sA} \in \mathbf{B}(X)$, and $t > 0$ so small that $e^{tA}\, L_1\, e^{-t_1 A} \in \mathbf{B}(X)$.
Similarly it follows that

$$\langle\!\langle \theta, L L_2 \rangle\!\rangle_1 = \langle\!\langle \theta\, L_2', L \rangle\!\rangle_1 \,. \qquad \square$$

(IV.2.26) <u>Corollary</u>

Let $\Omega \in \mathbf{T}(S_{X,A})$ and let $K \in \mathbf{L}(T_{X,A})$. Then we have

(i) $\quad \langle\!\langle \Omega, K \rangle\!\rangle_1 = \overline{\langle\!\langle \Omega', K' \rangle\!\rangle}_1 \,.$

(ii) $\quad \langle\!\langle \Omega, K_1\, K\, K_2 \rangle\!\rangle_1 = \langle\!\langle K_1'\, \Omega\, K_2', K \rangle\!\rangle_1 \quad$ for all fixed $K_1, K_2 \in \mathbf{L}(T_{X,A})$. $\square$

Finally, we devote some attention to the nuclear case. If $T_{X,A}$ is nuclear, than all operators e^{-tA}, $t > 0$, are Hilbert-Schmidt. From the kernel theorem for $\mathbf{L}(T_{X,A})$ it follows that $TS_{X\otimes X;A\otimes I,I\otimes A}$ comprises the whole algebra $\mathbf{L}(T_{X,A})$. For $\Omega \in \mathbf{T}(S_{X,A})$ we define tha mapping $\hat{\Omega}$ by

$$\hat{\Omega}(t,s) \mapsto e^{-sA}\,\Omega\, e^{-tA} \,, \qquad s,t > 0 \,.$$

From the properties of the elements of $\mathbf{T}(S_{X,A})$ it follows that each mapping $\hat{\Omega}$ can be extended to $(0,\infty) \times [-\sigma,\infty)$ for some $\sigma > 0$, in such a way that $\hat{\Omega}(t,s) \in X \otimes X$ for all $t > 0$ and all $s \geq -\sigma$. Hence $\hat{\Omega} \in ST_{X\otimes X;A\otimes I,I\otimes A}$.
On the other hand, let $\hat{\zeta} \in ST_{X\otimes X;A\otimes I,I\otimes A}$. Since $e^{-tA} \in X \otimes X$ for all $t > 0$, it follows that $e^{-tA} \in \mathbf{T}(X)$ for all $t > 0$. Now define the linear mapping ζ by

$$\zeta w = \hat{\zeta}(t,0)\, e^{tA}\, w = e^{-\sigma A}\, \hat{\zeta}(t,-\sigma)\, e^{tA}\, w$$

where $w \in S_{X,A}$ and $t > 0$ sufficiently small. Then for all $t > 0$

$$e^{\sigma A} \zeta e^{-tA} \in \mathbf{T}(X)$$

and hence $\zeta \in \mathbf{T}(S_{X,A})$. We thus obtain the following identification $\mathbf{T}(S_{X,A}) \triangleq ST_{X\otimes X;A\otimes I,I\otimes A}$. We conclude that in the nuclear case the pairing between $\mathbf{T}(S_{X,A})$ and $\mathbf{L}(T_{X,A})$ is just a representation of the usual duality between the spaces $ST_{X\otimes X;A\otimes I,I\otimes A}$ and $TS_{X\otimes X;A\otimes I,I\otimes A}$.

In (IV.2.5) we have defined the transposition ^{c} on $TT_{X\otimes X;A\otimes I,I\otimes A}$ as follows:

(IV.2.5) $Q^c(t,s) = Q(t,s)^*$, $t > 0,\ s > 0$.

We can restrict the mapping ^{c} to the spaces $TS_{X\otimes X;I\otimes A,A\otimes I}$, $TS_{X\otimes X;A\otimes I,I\otimes A}$, $ST_{X\otimes X;I\otimes A,A\otimes I}$ and $ST_{X\otimes X;A\otimes I,I\otimes A}$, because these spaces are all included in $TT_{X\otimes X;A\otimes I,I\otimes A}$. (Here we use the identification $TT_{X\otimes X;A\otimes I,I\otimes A} = TT_{X\otimes X;I\otimes A,A\otimes I}$.) We observed already, that ^{c} is an anti-linear homeomorphism from $TS_{X\otimes X;I\otimes A,A\otimes I}$ onto $TS_{X\otimes X;A\otimes I,I\otimes A}$ which satisfies $P^{cc} = P$, $P \in TS_{X\otimes X;I\otimes A,A\otimes I}$. Also, it is easy to check that the mapping ^{c} is an anti-linear homeomorphism from $ST_{X\otimes X;A\otimes I,I\otimes A}$ onto $ST_{X\otimes X;I\otimes A,A\otimes I}$.
We observe that for each $\theta \in \mathbf{T}(T_{X,A})$ with representant $\hat{\theta} \in ST_{X\otimes X;I\otimes A,A\otimes I}$ the dual $\theta' \in \mathbf{T}(S_{X,A})$ has representant $\hat{\theta}^c \in ST_{X\otimes X;A\otimes I,I\otimes A}$. Observe that $\hat{\theta}^{cc} = \hat{\theta}$.

The relations between the various topologies for $\mathbf{L}(T_{X,A})$ and $TS_{X\otimes X;A\otimes I,I\otimes A}$ are the following:

(IV.2.27) Theorem

Let $T_{X,A}$ be a nuclear space. Then the topology $\lambda\tau_{uw}$ for $\mathbf{L}(T_{X,A})$ is equivalent to the topology $\tau\sigma_w$ for $TS_{X\otimes X;A\otimes I,I\otimes A}$. Moreover, the topology $\lambda\tau_{us}$ is equivalent to the topology $\tau\sigma_{proj}$, and $\tau\sigma_{proj}$ is equivalent to $\lambda\tau_u$. So $\lambda\tau_{us}$ and $\lambda\tau_u$ are equivalent.

Proof. Cf. Theorem (IV.1.23). □

IV.3. The algebra $\mathbf{E}(S_{X,A})$

The space $\mathbf{E}(S_{X,A})$ consists of all continuous linear mappings from $S_{X,A}$ into $S_{X,A}$ which are extendible to a continuous linear mapping from $T_{X,A}$ into $T_{X,A}$. From Theorem (I.4.10) we obtain the following characterization of the space $\mathbf{E}(S_{X,A})$.

(IV.3.1) Characterization

Let $E \in \mathbf{L}(S_{X,A})$. Then $E \in \mathbf{E}(S_{X,A})$ iff $E^{*}\restriction_{S_{X,A}} \in \mathbf{L}(S_{X,A})$. Its extension $\bar{E}$ to $\mathbf{L}(T_{X,A})$ is given by $\bar{E} = (E^{*}\restriction_{S_{X,A}})'$.

Remark. If no confusion is likely to arise we shall denote $E^{*}\restriction_{S_{X,A}}$ by E^{*}. Here E^{*} denotes the usual adjoint of the operator E in X.

(IV.3.2) Lemma

(i) Let $E_1, E_2 \in \mathbf{E}(S_{X,A})$. Then $E_1 \cdot E_2 \in \mathbf{E}(S_{X,A})$.

(ii) Let $E \in \mathbf{E}(S_{X,A})$. Then $E^{*} \in \mathbf{E}(S_{X,A})$.

Proof

(i) We have $(E_1E_2)^{*} \supset E_2^{*}E_1^{*}$, from which the result follows.

(ii) The linear mapping $E' \in \mathbf{L}(T_{X,A})$ extends E^{*} because for all $u,w \in S_{X,A}$ we have

$$\langle u,E^{*}w\rangle = (Eu,w) = \langle u,E'w\rangle \;. \qquad \square$$

So $\mathbf{E}(S_{X,A})$ can be regarded as a $*$-algebra of unbounded operators with common dense invariant domain $S_{X,A}$. Algebras of this type are called Op*-algebras by Lassner et al. (Cf. [La], [An Ma].) Furthermore, $\mathbf{E}(S_{X,A})$ can be regarded as a subalgebra of $\mathbf{L}(S_{X,A})$, and, identifying $\mathbf{E}(S_{X,A})$ and $\bar{\mathbf{E}}(T_{X,A})$,

$$\bar{\mathbf{E}}(T_{X,A}) = \{\bar{E} = (E^{*})' \mid E \in \mathbf{E}(S_{X,A})\} \;,$$

as a subalgebra of $\mathbf{L}(T_{X,A})$. Since both $\mathbf{L}(S_{X,A}) \hat{\subset} \mathbf{L}(S_{X,A},T_{X,A})$ and $\mathbf{L}(T_{X,A}) \hat{\subset} \mathbf{L}(S_{X,A},T_{X,A})$, it makes sense to write

$$\mathbf{E}(S_{X,A}) = \mathbf{L}(S_{X,A}) \cap \mathbf{L}(T_{X,A}) \;.$$

Introduced in this way $\mathbf{E}(S_{X,A})$ can be written as $\mathbf{L}^+(S_{X,A})$ in the sense of [La].

In this section we work out the same scheme as has already been accomplished for the spaces $\mathbf{L}(S_{X,A})$ and $\mathbf{L}(T_{X,A})$.

(IV.3.3) Definition

The uniform topology $\varepsilon\sigma_u$ for $\mathbf{E}(S_{X,A})$ is the locally convex topology generated by the seminorms $e_{\psi,t}$,

$$e_{\psi,t}(E) = \max\{\|\psi(A)\,E\,e^{-tA}\|, \|\psi(A)\,E^{*}e^{-tA}\|\}$$

where $\psi \in \mathbf{B}_+(\mathbb{R})$ and $t > 0$.

Let $E \in \mathbf{E}(S_{X,A})$. Since $(\psi(A)\,E^{\star}e^{-tA}x,y) = \langle e^{-tA}x,\bar{E}\psi(A)y\rangle$ for all $x,y \in X$, we have

$$\|\psi(A)\,E^{\star}e^{-tA}\| = \|e^{-tA}\,\bar{E}\,\psi(A)\| = t_{\psi,t}(E)\ .$$

Further, by definition, $s_{\psi,t}(E) = \|\psi(A)\,E\,e^{-tA}\|$. We recall that the seminorms $s_{\psi,t}$ and $t_{\psi,t}$, $\psi \in \mathbf{B}_+(\mathbb{R})$, $t > 0$, generate the topology $\lambda\sigma_u$ and $\lambda\tau_u$, respectively. Cf. Definitions (IV.1.2) and (IV.2.2). We get the following characterization of $\varepsilon\sigma_u$.

(IV.3.3) Lemma

The topology $\varepsilon\sigma_u$ is the coarsest locally convex topology for $\mathbf{E}(S_{X,A})$ for which the embeddings $\mathbf{E}(S_{X,A}) \subset \mathbf{L}(S_{X,A})$ and $\mathbf{E}(S_{X,A}) \subset \mathbf{L}(T_{X,A})$ are continuous where $\mathbf{L}(S_{X,A})$ and $\mathbf{L}(T_{X,A})$ carry their uniform topology.

(IV.3.4) Lemma

Let $S_{X,A}$ be a nuclear space. Then $TS_{X\otimes X;A \,\wedge\, A,A \,\vee\, A}$ comprises all elements of $\mathbf{E}(S_{X,A})$. The topology $\tau\sigma_{proj}$ for $TS_{X\otimes X;A \,\wedge\, A,A \,\vee\, A}$ is equivalent to the topology $\varepsilon\sigma_u$ for $\mathbf{E}(S_{X,A})$.

Proof. Each of the operators e^{-tA}, $t > 0$, is Hilbert-Schmidt. So following the kernel theorem for $\mathbf{E}(S_{X,A})$ each element of $\mathbf{E}(S_{X,A})$ arises from an element of $TS_{X\otimes X;A \,\wedge\, A,A \,\vee\, A}$.

From the definition of the operators $A \circledwedge A$ and $A \circledvee A$ it follows that the topology $\tau\sigma_{proj}$ for $TS_{X\otimes X;A \circledwedge A, A \circledvee A}$ is the coarsest topology for which the embeddings

$$TS_{X\otimes X;A \circledwedge A, A \circledvee A} \hookrightarrow TS_{X\otimes X;I\otimes A, A\otimes I}$$

and

$$TS_{X\otimes X;A \circledwedge A, A \circledvee A} \hookrightarrow TS_{X\otimes X;A\otimes I, I\otimes A}$$

are continuous, where both spaces on the right hand side carry their topology $\tau\sigma_{proj}$. Further, from Lemma (IV.1.3) and Lemm (IV.2.6) it follows that the topology $\tau\sigma_{proj}$ for $TS_{X\otimes X;I\otimes A, A\otimes I}$ is equivalent to the topology $\lambda\sigma_u$ for $\mathbf{L}(S_{X,A})$ and that the topology $\tau\sigma_{proj}$ for $TS_{X\otimes X;A\otimes I, I\otimes A}$ is equivalent to the topology $\lambda\tau_u$ for $\mathbf{L}(T_{X,A})$. All these observations together with Lemma (IV.3.3) yield the wanted result. □

With Lemma (IV.1.4), Lemma (IV.2.3) and the characterization of the topology $\varepsilon\sigma_u$ given in Lemma (IV.3.3) we obtain

(IV.3.5) Theorem

$\mathbf{E}(S_{X,A})$ with topology $\varepsilon\sigma_u$ is a complete locally convex algebra.

In a natural way the topologies $\lambda\sigma_s$, $\lambda\sigma_w$ for $\mathbf{L}(S_{X,A})$ and $\lambda\tau_s$, $\lambda\tau_w$ for $\mathbf{L}(T_{X,A})$ induce locally convex topologies on $\mathbf{E}(S_{X,A})$.

(IV.3.6) Definition

The strong topology $\varepsilon\sigma_s$ for $\mathbf{E}(S_{X,A})$ is the coarsest locally convex topology for which the natural embeddings $\mathbf{E}(S_{X,A}) \hookrightarrow \mathbf{L}(S_{X,A})$ and $\mathbf{E}(S_{X,A}) \hookrightarrow \mathbf{L}(T_{X,A})$ are continuous, where $\mathbf{L}(S_{X,A})$ and $\mathbf{L}(T_{X,A})$ carry their strong topologies $\lambda\sigma_s$ and $\lambda\tau_s$.

The net (E_α) in $\mathbf{E}(S_{X,A})$ is $\varepsilon\sigma_s$-convergent to E if and only if for all $w \in S_{X,A}$ the net $(E_\alpha w)$ is σ_{ind}-convergent to Ew, and for all $F \in T_{X,A}$ the net $(\bar{E}_\alpha F)$ is τ_{proj}-convergent to $\bar{E}F$. The topology $\varepsilon\sigma_s$ is the coarsest topology for which the mappings

$$E \mapsto Ew \quad \text{and} \quad E \mapsto \bar{E}G , \qquad E \in \mathbf{E}(S_{X,A}) ,$$

for any $w \in S_{X,A}$ and $G \in T_{X,A}$ are continuous from $\mathbf{E}(S_{X,A})$ into $S_{X,A}$ with topology σ_{ind} and from $\mathbf{E}(S_{X,A})$ into $T_{X,A}$ with topology τ_{proj}.

(IV.3.7) Definition

The weak topology $\varepsilon\sigma_w$ for $\mathbf{E}(S_{X,A})$ is the coarsest locally convex topology for which the embeddings $\mathbf{E}(S_{X,A}) \hookrightarrow \mathbf{L}(S_{X,A})$ and $\mathbf{E}(S_{X,A}) \hookrightarrow \mathbf{L}(T_{X,A})$ are continuous where $\mathbf{L}(S_{X,A})$ and $\mathbf{L}(T_{X,A})$ carry their weak topologies $\lambda\sigma_w$ and $\lambda\tau_w$.

The net (E_α) in $\mathbf{E}(S_{X,A})$ is $\varepsilon\sigma_w$-convergent to E if and only if for all $w \in S_{X,A}$ and all $G \in T_{X,A}$ the net $(\langle E_\alpha w,G\rangle)$ tends to $\langle Ew,G\rangle$ and the net $(\langle E_\alpha^* w,G\rangle)$ tends to $\langle E^*w,G\rangle$. The topology $\varepsilon\sigma_w$ is the coarsest topology for which the functionals

$$E \mapsto \langle Ew,G\rangle \quad \text{and} \quad E \mapsto \langle E^*w,G\rangle \;, \qquad E \in \mathbf{E}(S_{X,A})$$

are continuous for any $w \in S_{X,A}$ and $G \in T_{X,A}$.

The so-called principle of uniform boundedness follows immediately from Theorem (IV.1.9) and Theorem (IV.2.10). So we have

(IV.3.8) Theorem

Let B be a bounded subset of $\mathbf{E}(S_{X,A})$. The following statements are equivalent:

(i) B is $\varepsilon\sigma_u$-bounded.

(ii) B is $\varepsilon\sigma_s$-bounded.

(iii) B is $\varepsilon\sigma_w$-bounded.

(IV.3.9) Theorem

$\mathbf{E}(S_{X,A})$ is sequentially $\varepsilon\sigma_s$- and $\varepsilon\sigma_w$-complete.

Proof. Cf. Corollary (IV.1.11) and Corollary (IV.2.12). □

From the two previous sections we derive the following characterizations of bounded subsets and null sequences in $\mathbf{E}(S_{X,A})$.

(IV.3.10) Theorem

(i) A set $B \subset \mathbf{E}(S_{X,A})$ is bounded iff for all $t > 0$ there exists $\tau > 0$ such that the set $\{e^{\tau A}\, E\, e^{-tA}, e^{\tau A}\, E^*\, e^{-tA} \mid E \in B\}$ is bounded in $\mathbf{B}(X)$.

(ii) A sequence $(E_n)_{n\in\mathbb{N}} \subset \mathbf{E}(S_{X,A})$ is an $\varepsilon\sigma_u$-null sequence iff for all $t > 0$ there exists $\tau > 0$ such that both sequences $(e^{\tau A} E_n e^{-tA})_{n\in\mathbb{N}}$ and $(e^{\tau A} E_n^* e^{-tA})_{n\in\mathbb{N}}$ tend to zero in $\mathbf{B}(X)$.

We note that multiplication is jointly sequentially $\varepsilon\sigma_u$-continuous in $\mathbf{E}(S_{X,A})$. Moreover, the transposition $E \to E^*$ is weakly continuous but not strongly continuous, in general.
Next, we introduce the ultra-strong and the ultra-weak topology for $\mathbf{E}(S_{X,A})$

(IV.3.11) Definition

The ultra-strong topology $\varepsilon\sigma_{us}$ for $\mathbf{E}(S_{X,A})$ is the coarsest locally convex topology for which the embeddings $\mathbf{E}(S_{X,A}) \subset \mathbf{L}(S_{X,A})$ and $\mathbf{E}(S_{X,A}) \subset \mathbf{L}(T_{X,A})$ are continuous, where both $\mathbf{L}(S_{X,A})$ and $\mathbf{L}(T_{X,A})$ carry their ultra-strong topology. So the topology $\varepsilon\sigma_{us}$ is generated by all seminorms $s_{(w_\ell),\psi}$ and $t_{(F_\ell),t}$ as defined in (IV.1.15) and (IV.2.16).

(IV.3.12) Definition

The ultra-weak topology $\varepsilon\sigma_{uw}$ for $\mathbf{E}(S_{X,A})$ is the coarsest locally convex topology for which the embeddings $\mathbf{E}(S_{X,A}) \subset \mathbf{L}(S_{X,A})$ and $\mathbf{E}(S_{X,A}) \subset \mathbf{L}(T_{X,A})$ are continuous, where both $\mathbf{L}(S_{X,A})$ and $\mathbf{L}(T_{X,A})$ carry their ultra-weak topology.

It follows that $\varepsilon\sigma_{uw}$ is the coarsest topology for which the functionals

$$E \mapsto \sum_{\ell=1}^{\infty} \langle Ew_\ell, F_\ell\rangle \quad \text{and} \quad E \mapsto \sum_{\ell=1}^{\infty} \langle E^* w_\ell, F_\ell\rangle$$

are continuous for each couple $(w_\ell) \in S_{d(X),d(A)}$ and $(F_\ell) \in T_{d(X),d(A)}$.
We observe that the transposition $E \mapsto E^*$ is in general not $\varepsilon\sigma_{us}$-continuous. This mapping is $\varepsilon\sigma_{uw}$-continuous.
Since $\mathbf{L}(S_{X,A})$ and $\mathbf{L}(T_{X,A})$ can be considered as linear subspaces of the space $\mathbf{L}(S_{X,A}, T_{X,A})$ of all continuous linear mappings from $S_{X,A}$ into $T_{X,A}$, it makes sense to write down $\mathbf{L}(S_{X,A}) + \mathbf{L}(T_{X,A})$ where + denotes the vector sum in $\mathbf{L}(S_{X,A}, T_{X,A})$. As a subspace of $\mathbf{L}(S_{X,A}) + \mathbf{L}(T_{X,A})$, we introduce the space $\mathbf{T}(S_{X,A}) + \mathbf{T}(T_{X,A})$.

(IV.3.13) Lemma

Let $\Phi \in \mathbf{T}(S_{X,A}) + \mathbf{T}(T_{X,A})$. Then there exist $\theta_1,\theta_2 \in \mathbf{T}(T_{X,A})$ such that $\Phi = \theta_1' + \theta_2$.

Proof. There exists $\Omega \in \mathbf{T}(S_{X,A})$ and $\theta_2 \in \mathbf{T}(T_{X,A})$ such that $\Phi = \Omega + \theta_2$. Now put $\theta_1 = \Omega'$. □

The space $\mathbf{T}(S_{X,A}) + \mathbf{T}(T_{X,A})$ and $\mathbf{E}(S_{X,A})$ can be paired.

(IV.3.14) Definition

Let $E \in \mathbf{E}(S_{X,A})$ and let $\theta_1,\theta_2 \in \mathbf{T}(T_{X,A})$, $\Phi = \theta_1' + \theta_2$. Then

$$\langle\!\langle \Phi, E \rangle\!\rangle_e := \overline{\langle\!\langle \theta_1, E^* \rangle\!\rangle_1} + \langle\!\langle \theta_2, E \rangle\!\rangle_1 \ .$$

Remark. The way of decomposing Φ does not matter. To see this, assume $0 = \theta_1' + \theta_2$. Then $\theta_1,\theta_2 \in \mathbf{T}(T_{X,A}) \cap \mathbf{T}(S_{X,A})$. Hence,

$$\overline{\langle\!\langle \theta_1, E^* \rangle\!\rangle_1} + \langle\!\langle \theta_2, E \rangle\!\rangle_1 = \langle\!\langle \theta_1' + \theta_2, E \rangle\!\rangle_e = 0$$

where $\theta_1' + \theta_2 \in \mathbf{T}(T_{X,A})$. (Cf. Theorem (IV.2.25.)

Remark. Let $E \in \mathbf{E}(S_{X,A})$. Let $\bar{E} \in \mathbf{L}(T_{X,A})$ denote its extension. Then $\bar{E} = (E^*)'$. We have

$$\overline{\langle\!\langle \theta_1, E^* \rangle\!\rangle_1} = \langle\!\langle \theta_1', \bar{E} \rangle\!\rangle_1 \ .$$

From the definition of the topology $\varepsilon\sigma_{uw}$ and from Theorem (IV.1.20) we obtain

(IV.3.15) Theorem

The ultra-weak topology $\varepsilon\sigma_{uw}$ for $\mathbf{E}(S_{X,A})$ is the coarsest locally convex topology for which the linear functionals

$$E \mapsto \overline{\langle\!\langle \Phi, E \rangle\!\rangle_e} \ , \qquad E \in \mathbf{E}(S_{X,A})$$

are continuous for each $\Phi \in \mathbf{T}(S_{X,A}) + \mathbf{T}(T_{X,A})$.

(IV.3.16) Theorem

Let $B \subset \mathbf{E}(S_{X,A})$. The following statements are equivalent:

(i) B is $\varepsilon\sigma_u$-bounded.

(ii) B is $\varepsilon\sigma_{us}$-bounded.

(iii) B is $\varepsilon\sigma_{uw}$-bounded.

Proof. Cf. Theorem (IV.1.22). □

(IV.3.17) Corollary

The algebra $\mathbf{E}(S_{X,A})$ is $\varepsilon\sigma_{us}$- and $\varepsilon\sigma_{uw}$-sequentially complete. □

It is clear that the transposition mapping $L \mapsto L'$ maps $\mathbf{L}(S_{X,A},T_{X,A})$ onto $\mathbf{L}(S_{X,A},T_{X,A})$. We can restrict this mapping to the space $\mathbf{T}(S_{X,A}) + \mathbf{T}(T_{X,A})$. To see how it operates, let $\Phi = \Theta_1' + \Theta_2$ with $\Theta_1,\Theta_2 \in \mathbf{T}(T_{X,A})$. Then we get

$$\Phi' = (\Theta_1' + \Theta_2)' = \Theta_1 + \Theta_2'$$

and hence $\Phi' \in \mathbf{T}(S_{X,A}) + \mathbf{T}(T_{X,A})$, also.
From Theorem (IV.2.25) we obtain

(IV.3.18) Theorem

Let $\Theta_1,\Theta_2 \in \mathbf{T}(T_{X,A})$ and let $E \in \mathbf{E}(S_{X,A})$. Put $\Phi = \Theta_1' + \Theta_2$.

(i) $\langle\Phi,E\rangle_e = \overline{\langle\Theta_1,E^*\rangle_1} + \langle\Theta_2,E\rangle_1 = \langle\Theta_1',\bar{E}\rangle_1 + \overline{\langle\Theta_2',\overline{E^*}\rangle}$.

(ii) $\langle\Phi,E\rangle_e = \langle\Phi',E^*\rangle_e$.

(iii) $\langle\Phi,E_1\,E\,E_2\rangle_e = \langle E_1^*\,\Phi\,E_2^*,E\rangle_e\ ,\quad E_1,E_2 \in \mathbf{E}(S_{X,A})$.

Here $E_1^*\,\Phi\,E_2^* = \bar{E}_2\,\Theta_1'\,\bar{E}_1 + E_1'\,\Theta_2\,E_2'$.

We recall that $\mathbf{T}(S_{X,A})$ is a two-sided ideal in $\mathbf{L}(S_{X,A})$ and, similarly, $\mathbf{T}(T_{X,A})$ is a two-sided ideal in $\mathbf{L}(T_{X,A})$. So the intersection $\mathbf{T}(S_{X,A}) \cap \mathbf{T}(T_{X,A})$ is a two-sided ideal in $\mathbf{E}(S_{X,A})$. Now observe that $\mathbf{T}(T_{X,A}) \subset ST_{X\otimes X;I\otimes A,A\otimes I}$ and $\mathbf{T}(S_{X,A}) \subset ST_{X\otimes X;A\otimes I,I\otimes A}$. From Section III.5 it follows that

$$SS_{X\otimes X;I\otimes A,A\otimes I} \triangleq ST_{X\otimes X;I\otimes A,A\otimes I} \cap ST_{X\otimes X;A\otimes I,I\otimes A}$$

and hence $\mathbf{T}(S_{X,A}) \cap \mathbf{T}(T_{X,A}) \subset SS_{X\otimes X;I\otimes A,A\otimes I}$. We conjecture that the intersection $\mathbf{T}(S_{X,A}) \cap \mathbf{T}(T_{X,A})$ consists of all operators of the form $e^{-tA}\Lambda e^{-tA}$ with $\Lambda \in T(X)$ and $t > 0$.

Finally, we shortly discuss the nuclear case, i.e. the operator e^{-tA} is Hilbert-Schmidt for all $t > 0$. From Section III.6 it follows that in this case $TS_{X\otimes X;A \textcircled{\scriptsize$\wedge$} A,A \textcircled{\scriptsize$\vee$} A}$ comprises the whole $\mathbf{E}(S_{X,A})$. Observe that

$$TS_{X\otimes X;A \textcircled{\scriptsize$\wedge$} A,A \textcircled{\scriptsize$\vee$} A} = TS_{X\otimes X;I\otimes A,A\otimes I} \cap TS_{X\otimes X;A\otimes I,I\otimes A} .$$

In sections IV.1 and IV.2 we have seen that

$$\mathbf{T}(T_{X,A}) \triangleq ST_{X\otimes X;I\otimes A,A\otimes I}$$

and

$$\mathbf{T}(S_{X,A}) \triangleq ST_{X\otimes X;A\otimes I,I\otimes A} .$$

Hence

$$\mathbf{T}(S_{X,A}) + \mathbf{T}(T_{X,A}) \triangleq ST_{X\otimes X;A \textcircled{\scriptsize$\wedge$} A,A \textcircled{\scriptsize$\vee$} A} .$$

The mapping ^{c} on $TT_{X\otimes X;A\otimes I,I\otimes A}$, defined in (IV.2.5), maps $TS_{X\otimes X;A \textcircled{\scriptsize$\wedge$} A,A \textcircled{\scriptsize$\vee$} A}$ homeomorphically onto itself. Similarly, it is a homeomorphism from $ST_{X\otimes X;A \textcircled{\scriptsize$\wedge$} A,A \textcircled{\scriptsize$\vee$} A}$ onto itself.
Let $E \in \mathbf{E}(S_{X,A})$. The involution $*$: $E \mapsto E^{*}$ is represented by the mapping ^{c}. The relations between the locally convex topologies of $\mathbf{E}(S_{X,A})$ and of the tensor product $TS_{X\otimes X;A \textcircled{\scriptsize$\wedge$} A,A \textcircled{\scriptsize$\vee$} A}$ are the following.

(IV.3.19) <u>Theorem</u>

Let $S_{X,A}$ be a nuclear space. Then the topology $\varepsilon\sigma_{uw}$ for $\mathbf{E}(S_{X,A})$ is equivalent to the topology $\tau\sigma_{w}$ for $TS_{X\otimes X;A \textcircled{\scriptsize$\wedge$} A,A \textcircled{\scriptsize$\vee$} A}$. Moreover, the topology $\varepsilon\sigma_{us}$ is equivalent to the topology $\tau\sigma_{proj}$ and $\tau\sigma_{proj}$ is equivalent to $\varepsilon\sigma_{u}$.
So the topologies $\varepsilon\sigma_{u}$ and $\varepsilon\sigma_{us}$ are equivalent.

<u>Proof</u>. Cf. Theorems (IV.1.23) and (IV.2.27). □

IV.4. Operator ideals in $\mathbf{L}(S_{X,A})$ and $\mathbf{L}(T_{X,A})$

In this section we show that the operator ideals in $\mathbf{B}(X)$ give rise to ideals in the algebras $\mathbf{L}(S_{X,A})$ and $\mathbf{L}(T_{X,A})$. Worth mentioning are the ideal $\mathbf{T}(X)$ of the trace class operators, the ideal $\mathbf{H}(X)$ of the Hilbert-Schmidt operators and the ideal $\mathbf{C}(X)$ of the compact operator. In his monograph [Scha], Schatten gives an extensive discussion of the so-called Von Neumann-Schatten classes of completely continuous (= compact) operators. Each class is a two-sided ideal in $\mathbf{B}(X)$. In this monograph, Schatten also considers the chains

$$\mathbf{C}(X) \subset \mathbf{C}^{(2)}(X) \subset \mathbf{C}^{(4)}(X) \subset \ldots$$

$$\mathbf{T}(X) \subset \mathbf{T}^{(2)}(X) \subset \mathbf{T}^{(4)}(X) \subset \ldots .$$

We note that $\mathbf{T}(X)$ is a representant of the dual of $\mathbf{C}(X)$. Further, $\mathbf{C}^{(2)} = \mathbf{C}^{**}(X) \triangleq \mathbf{B}(X)$, i.e. $\mathbf{T}(X)$ is a representant of the pre-dual of $\mathbf{B}(X)$. Possibly, a similar scheme can be developed in our operator theory.

(IV.4.1) Definition

Let $\mathbf{I}(X)$ be a two-sided operator ideal in $\mathbf{B}(X)$. Then the subspace $\mathbf{I}_{ts}(S_{X,A})$ of $\mathbf{L}(S_{X,A})$ consists of all $L \in \mathbf{L}(S_{X,A})$ with the property

$$\forall_{s>0}\ \exists_{t>0}:\ e^{tA}\, L e^{-sA} \in \mathbf{I}(X)\ .$$

(IV.4.2) Lemma

$\mathbf{I}_{ts}(S_{X,A})$ is a two-sided ideal in $\mathbf{L}(S_{X,A})$.

Proof. Let $L_1, L_2 \in \mathbf{L}(S_{X,A})$, and let $L \in \mathbf{I}_{ts}(S_{X,A})$. Let $s > 0$. Then there exists $\sigma_1 > 0$, $\sigma_2 > 0$ and $t > 0$ such that

$$e^{\sigma_1 A}\, L_1\, e^{-sA} \in \mathbf{B}(X)\ ,$$

$$e^{\sigma_2 A}\, L e^{-\sigma_1 A} \in \mathbf{I}(X)\ ,$$

and

$$e^{tA}\, L_2\, e^{-\sigma_2 A} \in \mathbf{B}(X)\ .$$

Therefore we obtain $e^{tA}(L_2\, L\, L_1)e^{-sA} \in \mathbf{I}(X)$. □

For $\mathbf{I}(X)$ we can take any operator ideal introduced by Schatten, in particular the ideals $\mathbf{T}(X)$, $\mathbf{H}(X)$ and $\mathbf{C}(X)$. In this way, we get the ideals $\mathbf{T}_{ts}(S_{X,A})$, $\mathbf{H}_{ts}(S_{X,A})$ and $\mathbf{C}_{ts}(S_{X,A})$.

(IV.4.3) Lemma

$$\mathbf{T}_{ts}(S_{X,A}) \subset \mathbf{H}_{ts}(S_{X,A}) \subset \mathbf{C}_{ts}(S_{X,A}) \subset \mathbf{B}_{ts}(S_{X,A}) = \mathbf{L}(S_{X,A}) .$$

Proof. The assertion follows from the inclusion chain

$$\mathbf{T}(X) \subset \mathbf{H}(X) \subset \mathbf{C}(X) \subset \mathbf{B}(X) . \qquad \square$$

Another way of introducing operator ideals in $\mathbf{L}(S_{X,A})$ is the following one.

(IV.4.4) Definition

Let $\mathbf{I}(X)$ be a two-sided ideal in $\mathbf{B}(X)$. Then the subspace $\mathbf{I}_{st}(S_{X,A})$ of $\mathbf{L}(S_{X,A})$ consists of all $L \in \mathbf{L}(S_{X,A})$ withe the property

$$\exists_{s>0} \forall_{t>0}: e^{sA} L e^{-tA} \in \mathbf{I}(X) .$$

(IV.4.5) Lemma

$\mathbf{I}_{st}(S_{X,A})$ is a two-sided ideal in $\mathbf{L}(S_{X,A})$.

Proof. Let $L_1, L_2 \in \mathbf{L}(S_{X,A})$ and let $\theta \in \mathbf{I}_{st}(S_{X,A})$. Then there exists a fixed $\sigma_2 > 0$ such that for all $t > 0$

$$e^{\sigma_2 A} \theta e^{-tA} \in \mathbf{I}(X) .$$

Now take $s > 0$ so small that $e^{sA} L_2 e^{-\sigma_2 A} \in \mathbf{L}(X)$, and for each $t > 0$ take $\tau > 0$ so small that $e^{\tau A} L_1 e^{-tA} \in \mathbf{L}(X)$. Then we write for all $t > 0$

$$e^{sA} L_2 \theta L_1 e^{-tA} = (e^{sA} L_2 e^{-\sigma_2 A})(e^{\sigma_2 A} \theta e^{-\tau A})(e^{\tau A} L_1 e^{-tA}) .$$

From the latter expression it follows that

$$\forall_{t>0}: e^{sA} L_2 \theta L_1 e^{-tA} \in \mathbf{I}(X) . \qquad \square$$

From the definition of $\mathbf{I}_{st}(S_{X,A})$ the following observation is obvious.

(IV.4.6) Corollary

$\mathbf{I}_{st}(S_{X,A})$ is a subideal of $\mathbf{I}_{ts}(S_{X,A})$.

As in Lemma (IV.4.3) we have

(IV.4.7) Lemma

$$\mathbf{T}_{st}(S_{X,A}) \subset \mathbf{H}_{st}(S_{X,A}) \subset \mathbf{C}_{st}(S_{X,A}) \subset \mathbf{B}_{st}(S_{X,A}) .$$

Remark. $\mathbf{T}(S_{X,A}) = \mathbf{T}_{st}(S_{X,A})$, see Definition (IV.1.18).

Let $P \in \mathbf{H}_{ts}(S_{X,A})$. Then the mapping $\hat{P}$ defined by

$$\hat{P}(t,s) = e^{-tA} P e^{-sA} , \qquad t \geq 0,\ s > 0$$

belongs to $TS_{X\otimes X;I\otimes A,A\otimes I}$. Conversely, each element of $TS_{X\otimes X;I\otimes A,A\otimes I}$ gives rise to a linear mapping of $\mathbf{L}(S_{X,A})$ which belongs to $\mathbf{H}_{ts}(S_{X,A})$ (see section III.6, case (c)). So we can identify $\mathbf{H}_{ts}(S_{X,A})$ and $TS_{X\otimes X;I\otimes A,A\otimes I}$, i.e. $\mathbf{H}_{ts}(S_{X,A}) \triangleq TS_{X\otimes X;I\otimes A,A\otimes I}$. Similarly, it can be seen that $\mathbf{H}_{st}(S_{X,A}) \triangleq$ $\triangleq ST_{X\otimes X;A\otimes I,I\otimes A}$.

Next we introduce operator ideals in $\mathbf{L}(T_{X,A})$.

(IV.4.8) Definition

Let $\mathbf{I}(X)$ be an operator ideal in $\mathbf{B}(X)$. We introduce $\mathbf{I}_{ts}(T_{X,A})$ and $\mathbf{I}_{st}(T_{X,A})$ as follows:
Let $K \in \mathbf{L}(T_{X,A})$. Then

(i) $\qquad K \in \mathbf{I}_{ts}(T_{X,A}) \ :\Leftrightarrow K' \in \mathbf{I}_{ts}(S_{X,A})$.

(ii) $\qquad K \in \mathbf{I}_{st}(T_{X,A}) \ :\Leftrightarrow K' \in \mathbf{I}_{st}(S_{X,A})$.

From the properties of the transposition mapping $K \mapsto K'$ it follows that the spaces $\mathbf{I}_{ts}(T_{X,A})$ and $\mathbf{I}_{st}(T_{X,A})$ are well-defined operator ideals of $\mathbf{L}(T_{X,A})$. The mentioned inclusion relations carry over to these ideals.

(IV.4.9) Lemma

Let $\mathbf{I}(X)$ be a two-sided ideal in $\mathbf{B}(X)$. Then we have

(i) $\mathbf{I}_{st}(T_{X,A}) \subset \mathbf{I}_{ts}(T_{X,A})$.

(ii) $\mathbf{T}_{ts}(T_{X,A}) \subset \mathbf{H}_{ts}(T_{X,A}) \subset \mathbf{C}_{ts}(T_{X,A}) \subset \mathbf{B}_{ts}(T_{X,A}) = \mathbf{L}(T_{X,A})$.

(iii) $\mathbf{T}(T_{X,A}) = \mathbf{T}_{st}(T_{X,A}) \subset \mathbf{H}_{st}(T_{X,A}) \subset \mathbf{C}_{st}(T_{X,A}) \subset \mathbf{B}_{st}(T_{X,A})$.

From the properties of the mapping $K \mapsto K'$ as discussed in Section IV.2 we have the identifications:

$$\mathbf{H}_{ts}(T_{X,A}) \triangleq TS_{X\otimes X;A\otimes I,I\otimes A} \quad \text{and} \quad \mathbf{H}_{st}(T_{X,A}) \triangleq ST_{X\otimes X;I\otimes A,A\otimes I} .$$

In the next theorem we give a condition on the semigroup $(e^{-tA})_{t>0}$ such that the ideals $\mathbf{I}_{ts}(S_{X,A})$ and $\mathbf{I}_{ts}(T_{X,A})$ coincide with $\mathbf{L}(S_{X,A})$ and $\mathbf{L}(T_{X,A})$, respectively. So we present a kind of extended kernel theorem.

(IV.4.10) Theorem

Let $\mathbf{I}(X)$ be a two-sided ideal in $\mathbf{L}(X)$. Then the following statements are equivalent:

(i) For all $t > 0$ the operator e^{-tA} belongs to $\mathbf{I}(X)$.

(ii) $\mathbf{I}_{ts}(S_{X,A}) = \mathbf{L}(S_{X,A})$.

(iii) $\mathbf{I}_{st}(S_{X,A}) = \mathbf{B}_{st}(S_{X,A})$.

(iv) $\mathbf{I}_{ts}(T_{X,A}) = \mathbf{L}(T_{X,A})$.

(v) $\mathbf{I}_{st}(T_{X,A}) = \mathbf{B}_{st}(T_{X,A})$.

Proof

(i) ⇒ (ii): We only have to prove that $\mathbf{L}(S_{X,A}) \subset \mathbf{I}_{ts}(S_{X,A})$. So let $L \in \mathbf{L}(S_{X,A})$. Let $s > 0$. Then there exists $t > 0$ such that $e^{tA} L e^{-sA} \in \mathbf{B}(X)$. We get

$$e^{\frac{1}{2}tA} L e^{-sA} = e^{-\frac{1}{2}tA}(e^{tA} L e^{-sA}) \in \mathbf{I}(X) .$$

It follows that $L \in \mathbf{I}_{ts}(S_{X,A})$.

(ii) ⇒ (i): Consider the identity I. In order that $I \in \mathbf{I}_{ts}(S_{X,A})$ it follows that $e^{-tA} \in \mathbf{I}(X)$ for all $t > 0$.

(i) ⇒ (iii): We only have to prove that $\mathbf{B}_{st}(S_{X,A}) \subset \mathbf{I}_{st}(S_{X,A})$. So, let $\theta \in \mathbf{B}_{st}(S_{X,A})$. Then there exists $s > 0$ such that for all $t > 0$,

$$e^{sA}\,\theta\, e^{-tA} \in \mathbf{B}(X) .$$

It follows that

$$e^{\frac{1}{2}sA}\,\theta\, e^{-tA} \in \mathbf{I}(X)$$

for all $t > 0$, and hence $\theta \in \mathbf{I}_{st}(S_{X,A})$.

(iii) ⇒ (i): The operator e^{-tA} belongs to $\mathbf{B}_{st}(S_{X,A})$. In order that $e^{-tA} \in \mathbf{I}_{st}(S_{X,A})$ we must have $e^{-(s-\varepsilon)A} \in \mathbf{I}(X)$ for all ε with $0 < \varepsilon < s$.

(i) ⇔ (iv): follows from the equivalence (i) ⇔ (ii) by transposition.

(i) ⇔ (v) : follows from the equivalence (i) ⇔ (iii) by transposition. □

(IV.4.11) Corollary

We have the following equivalences:

(i) $\forall_{t>0}\colon e^{-tA} \in \mathbf{H}(X)$

$$\Leftrightarrow \mathbf{T}_{ts}(S_{X,A}) = \mathbf{H}_{ts}(S_{X,A}) = \mathbf{L}(S_{X,A})$$

$$\Leftrightarrow \mathbf{T}(S_{X,A}) = \mathbf{H}_{st}(S_{X,A}) = \mathbf{B}_{st}(S_{X,A})$$

$$\Leftrightarrow \mathbf{T}_{ts}(T_{X,A}) = \mathbf{H}_{ts}(T_{X,A}) = \mathbf{L}(T_{X,A})$$

$$\Leftrightarrow \mathbf{T}(T_{X,A}) = \mathbf{H}_{st}(T_{X,A}) = \mathbf{B}_{st}(T_{X,A})$$

(ii) $\forall_{t>0}\colon e^{-tA} \in \mathbf{C}(X)$

$$\Leftrightarrow \mathbf{C}_{ts}(S_{X,A}) = \mathbf{L}(S_{X,A})$$

$$\Leftrightarrow \mathbf{C}_{st}(S_{X,A}) = \mathbf{B}_{st}(S_{X,A})$$

$$\Leftrightarrow \mathbf{C}_{ts}(T_{X,A}) = \mathbf{L}(T_{X,A})$$

$$\Leftrightarrow \mathbf{C}_{st}(T_{X,A}) = \mathbf{B}_{st}(T_{X,A}) .$$

Proof. The proof follows from the previous theorem and the equivalence

$$(\forall_{t>0}\colon e^{-tA} \in \mathbf{T}(X)) \Leftrightarrow (\forall_{t>0}\colon e^{-tA} \in \mathbf{H}(X)) .$$ □

Remark. In [Tr], Treves shows that the ε-topological tensor product $X \otimes_\varepsilon X$ of X with itself can be identified with the Banach algebra $\mathbf{C}(X)$ of all compact operators on X. It seems likely that $\mathbf{C}_{ts}(S_{X,A})$ represents the ε-topological tensor product $T_{X,A} \otimes_\varepsilon S_{X,A}$, and $\mathbf{C}_{ts}(T_{X,A})$ represents $S_{X,A} \otimes_\varepsilon T_{X,A}$. Then Corollary (IV.4.11) gives two kernel theorems in terms of ε-topological tensor products, where Section III.6 presents the kernel theorems in terms of the so-called π-topological products. We do not intend to go into more detail here. However, it is an interesting subject for further study.

IV.5. The nuclear case, a matrix calculus

In the present section we assume that $S_{X,A}$ is a nuclear space. It follows that there exists an orthonormal basis $(v_\ell)_{\ell\in\mathbb{N}}$ in X which consists of eigenvectors of A with eigenvalues λ_ℓ, $\ell \in \mathbb{N}$, satisfying $\sum_{\ell=1}^\infty e^{-\lambda_\ell t} < \infty$ for all $t > 0$. For convenience we take the eigenvalues ordered, i.e. $0 \leq \lambda_1 \leq \leq \lambda_2 \leq \dots$.

The mapping V defined by

$$V\colon x \mapsto ((x,v_\ell))_{\ell\in\mathbb{N}}\,, \qquad x \in X$$

is a unitary operator from X onto ℓ_2. For any ℓ_2-sequence $(\varphi_\ell)_{\ell\in\mathbb{N}}$ we have

$$V^*(\varphi_\ell)_{\ell\in\mathbb{N}} = \sum_{\ell=1}^\infty \varphi_\ell\, v_\ell \in X\,.$$

Consequently, X can be identified with the Hilbert space ℓ_2 in this setting. Similarly, the analyticity space $S_{X,A}$ and the trajectory space $T_{X,A}$ can be identified with sequence spaces of a special type.

(IV.5.1) Theorem

(i) $w \in S_{X,A}$ iff there exists a sequence $(\omega_\ell)_{\ell\in\mathbb{N}}$ satisfying

$$\exists_{t>0}\colon \sup_{\ell\in\mathbb{N}} |\omega_\ell| e^{\lambda_\ell t} < \infty$$

such that $(w,v_\ell) = \omega_\ell$. The series $\sum_{\ell=1}^\infty \omega_\ell v_\ell$ converges absolutely in each Hilbert space $e^{-(t-\varepsilon)A}(X)$, $0 < \varepsilon < t$.

(ii) $G \in T_{X,A}$ iff there exists a sequence $(\gamma_\ell)_{\ell\in\mathbb{N}}$ satisfying

$$\forall_{t>0}: \sup_{\ell\in\mathbb{N}} |\gamma_\ell| e^{-\lambda_\ell t} < \infty$$

such that

$$\gamma_\ell = \overline{\langle v_\ell, G\rangle} \quad \text{and} \quad G(t) = \sum_{\ell=1}^{\infty} e^{-\lambda_\ell t} \gamma_\ell v_\ell \,, \qquad t > 0 \,.$$

Proof

(i) Let $w \in e^{-tA}(X)$ for some $t > 0$. Then the series $\omega_\ell = (w, v_\ell)$ satisfies

$$\sum_{\ell=1}^{\infty} e^{2t\lambda_\ell} |\omega_\ell|^2 < \infty \,.$$

It follows that

$$\sup_{\ell\in\mathbb{N}} (|\omega_\ell| e^{t\lambda_\ell}) < \infty \,.$$

Conversely, let $(\omega_\ell)_{\ell\in\mathbb{N}}$ be a sequence with the indicated property for some $t > 0$. Let $0 < \varepsilon < t$. Then we find

$$\sum_{\ell=1}^{\infty} e^{2(t-\varepsilon)\lambda_\ell} (\omega_\ell)^2 < \sup_{\ell\in\mathbb{N}} (e^{2t\lambda_\ell} |\omega_\ell|^2) \sum_{\ell=1}^{\infty} e^{-2\varepsilon\lambda_\ell} \,.$$

and hence $w = \sum_{\ell=1}^{\infty} \omega_\ell v_\ell$ belongs to $e^{-(t-\varepsilon)A}(X)$.

(ii) Let $G \in T_{X,A}$. Then for all $t > 0$, $(v_\ell, G(t)) = e^{-\lambda_\ell t} \langle v_\ell, G\rangle$. Since $((v_\ell, G(t)))_{\ell\in\mathbb{N}} \in \ell_2 \subset \ell_\infty$ the result follows.
On the other hand, let $(\gamma_\ell)_{\ell\in\mathbb{N}}$ be as indicated. Then for all $t > 0$

$$\sum_{\ell=1}^{\infty} e^{-2\lambda_\ell t} |\gamma_\ell|^2 < \infty$$

and hence $G: t \mapsto \sum_{\ell=1}^{\infty} e^{-\lambda_\ell t} \gamma_\ell v_\ell$, $t > 0$ maps $(0,\infty)$ into X. Further we have for all $t, \tau > 0$,

$$G(t+\tau) = \sum_{\ell=1}^{\infty} e^{-\lambda_\ell (t+\tau)} \gamma_\ell v_\ell = \sum_{\ell=1}^{\infty} e^{-\lambda_\ell t} \gamma_\ell (e^{-\tau A} v_\ell) = e^{-\tau A} G(t) \,. \quad \square$$

Since $S_{X,A}$ and $T_{X,A}$ can be regarded as sequence spaces, the elements of $L(S_{X,A})$ and $L(T_{X,A})$ can be represented by infinite matrices. In this section we give complete characterizations of the corresponding types of infinite matrices.

Let $L \in L(S_{X,A})$. Then to L corresponds the matrix $(L_{k\ell})$ defined by

$$L_{k\ell} = (Lv_\ell, v_k) \ , \quad k,\ell \in \mathbb{N} \ .$$

Since L is a continuous linear mapping from $S_{X,A}$ into $S_{X,A}$, it follows that

$$\forall_{s>0} \ \exists_{\tau>0} \ \exists_{C>0}: \ \|e^{\tau A} L e^{-sA}\|_{X\otimes X} \leq C \ .$$

This condition implies that the column vectors Lv_ℓ, $\ell \in \mathbb{N}$, of the matrix $(L_{k\ell})$ satisfy

$$\text{(IV.5.2)} \quad \forall_{s>0} \ \exists_{\tau>0} \ \exists_{C>0} \ \forall_{\ell\in\mathbb{N}}: \ \|e^{\tau A} L v_\ell\| \leq Ce^{\lambda_\ell s} \ .$$

The vectors Lv_ℓ, $\ell \in \mathbb{N}$, span the range $\{Lw \mid w \in S_{X,A}\}$. From (IV.5.2) it follows that there exists $\tau > 0$ such that $Lv_\ell \in e^{-\tau A}(X)$ for all $\ell \in \mathbb{N}$. Define the mapping $\hat{L}: [0,\infty) \times (0,\infty) \to X \otimes X$ by

$$\hat{L}(t,s) = \sum_{\ell=1}^{\infty} e^{-\lambda_\ell s} (v_\ell \otimes (e^{-tA} L v_\ell)) \ .$$

Then $\hat{L}(0,s) \in S_{X\otimes X, I\otimes A}$. To show this assertion, let $0 < \sigma < s$ and choose $\tau > 0$ and $C > 0$ such that

$$\|e^{\tau A} L v_\ell\| \leq Ce^{\lambda_\ell \sigma} \ , \quad \ell \in \mathbb{N} \ .$$

Then we derive

$$\sum_{\ell=1}^{\infty} e^{-\lambda_\ell s} \|e^{\tau A} L v_\ell\| \leq C \sum_{\ell=1}^{\infty} e^{-\lambda_\ell (s-\sigma)} < \infty \ ,$$

and hence

$$\sum_{\ell=1}^{\infty} \|e^{-\lambda_\ell s} (v_\ell \otimes (e^{\tau A} L v_\ell))\|_{X\otimes X} < \infty \ .$$

Thus we have shown that $e^{\tau A} \hat{L}(0,s) \in S_{X\otimes X, I\otimes A}$, and consequently $\hat{L} \in TS_{X\otimes X; I\otimes A, A\otimes I}$. Since for each $w \in S_{X,A}$

$$\hat{L}w = \hat{L}(0,s)(e^{sA}\, w) = \sum_{\ell=1}^{\infty} (w,v_\ell)Lv_\ell = Lw$$

the linear mapping L is represented by the series

$$\sum_{\ell=1}^{\infty} v_\ell \otimes Lv_\ell \in TS_{X\otimes X;I\otimes A,A\otimes I}$$

where for each $\ell \in \mathbb{N}$, $v_\ell \otimes Lv_\ell$ denotes the mapping

$$(v_\ell \otimes Lv_\ell)(t,s) = e^{-\lambda_\ell s}(v_\ell \otimes (e^{-tA}\, Lv_\ell)) \,, \qquad s > 0,\ t \geq 0 \,.$$

On the other hand, we can start with given $b_1, b_2, \ldots$ in $S_{X,A}$ which satisfy

$$\forall_{s>0}\, \exists_{\tau>0}\, \exists_{C>0}\, \forall_{\ell\in\mathbb{N}}:\ \|e^{\tau A}\, b_\ell\| < Ce^{\lambda_\ell s} \,.$$

Then the mapping

$$(t,s) \mapsto \sum e^{-\lambda_\ell s}(v_\ell \otimes e^{-tA}\, b_\ell) \,, \qquad t \geq 0,\ s > 0$$

belongs to $TS_{X\otimes X;I\otimes A,A\otimes I}$. It represents the linear operator

$$w \mapsto \sum_{\ell=1}^{\infty} (w,v_\ell)b_\ell \in \mathbf{L}(S_{X,A}) \,.$$

We arrive at the following characterization.

(IV.5.3) <u>Theorem</u>

Let P be a linear operator in X, the domain of which contains the span $<\{v_\ell \mid \ell \in \mathbb{N}\}>$. Then P extends to an element of $\mathbf{L}(S_{X,A})$ iff the vectors Pv_ℓ, $\ell \in \mathbb{N}$, belong to $S_{X,A}$ and satisfy the condition stated in (IV.5.2). Its extension in $\mathbf{L}(S_{X,A})$ is given by the series $\sum_{\ell=1}^{\infty} v_\ell \otimes Pv_\ell$.

The dual mapping L' of $L \in \mathbf{L}(S_{X,A})$ is an element of $\mathbf{L}(T_{X,A})$. So it satisfies the following condition

$$\forall_{s>0}\, \exists_{\tau>0}\, \exists_{C>0}:\ \|e^{-sA}\, L'\, e^{\tau A}\|_{X\otimes X} \leq C \,.$$

If we put $B_k = L'v_k \in T_{X,A}$, we get

(IV.5.4) $\forall_{s>0}\, \exists_{\tau>0}\, \exists_{C>0}\, \forall_{k\in\mathbb{N}}:\ \|B_k(s)\| \leq Ce^{-\lambda_k \tau}$.

The trajectories B_k span $\{L'G \mid G \in \mathcal{T}_{X,A}\}$, and

$$B_k(s) = e^{-sA} L' v_k = \sum_{\ell=1}^{\infty} e^{-\lambda_\ell s} \overline{(Lv_\ell, v_k)} v_\ell = \sum_{\ell=1}^{\infty} e^{-\lambda_\ell s} \bar{L}_{k\ell} v_\ell .$$

Hence B_k is the k-th row vector of the matrix $(\bar{L}_{k\ell})$.
Define the mapping $\tilde{L}$: $[0,\infty) \times (0,\infty) \to X \otimes X$ by

$$\tilde{L}(t,s) = \sum_{k=1}^{\infty} e^{-\lambda_k t} (B_k(s) \otimes v_k) , \quad t \geq 0, s > 0 .$$

Then for each $s > 0$ and for $\tau > 0$, $C > 0$ as indicated in (IV.5.4) we have

$$\sum_{\ell=1}^{\infty} \| (B_k(s) \otimes (e^{\lambda_k \tau_0} v_k)) \|_{X \otimes X} \leq C \sum_{k=1}^{\infty} e^{-\lambda_k(\tau-\tau_0)} < \infty$$

for any τ_0 with $0 < \tau_0 < \tau$. It follows that $e^{\tau_0 A} \tilde{L}(0,s) \in X \otimes X$, and hence $\tilde{L} \in \mathcal{TS}_{X\otimes X; I\otimes A, A\otimes I}$.
Since

$$\tilde{L}w = \tilde{L}(0,s) e^{sA} w = \sum_{k=1}^{\infty} \langle f, B_k \rangle v_k = \sum_{k=1}^{\infty} (Lw, v_k) v_k ,$$

the operator L is represented by the series $\sum_{k=1}^{\infty} B_k \otimes v_k$. The latter series converges absolutely in the topology $\lambda\sigma_u$ for $\mathbf{L}(\mathcal{S}_{X,A})$.
If we start with given $B_1, B_2, \ldots$ in $\mathcal{T}_{X,A}$ which satisfy the condition stated in (IV.5.4), then it follows similarly, that the series $\sum_{k=1}^{\infty} B_k \otimes v_k$ represents the linear operator

$$w \mapsto \sum_{k=1}^{\infty} \langle w, B_k \rangle v_k , \quad w \in \mathcal{S}_{X,A}$$

which belongs to $\mathbf{L}(\mathcal{S}_{X,A})$. We obtain a second characterization of the elements of $\mathbf{L}(\mathcal{S}_{X,A})$.

(IV.5.5) Theorem

Let P be a linear operator in X, the domain of which contains the linear span $\langle\{v_\ell \mid \ell \in \mathbb{N}\}\rangle$. Put $B_k = \sum_{\ell=1}^{\infty} \overline{(Lv_\ell, v_k)} v_\ell$. Then L can be extended to an element of $\mathbf{L}(\mathcal{S}_{X,A})$ iff the formal series B_k, $k \in \mathbb{N}$, belong to $\mathcal{T}_{X,A}$ and fulfill the condition stated in (IV.5.4). We have $L = \sum_{k=1}^{\infty} B_k \otimes v_k$.

In the following theorem we give a useful characterization of the entries $L_{k\ell}$ of the matrix $(L_{k\ell})$ belonging to $L \in \mathbf{L}(S_{X,A})$.

(IV.5.6) Theorem

(i) Let the infinite matrix $(L_{k\ell})$ satisfy

$$\text{(IV.5.7)}\quad \forall_{s>0}\ \exists_{\tau>0}: \sup_{\ell,k\in\mathbb{N}} \left(e^{-\lambda_\ell s}\, e^{\lambda_k \tau}\, |L_{k\ell}|\right) < \infty .$$

Then the series

$$\sum_{k,\ell=1}^{\infty} L_{k\ell}(v_\ell \otimes v_k)$$

converges in $\mathbf{L}(S_{X,A})$ with respect to the topology $\lambda\sigma_u$. This series represents the mapping $L \in \mathbf{L}(S_{X,A})$ defined by

$$Lw = \sum_{k,\ell=1}^{\infty} L_{k\ell}(w,v_\ell)v_k\ ,\qquad w \in S_{X,A}\ .$$

(ii) Conversely, let $L \in \mathbf{L}(S_{X,A})$. Then its matrix $(L_{k\ell})$ satisfies (IV.5.7).

Proof

(i) Let $s_0 > 0$. Then there are $\tau_0 > 0$ and $C > 0$ such that

$$e^{-\lambda_\ell s_0}\, e^{\lambda_k \tau_0}\, |L_{k\ell}| \leq C\ ,\qquad k,\ell \in \mathbb{N}\ .$$

We obtain the following estimate for all $s > s_0$ and $0 < \tau < \tau_0$

$$\sum_{k,\ell=1}^{\infty} \left(e^{-2\lambda_\ell s}\, e^{2\lambda_k \tau}\, |L_{k\ell}|^2\right) \leq C^2 \sum_{k,\ell=1}^{\infty} e^{-2\lambda_\ell (s-s_0)}\, e^{-2\lambda_k(\tau_0-\tau)}$$

and hence

$$\sum_{k,\ell=1}^{\infty} e^{-\lambda_\ell s}\, e^{\lambda_k \tau}\, L_{k\ell}(v_\ell \otimes v_k) \in X \otimes X\ .$$

Since $s > s_0$ and $s_0 > 0$ have been taken arbitrarily, it follows that the series $\sum_{k,\ell=1}^{\infty} L_{k\ell}(v_\ell \otimes v_k)$ converges in $\lambda\sigma_u$-sense. Moreover,

$$\left(\sum_{k,\ell=1}^{\infty} L_{k\ell}(v_\ell \otimes v_k)\right)(w) = \sum_{k,\ell=1}^{\infty} L_{k\ell}(w,v_\ell)v_k\ ,\qquad w \in S_{X,A}\ .$$

(ii) Since $L \in \mathbf{L}(S_{X,A})$ we have

$$\forall_{s>0}\ \exists_{\tau>0}:\ \sum_{k,\ell=1}^{\infty} \left(e^{-2s\lambda_\ell}\, e^{2\tau\lambda_k}\, |L_{k\ell}|^2\right) < \infty .$$

Thus the wanted result follows. □

Condition (IV.5.7) enables us to construct a lot of examples of elements in $\mathbf{L}(S_{X,A})$. This condition has been used already in several applications of our theory. We refer to [EG 1-3]. In the next section it is of great help in the construction of analyticity spaces on which a given set of operators act continuously.

(IV.5.8) Lemma

Let $P,Q \in \mathbf{L}(S_{X,A})$ with matrices $(P_{k\ell})$ and $(Q_{k\ell})$. Then the matrix of the product PQ is given by

$$(PQ)_{k\ell} = \sum_{j=1}^{\infty} P_{kj}\, Q_{j\ell}\,, \qquad k,\ell \in \mathbb{N}$$

where each series converges absolutely.

Proof. Let $s > 0$ and let $k,\ell \in \mathbb{N}$. Then there exist $\tau,\tau_0 > 0$ such that

$$|P_{kj}| \leq C_P\, e^{\frac{1}{2}\lambda_j\tau_0}\, e^{-\lambda_k\tau}$$

and

$$|Q_{j\ell}| \leq C_Q\, e^{-\lambda_j\tau_0}\, e^{\lambda_\ell s}$$

for some $C_P, C_Q > 0$. It leads to the following estimate:

$$\sum_{j=1}^{\infty} |e^{\lambda_k\tau}\, P_{kj}\, e^{-\frac{1}{2}\lambda_j\tau_0}|\,|e^{\lambda_j\tau_0}\, Q_{j\ell}\, e^{-\lambda_\ell s}|\, e^{-\frac{1}{2}\lambda_j\tau_0} \leq$$

$$\leq C_P\, C_Q \left(\sum_{j=1}^{\infty} e^{-\frac{1}{2}\lambda_j\tau_0}\right).$$

This estimate shows that the series

$$\sum_{j=1}^{\infty} P_{kj}\, Q_{j\ell}$$

converges absolutely for each $k,\ell \in \mathbb{N}$, and also that the matrix

$$\sum_{j=1}^{\infty} P_{kj}\, Q_{j\ell}$$

satisfies condition (IV.5.7). So it represents an element of $\mathsf{L}(S_{X,A})$. We have

$$PQ = \left(\sum_{k,i=1}^{\infty} P_{ki}(v_i \otimes v_k)\right)\left(\sum_{j,\ell=1}^{\infty} Q_{j\ell}(v_\ell \otimes v_j)\right) =$$

$$= \sum_{k,\ell=1}^{\infty} \sum_{i,j=1}^{\infty} P_{ki}\, Q_{j\ell}(v_i \otimes v_k)(v_\ell \otimes v_j) =$$

$$= \sum_{k,\ell=1}^{\infty} \left(\sum_{j=1}^{\infty} P_{kj}\, Q_{j\ell}\right) v_\ell \otimes v_k \,. \qquad \square$$

As can be expected, the transposition mapping $\mathsf{L} \mapsto \mathsf{L}'$ from $\mathsf{L}(S_{X,A})$ onto $\mathsf{L}(T_{X,A})$ can be translated into the usual hermitean conjugation of matrices. First we need the following definition.

(IV.5.9) Definition

Let $\mathsf{K} \in \mathsf{L}(T_{X,A})$, and put $\mathsf{K}_{k\ell} = \overline{\langle v_k, \mathsf{K}v_\ell\rangle}$. Then $(\mathsf{K}_{k\ell})$ is called the matrix of K.

(IV.5.10) Lemma

Let $\mathsf{L} \in \mathsf{L}(S_{X,A})$ with matrix $(\mathsf{L}_{k\ell})$. Then the mapping

$$\hat{\mathsf{L}}^c : (t,s) \mapsto \sum_{k,\ell=1}^{\infty} \overline{\mathsf{L}_{\ell k}}\, e^{-\lambda_k s}\, e^{-\lambda_\ell t}\, (v_\ell \otimes v_k)\,, \qquad t \geq 0,\ s > 0$$

is well-defined and $\hat{\mathsf{L}}^c \in TS_{X\otimes X;\, A\otimes I,\, I\otimes A}$.
It follows that the dual mapping $\mathsf{L}' \in \mathsf{L}(T_{X,A})$ is represented by $\hat{\mathsf{L}}^c \in TS_{X\otimes X;\, A\otimes I,\, I\otimes A}$.

Proof. From Theorem (IV.5.6) we derive that the trajectory

$$\hat{\mathsf{L}} : (t,s) \to \sum_{k,\ell=1}^{\infty} e^{-\lambda_k t}\, e^{-\lambda_\ell s}\, \mathsf{L}_{k\ell}(v_\ell \otimes v_k)\,, \qquad t \geq 0,\ s > 0$$

belongs to $TS_{X\otimes X;\, I\otimes A,\, A\otimes I}$. From the definition of the operation ^{c} in Section IV.2 we obtain

$$\hat{L}^c(t,s) = L(t,s)^* = \sum_{k,\ell=1}^{\infty} e^{-\lambda_k t} e^{-\lambda_\ell s} \bar{L}_{k\ell}(v_k \otimes v_\ell) .$$

Hence $\hat{L}^c \in TS_{X\otimes X;A\otimes I,I\otimes A}$.

(IV.5.11) Corollary

(i) Let $L \in \mathbf{L}(S_{X,A})$ with matrix $(L_{k\ell})$. Then the matrix of L' equals $(\bar{L}_{\ell k})$.

(ii) Let $K \in \mathbf{L}(T_{X,A})$ with matrix $(K_{k\ell})$. Then the matrix of K' equals $(\bar{K}_{\ell k})$.

(IV.5.12) Theorem

Let $(M_{k\ell})$ be an infinite matrix. Then

$$M = \sum_{k,\ell=1}^{\infty} M_{k\ell}(v_\ell \otimes v_k)$$

represents an element of $\mathbf{L}(T_{X,A})$ iff

(IV.5.13) $$\forall_{s>0}\ \exists_{\tau>0}\ \sup_{k,\ell\in\mathbb{N}} (e^{-\lambda_k s} e^{\lambda_\ell \tau} |M_{k\ell}|) < \infty .$$

(IV.5.14) Theorem

Let $(E_{k\ell})$ be an infinite matrix. Then

$$E = \sum_{k,\ell=1}^{\infty} E_{k\ell}(v_\ell \otimes v_k)$$

represents an element of $\mathbf{E}(S_{X,A})$ iff the entries $E_{k\ell}$, $k,\ell \in \mathbb{N}$ satisfy conditions (IV.5.7) and (IV.5.13).

At the end of this section we introduce the class of weighted shift operators. For reasons of elegancy we impose the following extra condition on the ordered eigenvalues λ_j, $j \in \mathbb{N}$:

(IV.5.15) $$\exists_{M>0}\ \forall_{j\in\mathbb{N}}: \frac{\lambda_{j+1}}{\lambda_j} \leq M .$$

So the eigenvalues of A may not grow too rapidly. We admit $\lambda_j = e^j$ but not $\lambda_j = j!$.

(IV.5.16) Definition

By $\mathbf{D}(S_{X,A})$ we denote the set of all diagonal operators in $\mathbf{E}(S_{X,A})$, i.e. $\Delta \in \mathbf{D}(S_{X,A})$ iff $(\Delta v_\ell, v_k) = (\Delta v_\ell, v_\ell)\delta_{k\ell}$, $k,\ell \in \mathbb{N}$.

(IV.5.17) Lemma

$\Delta \in \mathbf{D}(S_{X,A})$ iff $\forall_{t>0}: \sup_{k\in\mathbb{N}} |\Delta_{kk}| e^{-\lambda_k t} < \infty$.

Proof. Cf. Theorem (IV.5.6). □

In order to introduce the class of weighted shift-operators we first prove the following lemma.

(IV.5.18) Lemma

Let U denote the unilateral shift defined by

$$U v_\ell = v_{\ell+1}\,, \qquad \ell \in \mathbb{N}\,.$$

Then U belongs to $\mathbf{E}(S_{X,A})$.

Proof. We have to check Conditions (IV.5.6) and (IV.5.13).
So let $s > 0$. First put $\tau = s$. Then we get

$$\sup_{k,\ell\in\mathbb{N}} e^{-\lambda_\ell s} e^{\lambda_k s} |(v_\ell, U v_k)| = \sup_{\ell\in\mathbb{N}} e^{-(\lambda_\ell - \lambda_{\ell-1})s} \leq 1\,.$$

Next, put $\tau = s/M$ with M as in (IV.5.15). Then we get

$$\sup_{k,\ell\in\mathbb{N}} e^{-\lambda_\ell s} e^{\lambda_k \tau} |(U v_\ell, v_k)| = \sup_{\ell\in\mathbb{N}} e^{-\lambda_\ell s} e^{\lambda_{\ell+1}\tau}$$

$$= \sup_{\ell\in\mathbb{N}} \exp\left(-\lambda_s\left(s - \frac{\lambda_{\ell+1}}{\lambda_\ell}\tau\right)\right) \leq 1\,.$$ □

(IV.5.19) Definition

The class of weighted shift operators $\mathbf{W}(S_{X,A}) \subset \mathbf{E}(S_{X,A})$ consists of all operators W for which there exists $\Delta \in \mathbf{D}(S_{X,A})$ and $n \in \mathbb{N}$ such that

$$W = U^n \Delta\,.$$

We show that $W(S_{X,A})$ contains precisely all weighted shift operators in $L(S_{X,A})$, i.e. operators with one nonzero co-diagonal.

(IV.5.20) Lemma

Let W be a weighted shift operator in $L(S_{X,A})$, i.e. there exists $n \in \mathbb{N}$ such that

$$\begin{cases} (Wv_\ell, v_k) = 0 \quad , & k \neq \ell+n \ , \ k,\ell \in \mathbb{N} \ , \\ (Wv_\ell, v_{\ell+n}) = \omega_\ell \ , & \ell \in \mathbb{N} \ , \ \omega_\ell \in \mathbb{C} \ . \end{cases}$$

Then $W = U^n \Delta$ with $\Delta \in D(S_{X,A})$, $\Delta_{\ell\ell} = \omega_\ell$.

Proof. Following Theorem (IV.5.6) the numbers ω_ℓ satisfy

$$(*) \qquad \forall_{s>0} \ \exists_{\tau>0}: \ \sup_{\ell\in\mathbb{N}} \left(|\omega_\ell| e^{-\lambda_\ell s} e^{\lambda_{\ell+n}\tau} \right) < \infty \ .$$

Because of assumption (IV.5.15) there exists $M > 0$ such that

$$\frac{\lambda_{j+n}}{\lambda_j} \leq M^n \ , \qquad j \in \mathbb{N} \ .$$

By taking $\tau = M^{-n} s$ we can replace (*) by

$$\forall_{s>0}: \ \sup_{\ell\in\mathbb{N}} \left(|\omega_\ell| e^{-\lambda_\ell s} \right) < \infty \ .$$

Define $\Delta \in D(S_{X,A})$ by $\Delta_{\ell\ell} = \omega_\ell$. Then $W = U^n \Delta$. □

The class $W(S_{X,A})$ 'spans' the algebra $L(S_{X,A})$:

(IV.5.21) Theorem

Let $L \in L(S_{X,A})$ with matrix $(L_{k\ell})$. Define the operators W_ν, $\nu \in \mathbb{Z}$, of $W(S_{X,A})$ as follows:

$$W_n v_\ell = L_{\ell+n,\ell} v_{\ell+n} \ , \qquad \ell \in \mathbb{N} \ , \ n \in \mathbb{N} \cup \{0\} \ ,$$

$$W_{-n} v_\ell = L_{\ell-n,\ell} v_{\ell+n} \ , \qquad \ell > -n \ , \ \ell,n \in \mathbb{N} \ .$$

Then $\sum_{n=0}^{\infty} W_n + \sum_{n=1}^{\infty} W_{-n}^*$ represents $\mathbf{L}$. The series converges absolutely in the topology $\lambda\sigma_u$ for $\mathbf{L}(S_{X,\mathbf{A}})$.

Proof. Since the eigenvalues of $\mathbf{A}$ are assumed to be ordered, $0 \leqq \lambda_1 \leqq \lambda_2 \leqq \ldots$, they satisfy the following estimates.

For $n \in \mathbb{N} \cup \{0\}$,

$$(*) \qquad e^{\lambda_{j+n} t} \leqq e^{-\lambda_n (t_0 - t)} e^{\lambda_{j+n} t_0}$$

where $j \in \mathbb{N}$, $t_0 > 0$ and $0 < t < t_0$.

For $n \in \mathbb{N}$,

$$(**) \qquad e^{-\lambda_j s} \leqq e^{-\lambda_n (s - s_0)} e^{-\lambda_j s_0}$$

with $j \in \mathbb{N}$, $j > n$ and $s > s_0 > 0$.

Let $s > 0$. Then there exists $\tau > 0$ such that

$$e^{2\tau \mathbf{A}} \mathbf{L} e^{-\frac{1}{2} s \mathbf{A}} \in X \otimes X .$$

First let $n \in \mathbb{N} \cup \{0\}$. Then by (*) we have

$$\| e^{\tau \mathbf{A}} W_n e^{-s\mathbf{A}} \|^2_{X \otimes X} \leqq e^{-2\lambda_n \tau} \sum_{\ell=1}^{\infty} | e^{2\tau \lambda_{\ell+n}} \mathbf{L}_{\ell+n,\ell} e^{-s\lambda_\ell} |^2$$

$$\leqq e^{-2\lambda_n \tau} \| e^{2\tau \mathbf{A}} \mathbf{L} e^{-\frac{1}{2} s} \|^2_{X \otimes X} .$$

Next, let $n \in \mathbb{N}$. Then by (**) we have

$$\| e^{\tau \mathbf{A}} W_{-n} e^{-s\mathbf{A}} \|^2_{X \otimes X} \leqq e^{-\lambda_n s} \sum_{\ell=n+1}^{\infty} | e^{\tau \lambda_{\ell-n}} \mathbf{L}_{\ell-n,\ell} e^{-\frac{1}{2} s \lambda_\ell} |^2$$

$$\leqq e^{-\lambda_n s} \| e^{2\tau \mathbf{A}} \mathbf{L} e^{-\frac{1}{2} s} \|^2_{X \otimes X} .$$

A combination of the above results yields for all $N_1, N_2 \in \mathbb{N}$

$$\sum_{n=0}^{N_1} \| e^{\tau \mathbf{A}} W_n e^{-s\mathbf{A}} \|_{X \otimes X} + \sum_{n=1}^{N_2} \| e^{\tau \mathbf{A}} W_{-n}^* e^{-s\mathbf{A}} \|_{X \otimes X}$$

$$\leqq \| e^{2\tau \mathbf{A}} \mathbf{L} e^{-\frac{1}{2} s \mathbf{A}} \|_{X \otimes X} \left(\sum_{n=0}^{N_1} e^{-\lambda_n \tau} + \sum_{n=0}^{N_2} e^{-\frac{1}{2} \lambda_n s} \right) .$$

So the series

$$\sum_{n=0}^{\infty} e^{\tau A} W_n e^{-sA} \quad \text{and} \quad \sum_{n=1}^{\infty} e^{\tau A} W_{-n}^{*} e^{-sA}$$

converge absolutely in $X \otimes X$. Comparing matrices it follows that

$$L = \sum_{n=0}^{\infty} W_n + \sum_{n=1}^{\infty} W_{-n}^{*} . \qquad \square$$

The operators in $W(S_{X,A})$, their adjoints and finite combinations thereof are important in applications. For instance, if the orthonormal basis $(v_k)_{k\in\mathbb{N}}$ consists of classical polynomials (or related special functions), then the differential operators with polynomial coefficients are finite linear combinations of elements of $W(S_{X,A})$ and their adjoints.

IV.6. The construction of nuclear analyticity spaces on which a given set of operators acts continuously

For a given analyticity space $S_{X,A}$, we have determined conditions on (unbounded) linear operators in X which ensure that they act continuously on $S_{X,A}$. However, it is also a natural question whether there exists a space $S_{X,A}$ such that a given set of linear operators in X becomes a subset of $L(S_{X,A})$. Sometimes the choice of the space $S_{X,A}$ is evident. E.g. for the algebra of the differential operators in $L_2(\mathbb{R})$ with polynomials coefficients we can take

$$X = L_2(\mathbb{R}) \quad \text{and} \quad A = \tfrac{1}{2}\left(-\frac{d^2}{dx^2} + x^2 + 1\right)$$

(cf. Section II.3). In Section II.5 we have seen that there is a natural choice for $S_{X,A}$, if we consider a set of linear operators in X which is part of a unitary representation of a Lie group or of the corresponding infinitesimal representation of its Lie algebra.
Because of their rich mathematical structure we only consider nuclear analyticity spaces in the sequel. We prove the existence of a nuclear analyticity space on which

- a given countable set of bounded operators
- a given finite set of commuting (unbounded) self-adjoint operators

acts continuously. Besides that we construct a nuclear trajectory space $T_{X,A}$ such that the operators of a given countable set of unbounded symmetric operators with common dense C^∞-domain, are closable as linear mappings in $T_{X,A}$.

We start with a countable set $\{L_k \mid k \in \mathbb{N}\}$ of densely defined linear operators in the separable Hilbert space X. The subspace $C^\infty(\{L_k \mid k \in \mathbb{N}\})$ of X is defined as follows:

$$C^\infty(\{L_k \mid k \in \mathbb{N}\}) = \{v \in X \mid \forall_{s\in\mathbb{N}} \forall_{p\in\mathbb{N}^{\mathbb{N}}} : v \in \mathcal{D}(L_{p(1)} L_{p(2)} \cdots L_{(s)})\} .$$

(IV.6.1) <u>Theorem</u>

Suppose $C^\infty(\{L_k \mid k \in \mathbb{N}\})$ is dense in X. Then there exists an orthonormal basis $(v_\ell)_{\ell\in\mathbb{N}} \in C^\infty(\{L_k \mid k \in \mathbb{N}\})$ with the property

$$\forall_{k\in\mathbb{N}} \forall_{\ell\in\mathbb{N}} : L_k v_\ell \in \langle v_1, \ldots, v_{n_{k+\ell-1}} \rangle$$

where $n_p \in \mathbb{N}$ satisfies $n_p \leq \frac{1}{2}(p+1)(p+2)$, $p \in \mathbb{N}$.
It follows that each operator L_k, $k \in \mathbb{N}$, has a column finite matrix representation with respect to the basis $(v_\ell)_{\ell\in\mathbb{N}}$.

<u>Proof</u>. There exists an orthonormal basis $(u_j)_{j\in\mathbb{N}} \subset C^\infty(\{L_k \mid k \in \mathbb{N}\})$ for X. We put $v_1 = u_1$. There is an orthonormal set $\{v_2, v_{n_1}\} \perp v_1$ with $n_1 \leq 3$ such that $\langle L_1 v_1, u_2 \rangle \subset \langle v_1, v_2, v_{n_1} \rangle$.
Next, there is an orthonormal set $\{v_{n_1+1}, \ldots, v_{n_2}\}$ with $n_2 \leq 6$ and $\{v_{n_1+1}, \ldots, v_{n_2}\} \perp \{v_1, \ldots, v_{n_1}\}$ such that

$$\langle L_1 v_2, L_2 v_1, u_3 \rangle \subset \langle v_1, \ldots, v_{n_2} \rangle .$$

Having obtained the orthonormal set $\{v_1, \ldots, v_{n_p}\}$ with the property

$$n_p \leq \tfrac{1}{2}(p+1)(p+2)$$

and

$$\langle L_1 v_p, L_2 v_{p-1}, \ldots, L_p v_1, u_{p+1} \rangle \subset \langle v_1, \ldots, v_{n_p} \rangle$$

we produce the orthonormal set $\{v_{n_p+1}, \ldots, v_{n_{p+1}}\}$ with

$$n_{p+1} \leq \tfrac{1}{2}(p+2)(p+3) \ , \qquad \{v_{n_p+1},\ldots,v_{n_{p+1}}\} \perp \{v_1,\ldots,v_{n_p}\}$$

and

$$\langle L_1 v_{p+1}, L_2 v_p, \ldots, L_{p+1} v_1, u_{p+2}\rangle \subset \langle v_1,\ldots,v_{n_{p+1}}\rangle \ .$$

In this way we inductively obtain an orthonormal system $(v_\ell)_{\ell\in\mathbb{N}}$. Since $u_j \in \langle v_1,\ldots,v_{n_{j-1}}\rangle$ for all $j \in \mathbb{N}$, the system $(v_\ell)_{\ell\in\mathbb{N}}$ is also an orthonormal basis for X. Because of our construction

$$\forall_{k\in\mathbb{N}}\ \forall_{\ell\in\mathbb{N}}:\ L_k v_\ell \in \langle v_1,\ldots,v_{n_{k+\ell-1}}\rangle \ , \qquad n_{k+\ell-1} \leq \tfrac{1}{2}(k+\ell)(k+\ell+1) \ .$$

So the length of the column $((L_k v_\ell, v_j))_{j\in\mathbb{N}}$ is at most $n_{k+\ell-1}$. □

<u>Remark</u>. In the above proof we assumed that the set $\{L_k \mid k \in \mathbb{N}\}$ is infinite. However, if this set is finite with cardinality $k_0 \in \mathbb{N}$, then we take $L_k = I$ (the identity) for $k > k_0$. Of course the estimate for the length of the columns n_p is much too rough now. In Theorem (IV.6.4) we give a sharper estimate for that case.

(IV.6.2) <u>Theorem</u>

Let $\{B_k \mid k \in \mathbb{N}\}$ be a countable set of bounded linear operators on X. Then there exists a nuclear analyticity space $S_{X,A}$ such that

$$\{B_k \mid k \in \mathbb{N}\} \subset \mathbf{L}(S_{X,A}) \ .$$

<u>Proof</u>. By Theorem (IV.6.1) there exists an orthonormal basis $(v_\ell)_{\ell\in\mathbb{N}}$ for X such that for all $k,\ell,j \in \mathbb{N}$

$$(B_k v_\ell, v_j) = 0 \quad \text{if} \quad j > \tfrac{1}{2}(k+\ell)(k+\ell+1) \ .$$

Define the positive self-adjoint operator A as follows:

$$A v_\ell = (\log(\ell))^2\, v_\ell \ , \qquad \ell \in \mathbb{N}$$

and usual self-adjoint extension. The space $S_{X,A}$ is nuclear because the series $\sum_{\ell=1}^{\infty} \exp(-\,t(\log(\ell))^2)$ converges for all $t > 0$.
Following Theorem (IV.5.6) we only have to prove

$$\forall_{k\in\mathbb{N}}\ \forall_{t>0}\ \exists_{\tau>0}:\ \sup_{\ell,k\in\mathbb{N}} \exp(-\,t(\log(\ell))^2 + \tau(\log(j))^2)\ |(B_k v_\ell, v_j)| < \infty \ .$$

Note first that $|(B_k v_\ell, v_j)| \leq \|B_k\|$, $k,\ell,j \in \mathbb{N}$. Next, let $k \in \mathbb{N}$ and let $t > 0$. Put $\tau = t/4$. Then for all $\ell, j \in \mathbb{N}$ with $1 \leq j \leq \frac{1}{2}(k+\ell)(k+\ell+1)$

$$\exp t(\tfrac{1}{4}(\log(j))^2 - (\log(\ell))^2) =$$

$$= \exp t(\tfrac{1}{2}\log(j) + \log(\ell))(\tfrac{1}{2}\log(j) - \log(\ell)) =$$

$$= \exp(t \log(\tfrac{1}{\ell}\sqrt{j}) \log \ell\sqrt{j}) \ .$$

Since

$$\frac{\sqrt{j}}{\ell} \leq \tfrac{1}{2}\sqrt{2}\sqrt{(\frac{k+\ell}{\ell})(\frac{k+\ell+1}{\ell})} \leq \tfrac{1}{2}\sqrt{2}\,(1 + \frac{k+1}{\ell})$$

we get

$$\log \frac{\sqrt{j}}{\ell} \leq \log(\tfrac{1}{2}\sqrt{2}) + \log(1 + \frac{k+1}{\ell}) \ .$$

Hence

$$\sup_{\substack{j,\ell \in \mathbb{N} \\ 1 \leq j \leq \frac{1}{2}(k+\ell)(k+\ell+1)}} [\exp t(\tfrac{1}{4}(\log(j))^2 - (\log(\ell))^2)] < \infty \ .$$

□

The result of Theorem (IV.6.1) can also be applied in the construction of a nuclear trajectory space $\mathcal{T}_{X,A}$ such that certain symmetric operators in X become closable linear mappings in $\mathcal{T}_{X,A}$. To this end, we first prove the following lemma.

(IV.6.3) Lemma

Let P be a densely defined linear mapping in $\mathcal{T}_{X,A}$. Suppose for all $t > 0$ and all $\psi \in B_+(\mathbb{R})$ that the operator $e^{-tA} P\psi(A)$ is closable in X. Then P is a closable linear mapping in $\mathcal{T}_{X,A}$. Its closure $\bar{P}$ is defined by

$$\mathrm{Dom}(\bar{P}) = \{F \in \mathcal{T}_{X,A} \mid F = \psi(A)x \wedge \forall_{t>0}: x \in \mathcal{D}((e^{-tA} P\psi(A))^{**})\} \ ,$$

$$\bar{P}F: t \mapsto (e^{-tA} P\psi(A))^{**} x \quad \text{where} \quad F = \psi(A)x \in \mathrm{Dom}(\bar{P}) \ .$$

Proof. First observe that $\bar{P}$ is well-defined. If $x \in \mathcal{D}((e^{-tA} P\psi(A))^{**})$ and $y \in \mathcal{D}((e^{-tA} P\chi(A))^{**})$ for all $t > 0$ and if $F = \psi(A)x = \chi(A)y$, then

$$\bar{P}F(t) = (e^{-tA} P\psi(A))^{**} x = (e^{-tA} P\chi(A))^{**} y \ ,$$

because

$$(e^{-tA}\,P\,\chi(A))^{**} \supset (e^{-tA}\,P\,\psi(A))^{**}\;\chi(A)\psi(A)^{-1}\;.$$

Next we show that $\bar{P}$ is closed.

Let $(F_n)_{n\in\mathbb{N}}$ be a sequence in $\mathrm{Dom}(\bar{P})$ with limit F, and suppose the sequence $(\bar{P}F_n)_{n\in\mathbb{N}}$ converges in $T_{X,A}$ to a limit G. Then there exists $\psi \in B_+(\mathbb{R})$ and a convergent sequence $(x_n)_{n\in\mathbb{N}} \subset X$ such that $F_n = \psi(A)x_n$, $F = \psi(A)x$. We have

$$\forall_{t>0}\colon\ (e^{-tA}\,P\,\psi(A))^{**}\;x_n \to G(t)$$

and

$$\|x_n - x\| \to 0\;.$$

Hence

$$x \in \mathcal{D}((e^{-tA}\,P\,\psi(A))^{**}) \quad \text{and} \quad G(t) = (e^{-tA}\,P\,\psi(A))^{**}\;x$$

for all $t > 0$. □

(IV.6.4) <u>Theorem</u>

Let $\{P_k \mid k \in \mathbb{N}\}$ be a countable set of unbounded symmetric operators in X such that $C^{\infty}(\{P_k \mid k \in \mathbb{N}\})$ is dense in X. Then there exists a nuclear trajectory space $T_{X,A}$ such that each operator P_k is a densely defined closable operator in $T_{X,A}$.

<u>Proof</u>. By Theorem (IV.6.1) there exists an orthonormal basis $(v_\ell)_{\ell\in\mathbb{N}}$ such that each operator has a column finite and hence row finite matrix with respect to this basis. Now let $(\lambda_\ell)_{\ell\in\mathbb{N}}$ be any sequence of positive numbers such that $\sum_{\ell=1}^{\infty} e^{-\lambda_\ell t} < \infty$ for all $t > 0$. Then we define the positive self-adjoint operator A by

$$Av_\ell = \lambda_\ell\,v_\ell\;, \qquad \ell \in \mathbb{N}\;,$$

and usual self-adjoint extension.

For each $k \in \mathbb{N}$ we take $\mathrm{Dom}(P_k) = \langle\{v_\ell \mid \ell \in \mathbb{N}\}\rangle$. Then for all $t > 0$ and all $\psi \in B_+(\mathbb{R})$ we have

$$(\psi(A)P_k\;e^{-tA}\;u,w) = (u,e^{-tA}\;P_k\;\psi(A)w)\;, \qquad u,w \in \mathrm{Dom}(P_k)$$

because each P_k is symmetric and column finite. It follows that $(e^{-tA}\;P_k\;\psi(A))^{*}$ is densely defined, and hence $e^{-tA}\;P_k\;\psi(A)$ is closable. The result follows from Lemma (IV.6.3). □

Remark. The previous corollary plays a role in Chapter V. There we prove the following result.
Let P be a self-adjoint operator in X, which is closable as a linear mapping in a nuclear space $T_{X,A}$. Then for almost each point in the spectrum of P with multiplicity m there are m generalized eigenvectors in $T_{X,A}$.

We now present another version of Theorem (IV.6.1) for a finite set $\{L_k \mid k = 1,\dots,n\}$.

(IV.6.5) Theorem

Suppose $C^\infty(\{L_k \mid k = 1,\dots,n\})$ is dense in X. Then there exists an orthonormal basis $(v_\ell)_{\ell\in\mathbb{N}} \subset C^\infty(\{L_k \mid k = 1,\dots,n\})$ with the property

$$L_k v_\ell \in \langle v_1,\dots,v_{p_\ell}\rangle , \qquad k = 1,\dots,n ,$$

where

$$p_\ell \leqq \ell(n+1) + 1 , \qquad \ell \in \mathbb{N} .$$

Proof. Let $(u_j)_{j\in\mathbb{N}} \subset C^\infty(\{L_k \mid k = 1,\dots,n\})$ be an orthonormal basis for X. Put $v_1 = u_1$. Then there exists an orthonormal set $\{v_2,\dots,v_{p_1}\} \perp \{v_1\}$ with $p_1 \leqq (n+1) + 1$ such that

$$L_k v_1 \in \langle v_1,\dots,v_{p_1}\rangle , \qquad k = 1,\dots,n ,$$

and

$$u_2 \in \langle v_1,\dots,v_{p_1}\rangle .$$

Similarly, there exists an orthonormal set $\{v_{p_1+1},\dots,v_{p_2}\} \perp \{v_1,\dots,v_{p_1}\}$ with $p_2 \leqq 2(n+1) + 1$ such that

$$L_k v_2 \in \langle v_1,\dots,v_{p_2}\rangle , \qquad k = 1,\dots,n ,$$

and

$$u_3 \in \langle v_1,\dots,v_{p_2}\rangle .$$

We thus inductively obtain orthonormal sets $\{v_{p_\ell+1},\dots,v_{p_{\ell+1}}\}$ with $p_{\ell+1} \leqq (\ell+1)(n+1) + 1$ and $\{v_{p_\ell+1},\dots,v_{p_{\ell+1}}\} \perp \{v_1,\dots,v_{p_\ell}\}$ such that

$$L_k v_{\ell+1} \in \langle v_1,\dots,v_{p_{\ell+1}}\rangle ,$$

and

$$u_{\ell+2} \in \langle v_1,\dots,v_{p_{\ell+1}}\rangle .$$

Since the orthonormal basis $(u_j)_{j\in\mathbb{N}}$ is contained in the linear span $\langle\{v_\ell \mid \ell \in \mathbb{N}\}\rangle$, the orthonormal system is a basis for X. By construction we have

$$L_k v_\ell \in \langle v_1,\dots,v_{p_\ell}\rangle , \qquad k = 1,\dots,n$$

with $p_\ell < \ell(n+1) + 1$. □

With the aid of the previous theorem we construct a nuclear analyticity space on which a given finite set of strongly commuting self-adjoint operators acts continuously.

So, let $P_1,\dots,P_n$ denote n strongly commuting self-adjoint operators in X with spectral resolutions $(F_\lambda^{(k)})_{\lambda\in\mathbb{R}}$, $k = 1,\dots,n$. Let Π_r, $r \in \mathbb{N}^n$, denote the projection

$$\Pi_r = F^{(1)}(\{\lambda \mid r_1-1 \leqq |\lambda| < r_1\}) \dots F^{(n)}(\{\lambda \mid r_n-1 \leqq |\lambda| < r_n\}) .$$

Then, for all $x_r \in \Pi_r(X)$,

$$P_k x_r \in \Pi_r(X) \quad \text{and} \quad \|P_k x_r\| \leqq r_k\|x_r\| .$$

Further we have

$$X = \bigoplus_{r\in\mathbb{N}^n} \Pi_r(X) .$$

Since each operator $P_k \restriction \Pi_r(X)$ is bounded, there exists an orthonormal basis $(v_\ell^{(r)})$ in $\Pi_r(X)$ such that for all $k = 1,\dots,n$

$$(P_k v_\ell^{(r)}, v_j^{(r)}) = 0 \quad \text{if} \quad j > \ell(n+1) + 1 .$$

In addition, for $r \neq s$, $r,s \in \mathbb{N}^n$ we have

$$(P_k v_\ell^{(r)}, v_j^{(s)}) = 0 .$$

Now define the self-adjoint operator A by

$$A v_\ell^{(r)} = (\ell+|r|) v_j^{(r)}$$

followed by the usual self-adjoint extension. (We take $|r| = r_1 + \dots + r_n$.)

Then the eigenvalues of A are the natural numbers n+p, p $\in \mathbb{N}$, with multiplicity $\binom{n+p-1}{n}$. So the analyticity space $S_{X,A}$ is nuclear. A renumbering of the orthonormal basis $(v_\ell^{(r)})_{\ell\in\mathbb{N},r\in\mathbb{N}^n}$ yields an orthonormal basis $(g_p)_{p\in\mathbb{N}}$.

(IV.6.5) Theorem

Each operator P_k, k = 1,...,n, is a member of $L(S_{X,A})$.

Proof. Let k = 1,...,n. Let t > 0 and let 0 < τ < t. Then we estimate as follows:

$$\sup_{p,q\in\mathbb{N}} |(e^{\tau A} P_k e^{-tA} g_p,g_q)| =$$

$$= \sup_{r\in\mathbb{N}^n} \sup_{j,\ell\in\mathbb{N}} e^{-(|r|+\ell)t} e^{(|r|+j)\tau} |(P_k v_\ell^{(r)}, v_j^{(r)})| \leq$$

$$\sup_{r\in\mathbb{N}^n} r_k e^{-|r|(t-\tau)} \sup_{j\in\mathbb{N}} (e^{-j(t-(n+1)\tau)}) < \infty . \qquad \square$$

IV.7. Some applications to quantum statistical mechanics

In this section we develop a formulation of quantum statistical mechanics in terms of the algebra $E(S_{X,A})$. As observed in the introduction to the present chapter, the algebra $E(S_{X,A})$ can be regarded as a *-algebra of unbounded operators with common dense invariant domain $S_{X,A}$. We restrict to nuclear analyticity spaces because of their rather simple mathematical structure.

Let us consider a quantum mechanical system in which the dynamics are determined by a hamiltonian operator H, i.e. a self-adjoint operator in an appropriate Hilbert space X. Let A be a positive self-adjoint operator such that $S_{X,A}$ is a nuclear space and such that H belongs to $E(S_{X,A})$ as well as the unitary operators $e^{i\alpha H}$, $\alpha \in \mathbb{R}$. For the states of the quantum system we take the one-dimensional subspaces of the trajectory space $T_{X,A}$. In the next chapter we prove that $T_{X,A}$ contains almost all generalized eigenvectors of H. Here we adopt both the terminology and the notations of Dirac. So $T_{X,A}$ is the ket space and the kets are denoted by $|F\rangle$. Conjugate to the ket space is the bra space, the elements of which are called bras and are denoted by $\langle F|$. The bra space is anti-linearly homeomorphic to the ket space $T_{X,A}$. One could say that the bra space consists of all densely defined (unbounded) linear functionals in $T_{X,A}$ with common domain $S_{X,A}$. Since Dirac does not bother about the domain of the functionals (bras), he introduces the bracket $\langle F|G\rangle$ without any assumption on $|F\rangle$ and $|G\rangle$. Dirac treats brackets as numbers. However, only the bras which come from an element of $S_{X,A}$, have the property that the bracket with any ket is a finite number. In fact, to each bra we have to adjoin its domain.

In order to keep in line with Dirac's ideas, we do not introduce the bracket as a number. In our view the bracket must be seen as a function in the following way. Let $\langle F|$ be the bra corresponding to the ket $|F\rangle$. Then $|F\rangle$ is a trajectory of $T_{X,A}$. It follows that for all $\alpha > 0$, $|F\rangle(\alpha) \in S_{X,A}$. So it makes sense to define

$$\langle F|G\rangle : \alpha \mapsto \langle \overline{|F\rangle(\alpha), |G\rangle}\rangle \ .$$

In this way, for any $\langle F|$ and any $|G\rangle$, the bracket is a well-defined mathematical object. If we want to regard $\langle F|$ as a linear functional in $T_{X,A}$, then for the natural domain of $\langle F|$ we can take the subspace of $T_{X,A}$ which consists of all $|G\rangle$ for which the limit

$$\lim_{\alpha\downarrow 0} \langle F|G\rangle(\alpha) =: \langle F|G\rangle_{\alpha\downarrow 0}$$

is finite. In the next chapter we discuss the consequences of this new bracket notion. We present an interpretation of Dirac's formalism which is based amongst others on this notion.

Let O be an observable. Let the system be in state $|S\rangle$. Then formally the expectation value is given by $\langle S|\mathsf{O}|S\rangle$. We say that the expectation value is detectable if $\langle S|\mathsf{O}|S\rangle_{\alpha\downarrow 0}$ exists. If the state $|S\rangle$ is represented by an element of $S_{X,A}$, then for all observables $\mathsf{O} \in \mathsf{E}(S_{X,A})$, the expectation value in the state $|S\rangle$ is detectable. Because of this property the states (kets) corresponding to the elements of $S_{X,A}$ are called test states (test kets).

At a certain instant the dynamical system is supposed to be in one or other of a number of possible states according to some given probability law. Following Dirac these states may establish a discrete set, a continuous range or both together. Here we only look at the discrete case.

Suppose that the possible states of the system are given by the orthonormal test kets $|m\rangle$, $m \in \mathbb{N}$. Let p_m denote the probability that the system is in the m-th state. Then we define the quantum probability ρ by

$$\text{(IV.7.1)}\quad \rho = \sum_{m=1}^{\infty} p_m \, |m\rangle\langle m| \; , \qquad \sum_{m=1}^{\infty} p_m = 1 \; , \qquad p_m \geqq 0$$

where, in Dirac's notation, $|m\rangle\langle m| = |m\rangle \otimes |m\rangle$.
In the so-called Schrödinger picture the kets will evolve in accordance with Schrödinger's equation

$$i\hbar \frac{d}{dt} |F\rangle = H|F\rangle \; .$$

The bras will evolve with the complex conjugate of this equation. Without disturbances the system remains in the same state. Hence, the p_m's are constant in time. This observation yields the following equation

$$\text{(IV.7.2)}\quad i\hbar\dot{\rho} = \sum_{m=1}^{\infty} p_m (H|m\rangle\langle m| - |m\rangle\langle m|H) = H\rho - \rho H = [H,\rho] \; .$$

For convenience's sake we take $\hbar = 1$ in the sequel.
In our interpretation, the observables of the quantum system are represented by symmetric elements of $\mathbf{E}(S_{X,A})$ which have a self-adjoint extension in X. Here occurs an ambiguity. If a symmetric $P \in \mathbf{E}(S_{X,A})$ has two self-adjoint extensions P_1 and P_2, the algebra $\mathbf{E}(S_{X,A})$ cannot see the difference between P_1 and P_2.
If the system is in the m-th state, the expectation value $\langle O\rangle$ of the observable O equals

$$\langle O\rangle = \langle m|O|m\rangle_{\alpha\downarrow 0} \; .$$

In the sequel we leave out the notation '$\alpha\downarrow 0$'.
We insert the distribution law of the system corresponding to the above introduced density operator ρ. Then the average expectation value $\langle O\rangle$ is given by

$$\text{(IV.7.3)}\quad \langle O\rangle = \sum_{m=1}^{\infty} p_m \, \langle m|O|m\rangle = \langle\!\langle \rho, O\rangle\!\rangle = \mathrm{tr}(\rho O)$$

whenever $\rho \in \mathbf{E}'(S_{X,A})$. Note that $\mathbf{E}'(S_{X,A}) \triangleq ST_{X\otimes X;A \,\wedge\, A, A \,\vee\, A}$. If we put $O = I$ then we get $\langle I\rangle = \sum_{m=1}^{\infty} p_m = 1$.

The solution of equation (IV.7.2) is given by

$$\rho(t) = e^{-itH} \rho_0 \, e^{itH} \; , \qquad t \geqq 0$$

where ρ_0 is the initial value of ρ at $t = 0$. Since the unitary operators e^{itH}, $t \in \mathbb{R}$, belong to $\mathbf{E}(S_{X,A})$ and since $\mathbf{E}'(S_{X,A})$ remains invariant under right and left multiplication by elements of $\mathbf{E}(S_{X,A})$, we get $\rho(t) \in \mathbf{E}'(S_{X,A})$ for all $t \geq 0$ if and only if $\rho_0 \in \mathbf{E}'(S_{X,A})$.

The average expectation value at time t equals

$$\langle O\rangle(t) = \langle\!\langle \rho(t), O(t)\rangle\!\rangle = \langle\!\langle \rho_0, e^{itH}\, O(t) e^{-itH}\rangle\!\rangle \; .$$

Here we write $O(t)$ to indicate that O can intrinsically depend on t. Now put $\underline{O}(t) = e^{itH}\, O(t) e^{-itH}$. Then we derive

(IV.7.4) $$\dot{\underline{O}} = i[H,\underline{O}] + \frac{\partial \underline{O}}{\partial t}$$

$$\frac{d}{dt}\,(\langle\underline{O}\rangle) = i\langle[H,\underline{O}]\rangle + \langle\frac{\partial \underline{O}}{\partial t}\rangle$$

where

$$\frac{\partial \underline{O}}{\partial t}\,(s) = e^{isH}\, \frac{dO}{dt}(s) e^{-isH} \; .$$

The differential equations (IV.7.4) determine the evolution of the observables in the Heisenberg picture.

Next we describe a quantum mechanical system in terms of observables out of a suitable algebra $\mathbf{E}(S_{X,A})$ and 'states' in the dual $\mathbf{E}'(S_{X,A})$. We emphasize that the notion of state differs from the one considered at the beginning of this section.

Remark. In the description of quantum statistical mechanics in terms of bounded operator algebras, the states are represented by trace class operators. Here we point at a great dissimilarity between the algebra of trace class operators and its substitute in our theory $\mathbf{E}'(S_{X,A})$. The algebra of trace class operators is a Banach algebra with the so-called trace norm. However, the set $\mathbf{E}'(S_{X,A})$ is not an algebra and it is complete only in the trivial case that X is a finite dimensional Hilbert space.

(IV.7.5) Definition

A symmetric element $P \in \mathbf{E}(S_{X,A})$ is called semi-bounded if $\langle w|P|w\rangle \geq a$ for some $a \in \mathbb{R}$ and all test kets $|w\rangle$. P is called positive if $a \geq 0$.

A semi-bounded element P of $\mathbf{E}(S_{X,A})$ is a semi-bounded symmetric operator in X. From [Wei] we obtain that each semi-bounded symmetric operator has self-adjoint extensions. So each semi-bounded element of $\mathbf{E}(S_{X,A})$ is an observable.

(IV.7.6) <u>Definition</u>

Let $\Sigma \in \mathbf{E}'(S_{X,A})$. Then Σ is called real if $\langle\Sigma,P\rangle \in \mathbb{R}$ for all symmetric $P \in \mathbf{E}(S_{X,A})$.

From the properties of the pairing $\langle\ ,\ \rangle$ we obtain the following characterization of the real functionals.

(IV.7.7) <u>Lemma</u>

The linear functional $\Sigma \in \mathbf{E}'(S_{X,A})$ is real iff $\Sigma' = \Sigma$.

<u>Proof</u>. With the aid of Theorem (IV.3.18) we derive the following equivalences

$$\langle\!\langle\Sigma,P\rangle\!\rangle \in \mathbb{R} \quad \text{for all symmetric } P \in \mathbf{E}(S_{X,A})$$

$$\Leftrightarrow \langle\!\langle\Sigma,P\rangle\!\rangle = \langle\!\langle\Sigma',P\rangle\!\rangle \quad \text{for all symmetric } P \in \mathbf{E}(S_{X,A})$$

$$\Leftrightarrow \Sigma = \Sigma' .$$

□

<u>Remark</u>. Let $\Sigma \in \mathbf{E}'(S_{X,A})$ with $\Sigma = \Sigma'$. Then $\Sigma = \Phi + \Omega$ with $\Phi \in \mathbf{T}(T_{X,A})$ and $\Omega \in \mathbf{T}(S_{X,A})$. Put $\Psi = (\Phi+\Omega')/2$. Then $\Psi \in \mathbf{T}(T_{X,A})$ and $\Sigma = \Psi + \Psi'$. Note that because of the assumed nuclearity of $S_{X,A}$,

$$\mathbf{T}(S_{X,A}) \triangleq ST_{X\otimes X;A\otimes I,I\otimes A}\ ,\quad \mathbf{T}(T_{X,A}) \triangleq ST_{X\otimes X;I\otimes A,A\otimes I}\ .$$

(IV.7.8) <u>Definition</u>

Let $\Sigma \in \mathbf{E}'(S_{X,A})$ be a real functional. Then Σ is called a normalized state if

- $\langle\!\langle\Sigma,P\rangle\!\rangle \geq 0$ for all positive $P \in \mathbf{E}(S_{X,A})$
- $\langle\!\langle\Sigma,I\rangle\!\rangle = 1$.

Instead of considering continuous linear functionals on $\mathbf{E}'(S_{X,A})$, one can consider densely defined linear functionals, of course. It leads to the notions of real densely defined functionals and unnormalizable states. However, this point of view is beyond the scope of this section.

We will characterize the states in $\mathsf{E}'(S_{X,A})$. Therefore we prove the following result.

(IV.7.9) Lemma

Let $\mathsf{E} \in \mathsf{E}(S_{X,A})$. Let Π_n denote the orthogonal projection onto the first n eigenvectors of A. Then the sequence $(\Pi_n \mathsf{E} \Pi_n)_{n\in\mathbb{N}}$ converges to E in the uniform topology of $\mathsf{E}(S_{X,A})$.

Proof. Let $t > 0$. Then there exists $\tau > 0$ such that both

$$e^{2\tau A} \mathsf{E} e^{-\frac{1}{2}tA} \in X \otimes X$$

and

$$e^{2\tau A} \mathsf{E}^{*} e^{-\frac{1}{2}tA} \in X \otimes X .$$

We proceed as follows:

$$\| e^{\tau A}(\mathsf{E} - \Pi_n \mathsf{E} \Pi_n) e^{-tA} \|_{X\otimes X} \leqq$$

$$\leqq \| (I - \Pi_n) e^{\tau A} \mathsf{E} e^{-tA} \Pi_n \|_{X\otimes X} + \| e^{\tau A} \mathsf{E}(I - \Pi_n) e^{-tA} \|_{X\otimes X} \leqq$$

$$\leqq (\| (I - \Pi_n) e^{-\tau A} \| + \| (I - \Pi_n) e^{-\frac{1}{2}tA} \|) \, \| e^{2\tau A} \mathsf{E} e^{-\frac{1}{2}tA} \|_{X\otimes X} .$$

Since both $\| (I - \Pi_n) e^{-\tau A} \|$ and $\| (I - \Pi_n) e^{-\frac{1}{2}tA} \| \to 0$ as $n \to \infty$, we get

$$\| e^{\tau A}(\mathsf{E} - \Pi_n \mathsf{E} \Pi_n) e^{-tA} \|_{X\otimes X} \to 0 \quad \text{as } n \to \infty .$$

Similarly, it can be proved that

$$\| e^{\tau A}(\mathsf{E}^{*} - \Pi_n \mathsf{E}^{*} \Pi_n) e^{-tA} \|_{X\otimes X} \to 0 \quad \text{as } n \to \infty . \qquad \square$$

Remark. Let P be a positive symmetric element of $\mathsf{E}(S_{X,A})$. Then for each $n \in \mathbb{N}$, $\Pi_n \mathsf{P} \Pi_n$ is positive and belongs to $\mathsf{E}(S_{X,A})$. In fact, $\Pi_n \mathsf{P} \Pi_n$ is a positive self-adjoint finite rank operator. So there exists $f_j^{(n)} \in \Pi_n(X)$, $j = 1,\dots,n$, such that

$$\Pi_n \mathsf{P} \Pi_n = \sum_{j=1}^{n} \mu_j^{(n)} \, | f_j^{(n)} \rangle\langle f_j^{(n)} |$$

with $\mu_j \geqq 0$. It leads to the following characterization.

(IV.7.10) Theorem

Let $\Sigma \in \mathbf{E}'(S_{X,A})$ be a real functional. Then Σ is a normalizable state iff

$$\langle \Sigma, |w\rangle\langle w| \rangle \geq 0$$

for all test kets $|w\rangle$.

Proof

$\Rightarrow$) All projections $P_{|w\rangle} = |w\rangle\langle w|$ with $|w\rangle$ a normalized test ket, are elements of $\mathbf{E}(S_{X,A})$. Hence $\langle \Sigma, P_{|w\rangle} \rangle \geq 0$.

$\Leftarrow$) Let P be a positive element of $\mathbf{E}(S_{X,A})$. Let the projections Π_n be defined as in the previous lemma. The functional Σ is $\varepsilon\sigma_u$-continuous on $\mathbf{E}(S_{X,A})$. Hence with the aid of the above remark we get

$$\langle \Sigma, \Pi_n P \Pi_n \rangle \geq 0 \ .$$

for all $n \in \mathbb{N}$. Further, because of Lemma IV.7.9,

$$\langle \Sigma, P \rangle = \lim_{n\to\infty} \langle \Sigma, \Pi_n P \Pi_n \rangle$$

and hence $\langle \Sigma, P \rangle \geq 0$. □

Remark. Since $\Sigma \in \mathbf{E}'(S_{X,A}) \hat{\subset} TT_{X\otimes X;I\otimes A,A\otimes I}$, and since $|w\rangle\langle w| \in SS_{X\otimes X;I\otimes A,A\otimes I}$ we get

$$\langle \Sigma, |w\rangle\langle w| \rangle = \langle w|\Sigma|w\rangle \ .$$

Special elements of $\mathbf{E}'(S_{X,A})$ are the pure normalized states. Here is the definition.

(IV.7.11) Definition

A normalized state Ξ is called pure if there exists a normalized test ket $|w\rangle$ such that $\Xi = |w\rangle\langle w|$.

Only test kets give rise to normalizable states. All other kets provide unbounded functionals in $\mathbf{E}(S_{X,A})$.

(IV.7.12) Lemma

Let $|x\rangle$ be a ket. Then we have

$$|x\rangle\langle x| \in \mathbf{E}'(S_{X,A}) \Leftrightarrow |x\rangle \text{ is a test ket.}$$

Proof

$\Leftarrow$) Trivial.

$\Rightarrow$) Suppose $|x\rangle \notin S_{X,A}$. Then there exists $\psi \in \mathbf{B}_+(\mathbb{R})$ such that $|x\rangle \notin \mathcal{D}(\psi(A))$. The operator $\psi(A)^2$ belongs to $\mathbf{E}(S_{X,A})$. However, $\psi(A)|x\rangle$ does not belong to X and hence $|x\rangle\langle x| \notin \mathbf{E}'(S_{X,A})$. □

The pure states can be characterized as follows.

(IV.7.13) Lemma

A normalized state Ξ is pure iff $\Xi \in \mathbf{T}(T_{X,A})$ (or $\mathbf{T}(S_{X,A})$) with $\Xi^2 = \Xi$.

Proof. If Ξ is pure, then $\Xi = |w\rangle\langle w|$ for some test ket $|w\rangle$. Hence $|w\rangle\langle w| \in SS_{X\otimes X;I\otimes A,A\otimes I} \mathrel{\hat=} \mathbf{T}(T_{X,A}) \cap \mathbf{T}(S_{X,A})$, and it is a projection. Conversely, if $\Xi \in \mathbf{T}(T_{X,A})$ is a state, then $\Xi = \Xi' \in \mathbf{T}(S_{X,A})$. Hence $\Xi \in \mathbf{T}(T_{X,A}) \cap \mathbf{T}(S_{X,A})$ and Ξ is a projection. It follows that Ξ is a projection with trace norm 1. So there exists a normalized $|w\rangle \in X$ with $\Xi = |w\rangle\langle w|$. By the previous lemma $|w\rangle$ is a test ket. □

(IV.7.14) Theorem

Every normalized pure state is an extreme point in the convex set of normalized states.

Proof. Let w be a normalized test ket, and let Π_n, $n \in \mathbb{N}$, denote the orthogonal projection as introduced in Lemma (IV.7.9). Suppose there exist normalized states $\Sigma_1, \Sigma_2 \in \mathbf{E}'(S_{X,A})$ and $0 < \alpha < 1$ such that

$$|w\rangle\langle w| = \alpha\Sigma_1 + (1-\alpha)\Sigma_2 \,.$$

Then for each $n \in \mathbb{N}$ so large that $\Pi_n|w\rangle \neq 0$, we have

$$\frac{\Pi_n\ |w\rangle\langle w|\ \Pi_n}{\|\Pi_n\ |w\rangle\|^2} = \alpha\, \frac{\langle\!\langle \Sigma_1,\Pi_n \rangle\!\rangle}{\|\Pi_n\ |w\rangle\|^2} \left[\frac{\Pi_n\ \Sigma_1\ \Pi_n}{\langle\!\langle \Sigma_1,\Pi_n \rangle\!\rangle}\right] +$$

$$+ (1-\alpha)\, \frac{\langle\!\langle \Sigma_2,\Pi_n \rangle\!\rangle}{\|\Pi_n\ |w\rangle\|^2} \left[\frac{\Pi_n\ \Sigma_1\ \Pi_n}{\langle\!\langle \Sigma_2,\Pi_n \rangle\!\rangle}\right] .$$

So we have written $\dfrac{\Pi_n\ |w\rangle\langle w|\ \Pi_n}{\|\Pi_n\ |w\rangle\|^2}$ as a convex combination of $\dfrac{\Pi_n\,\Sigma_1\,\Pi_n}{\langle\!\langle\Sigma_1,\Pi_n\rangle\!\rangle}$ and $\dfrac{\Pi_n\,\Sigma_2\,\Pi_n}{\langle\!\langle\Sigma_2,\Pi_n\rangle\!\rangle}$. Since these three expressions denote three elements of the unit sphere of the finite dimensional Hilbert space $\Pi_n(X) \otimes \Pi_n(X)$ we may assume

$$\Pi_n\ |w\rangle\langle w|\ \Pi_n = \Pi_n\ \Sigma_1\ \Pi_n\ .$$

Since $\Pi_k\ \Pi_\ell = \Pi_k$ for all $\ell \geq k$, we derive

$$\forall_{n\in\mathbb{N}} : \Pi_n\ |w\rangle\langle w|\ \Pi_n = \Pi_n\ \Sigma_1\ \Pi_n\ .$$

By Lemma (IV.7.9) the sequences $(\Pi_n\ |w\rangle\langle w|\ \Pi_n)_{n\in\mathbb{N}}$ and $(\Pi_n\ \Sigma_1\ \Pi_n)_{n\in\mathbb{N}}$ converge weakly to $|w\rangle\langle w|$ and Σ_1, respectively. Hence $\Sigma_1 = |w\rangle\langle w|$. □

The next theorem states that the pure normalized states are the only extreme points in the set of normalized states.

(IV.7.15) <u>Theorem</u>

Let Ξ be an extreme point in the set of normalized states. Then Ξ is a pure normalized state.

<u>Proof</u>. Since $\Xi \neq 0$, there exists a normalized test ket $|w\rangle$ such that

$$\langle\!\langle\Xi,|w\rangle\langle w|\rangle\!\rangle \neq 0\ .$$

Now Ξ can be written as

$$\Xi = \Xi \circ |w\rangle\langle w| + \Xi \circ (I - |w\rangle\langle w|)\ .$$

Here

$$\langle\!\langle\Xi \circ |w\rangle\langle w|,E\rangle\!\rangle = \langle\!\langle\Xi,|w\rangle\langle w| \circ E\rangle\!\rangle\ .$$

So

$$\langle\!\langle\Xi \circ |w\rangle\langle w|,I\rangle\!\rangle = \langle w|\Xi|w\rangle \neq 0\ .$$

Suppose that $\Xi \circ (I - |w\rangle\langle w|) \neq 0$, and consequently $\langle\!\langle\Xi,I - |w\rangle\langle w|\rangle\!\rangle \neq 0$. (See the remark after this proof.) Then we can write $\Xi = \alpha\Xi_1 + (1-\alpha)\Xi_2$ with

$$\Xi_1 = \frac{\Xi \circ |w\rangle\langle w|}{\langle w|\Xi|w\rangle}\ , \qquad \Xi_2 = \frac{\Xi \circ (I - |w\rangle\langle w|)}{1 - \langle w|\Xi|w\rangle}$$

and

$$\alpha = \langle w|\Xi|w\rangle .$$

The functionals Ξ_1 and Ξ_2 are normalized states. This follows from the following observation.

Let Π denote a projection in $\mathbf{E}(S_{X,A})$. Then for each normalized state Ξ, the functional $\Sigma = \frac{\Xi \circ \Pi}{\Xi(\Pi)}$ is a normalized state if $\Xi(\Pi) \neq 0$. To see this assertion, observe that

$$\langle\!\langle \Sigma, E \rangle\!\rangle = \Xi(\Pi)^{-1} \langle\!\langle \Xi, \Pi \circ E \rangle\!\rangle = \Xi(\Pi)^{-1} \langle\!\langle \Xi, \Pi E \Pi \rangle\!\rangle$$

and

$$\langle\!\langle \Sigma, I \rangle\!\rangle = \Xi(\Pi)^{-1} \Xi(\Pi) = 1 .$$

Thus it follows that $\langle\!\langle \Sigma, E \rangle\!\rangle \in \mathbb{R}$ if E is symmetric, and $\langle\!\langle \Sigma, E \rangle\!\rangle \geq 0$ if E is positive.

However, we now get a contradiction with the assumption that Ξ is extreme. Hence $\Xi \circ (I - |w\rangle\langle w|) = 0$ and consequently $\Xi = \Xi \circ |w\rangle\langle w|$, $\langle\!\langle \Xi, |w\rangle\langle w| \rangle\!\rangle = 1$. So for all test kets $|v\rangle$ we get

$$\langle\!\langle \Xi, |v\rangle\langle v| \rangle\!\rangle = \langle\!\langle \Xi, (|w\rangle\langle w|) \circ (|v\rangle\langle v|) \rangle\!\rangle = |\langle w|v\rangle|^2 .$$

In the final step of the proof we again employ the finite rank projection Π_n as introduced in Lemma (IV.7.9).

Let $E \in \mathbf{E}(S_{X,A})$ be symmetric. Then $\Pi_n E \Pi_n$ is a symmetric operator on the finite dimensional Hilbert space $\Pi_n(X)$. So for each $n \in \mathbb{N}$ there exist $|v_j^{(n)}\rangle \in \Pi_n(X)$, $j = 1,\dots,n$, and numbers $\mu_j^{(n)} \in \mathbb{R}$, $j = 1,\dots,n$, such that

$$\Pi_n E \Pi_n = \sum_{j=1}^{n} \mu_j^{(n)} |v_j^{(n)}\rangle\langle v_j^{(n)}| .$$

It follows that

$$\langle\!\langle \Xi, \Pi_n E \Pi_n \rangle\!\rangle = \sum_{j=1}^{n} \mu_j^{(n)} |\langle w|v_j^{(n)}\rangle|^2 = \langle w|\Pi_n E \Pi_n|w\rangle .$$

Letting $n \to \infty$ we obtain by Lemma (IV.7.9)

$$\langle\!\langle \Xi, E \rangle\!\rangle = \langle w|E|w\rangle$$

for all symmetric $E \in \mathbf{E}(S_{X,A})$.

This yields $\Xi = |w\rangle\langle w|$. □

Remark. Let $\Sigma \in \mathbf{E}'(S_{X,A})$ be a state. Let $n \in \mathbb{N}$ and let $E \in \mathbf{E}(S_{X,A})$. Then we have by Cauchy-Schwarz's inequality

$$|\langle\Sigma,\Pi_n E \Pi_n\rangle|^2 \leq \langle\Sigma,\Pi_n\rangle \langle\Sigma,\Pi_n E^* E \Pi_n\rangle .$$

So in the limit $n \to \infty$ we obtain

$$\langle\Sigma,E\rangle^2 \leq \langle\Sigma,I\rangle \langle\Sigma,E^*E\rangle .$$

Consequently:

$$\langle\Sigma,I\rangle = 0 \Leftrightarrow \Sigma = 0 .$$

(IV.7.16) Lemma

The linear span of the pure states is dense in $\mathbf{E}'(S_{X,A})$.

Proof. Let $E \in \mathbf{E}(S_{X,A})$. If $\langle w|E|w\rangle = 0$ for all test kets $|w\rangle$ then it easily follows that $E = 0$. □

We note that the states corresponding to the elements of $\mathbf{T}(T_{X,A}) \cap \mathbf{T}(S_{X,A})$ are positive self-adjoint trace class operators Λ with trace equal to one for which there exists $t > 0$ and $\Lambda_0 \in \mathbf{T}(X)$ such that $\Lambda = e^{-tA} \Lambda_0 e^{-tA}$.

As a rule the dynamical state of a quantum system at a certain instant cannot be represented by one single ket, but we need to consider a statistical mixture of kets. Therefore we introduced the quantum density operator ρ in the beginning of this section. According to the probability law determined by ρ, the quantum system is in one or other of a number of possible states. So it makes sense to call ρ the state of the quantum system at a given instant.

If at $t = 0$ the quantum system is in the state ρ_0, at $t = \tau$ the system is in the state $\rho(\tau)$ with

$$\rho(\tau) = e^{-i\tau H} \rho_0 e^{i\tau H} .$$

So ρ satisfies the evolution equation

$$\mathring{\rho} = - i[H,\rho] .$$

Here we consider $\rho_0 \in \mathbf{E}'(S_{X,A})$ because these states give rise to finite expectation values for every observable in $\mathbf{E}(S_{X,A})$. Then the expectation value of the observable O is given by

$$\langle O\rangle(t) = \langle\!\langle \rho(t), O\rangle\!\rangle \ .$$

Now, in general, we may assume that a normalized state in $\mathbf{E}'(S_{X,A})$ represents an initial state of the quantum system in the above indicated way. A state Σ_0 evolves according to

$$e^{-itH} \, \Sigma_0 \, e^{itH} \ , \qquad t > 0 \ .$$

So the statistical mixture determined by the quantum density operator ρ is a very special type of states; states like ρ have a genuine physical interpretation. The elements of $\mathbf{T}(T_{X,A}) \cap \mathbf{T}(S_{X,A})$ are of this type. The pure states represent states which the system occupies with probability 1.

We conclude this section with a short discussion of the three possible types of dynamical quantum systems.

(1) <u>The Hamiltonian operator H admits a purely discrete spectrum</u>

This case is the easiest one to treat and it probably contains the most useful results.

Let H be a Hamiltonian operator in X with eigenvalues $E_1 \leq E_2 \leq \ldots$, and corresponding normalized eigenkets $|E_1\rangle, |E_2\rangle, \ldots$. Then the eigenkets $|E_i\rangle$ of H establish a complete orthonormal basis for X. Define the positive numbers λ_n, $n \in \mathbb{N}$, as follows

$$\lambda_1 = E_1 \ , \quad \lambda_n = \max(\lambda_{n-1}+1, |E_n|) \ , \qquad n > 1 \ ,$$

and the self-adjoint operator A by

$$A|E_n\rangle = \lambda_n |E_n\rangle$$

followed by linear extension and unique self-adjoint extension to X. Then the analyticity space $S_{X,A}$ is nuclear because $\sum_{n=1}^{\infty} e^{-\lambda_n t} < \infty$ for all $t > 0$. Further, H is continuous on $S_{X,A}$ because

$$\sup_{n\in\mathbb{N}} \left(|E_n| e^{-\lambda_n t}\right) < \infty \ .$$

Hence, $H \in \mathbf{E}_A$. Similarly it follows that the unitary operators $e^{i\alpha H}$, $\alpha \in \mathbb{R}$, are elements of $\mathbf{E}_A$. So the space $S_{X,A}$ satisfies the required conditions.

An important example of a statistical mixture is given by the state

$$\rho_0 = \sum_{n=1}^{\infty} p_n \, |E_n\rangle\langle E_n| \, , \qquad p_n \geq 0 \, , \qquad \sum_{n=1}^{\infty} p_n = 1 \, .$$

Then ρ is represented by a diagonal matrix, and seen as a bounded operator on X, ρ clearly commutes with A and H. Since $\rho \in E_A'$, it satisfies

$$\exists_{\alpha>0} \, \forall_{a>0} \, \exists_{M>0} \, \forall_{n\in\mathbb{N}}: \, (p_n \, e^{-a\lambda_n} \, e^{\alpha\lambda_n}) < M \, .$$

Hence $p_n = O(e^{-\alpha\lambda_n})$, and $\rho \in S_{X\otimes X, A\otimes I + I\otimes A}$. It is obvious that without disturbance the state ρ does not depend on the time t. We note that it is obvious that every term $|E_n\rangle\langle E_n|$ of the series does not depend on t, i.e. the system remains in a stationary state as long as disturbances do not occur. In general, a state ρ is given by

$$\rho = \sum_{n,m} \rho_{nm} \, |E_n\rangle\langle E_m| \, .$$

However, in many physically realistic cases the nondiagonal elements can be neglected.

An example for class (1) is given by the one-dimensional harmonic oscillator where $H = \frac{1}{2}(-\frac{d^2}{dx^2} + x^2 + 1)$. Then H is self-adjoint in $L_2(\mathbb{R})$ with $E_n = n$, $n \in \mathbb{N}$, as its eigenvalues and the Hermite functions as its eigenfunctions. Hence, we can take $A = H$. We note that the space $S_{L_2(\mathbb{R}),H}$ is equal to the space $S_{\frac{1}{2}}^{\frac{1}{2}}$ of Gelfand-Shilov. Well-defined observables are the momentum operator $i(d/dx)$ and the position operator x.

(2) <u>The Hamiltonian operator H admits a purely continuous spectrum</u>

This is a harder case. In the previous section we have been able to construct a nuclear analyticity space $S_{X,A}$ such that H is continuous on $S_{X,A}$. Then to almost every point in the spectrum of H there corresponds an eigenket in the trajectory space $T_{X,A}$. However, it is not clear whether the unitary operators $e^{i\alpha H}$, $\alpha \in \mathbb{R}$, are continuous on $S_{X,A}$, and this problem has not been solved yet. Of course, we could weaken the conditions on $S_{X,A}$ and skip nuclearity. Then the analyticity space $S_{X,|H|}$ with $(H) = (H^2)^{\frac{1}{2}}$ would be ideal. But nuclearity seems to play an essential role both in the discussions of this section and in our interpretation of Dirac's formalism in the next chapter.

There is another approach. Sometimes iH is one of the skew-adjoint generators of a unitary Lie group representation on X with nuclear analyticity space. We shall explain this to some extent. Let G be a finite dimensional Lie group with Lie algebra $A(G)$. Let U be a representation of G into the space of unitary operators on X, and ∂U the corresponding infinitesimal representation of $A(G)$ in X. Then for every $a \in A(G)$ the operator $\partial U(a)$ is skew-adjoint in X, by Stone's theorem.
Our first assertion is the following one.

- There exists $a_1 \in A(G)$ such that $iH = \partial A(a_1)$.

Since G has dimension $d < \infty$ there are $a_2,\dots,a_d \in A(G)$ such that $\{a_1,\dots,a_d\}$ generates the Lie group G in the usual way. Following Nelson, [Ne], the analyticity space corresponding to the unitary representation U is equal to

$$S_{X,\Delta^{\frac{1}{2}}}$$

wher

$$\Delta = 1 - \left((\partial U(a_1))^2 + (\partial U(a_2))^2 + \dots + (\partial U(a_d))^2\right) .$$

Then our second assumption is

- $S_{X,\Delta^{\frac{1}{2}}}$ is nuclear.

In Section II.4, we have given several cases of unitary representations of Lie groups G with a nuclear analyticity space $S_{X,\Delta^{\frac{1}{2}}}$. Moreover, we have proved that both the unitary operators $U(g)$, $g \in G$, and the skew-adjoint operators $\partial U(a_j)$, $j = 1,\dots,d$, are all continuous on $S_{X,\Delta^{\frac{1}{2}}}$. So under the above-mentioned assumptions the nuclear analyticity space $S_{X,\Delta^{\frac{1}{2}}}$ has the desired properties.

An example for this type of operators is the Hamiltonian operator of the free particle in one dimension,

$$H = -\frac{d^2}{dx^2} .$$

An appropriate algebra is the six-dimensional algebra generated by

$$i\frac{d^2}{dx^2} , \; i\left(\frac{d}{dx}\,x + x\,\frac{d}{dx}\right) , \; ix^2 , \; ix , \; \frac{d}{dx} , \; i .$$

It corresponds to the infinitesimal representation belonging to the unitary representation of the Schrödinger groups on $L_2(\mathbb{R})$. The Schrödinger group is is obtained as a semi-direct product of $\mathbf{SL}(2,\mathbb{R})$ and of the Heisenberg group. We note that the Schrödinger group is the symmetry group of the Schrödinger equation of the free particle (see [Mi]).

(3) The Hamiltonian operator H admits a discrete/continuous spectrum

In many applications the interesting part of the spectrum of H is the discrete one. So we split X into the direct sum $X = X_d \oplus X_c$ such that H_d, the restriction of H to X_d, acts invariantly in X_d and H_d is a self-adjoint operator in X_d with discrete spectrum, and such that H_c, the restriction of H to X_c, acts invariantly in X_c and H_c is a self-adjoint operator in X_c with a purely continuous spectrum.
An example for this case is the Hamiltonian operator of the hydrogen atom.

CHAPTER V

A mathematical interpretation of Dirac's formalism for quantum mechanics

Introduction

Several attempts have been made by mathematicians to justify the various non-mathematical techniques of Dirac's formalism which has become a common language for theoretical physicists in the passed fifty years. One of the reasons for the furious search for a justification of this formalism is the strong feeling that the ideas provoked by Dirac fully appease the mathematical intuition. However, appararently mathematicians have not been sufficiently flexible in their attempts to rigorize this formalism and to turn it into a genuine mathematical apparatus. In the preface to his book on the foundations of quantum mechanics, John von Neumann says about the fomalism that it is scarcely to be surpassed in brevity and elegance but in no way satisfies the requirements of mathematical rigour. Of course, his observation is quite right. But is it the ultimate goal of mathematics to replace a mathematically non-rigorous but elegant and brief theory by a mathematical theory which is less elegant, not brief and incomplete?
In fact, the ideas of Dirac are simple. At the time he wrote his book the theory of finite dimensional vector spaces was known. So he was used to the concepts of hermitean operator, of spectrum of a hermitean operator and its orthonormal basis of eigenvectors, of multiplicity of an eigenvalue, of matrix of an operator with respect to a basis and so on. Regarded from a mathematical point of view, Dirac claims that all properties of hermitean operators on finite dimensional vector spaces remain valid for (unbounded) self-adjoint operators in infinite dimensional Hilbert spaces. This validity has to be interpreted in the way Dirac describes in his book. As far as we know, no paper on Dirac's formalism mathematically foundates the bold claims on which Dirac based his principles of quantum mechanics. Most papers on this subject only solve the so-called generalized eigenvalue problem in the

setting of rigged Hilbert space theory. The other aspects of the formalism such as no discrimination between kets and bras, continuous basis and corresponding continuous matrices, are almost completely neglected.

In the present chapter we describe a justification and a rigorification of Dirac's formalism in terms of our theory. We define the ket space and the bra space. Both are considered as trajectory spaces. The ket space is a linear vector space. Each ket is in one to one correspondence with a bra. As Dirac assumes, the bra space is the hermitean conjugate of the ket space. In fact, the ket space generalizes the space of 'column vectors' and the bra space the space of 'row vectors' from elementary linear algebra. There, the product of a row and a column is a scalar. Similarly, Dirac introduces the product of a bra and a ket, and he calls this product the bracket. In order to describe Dirac's bracket notion we adjoin to each ket and to each bra a function from $(0,\infty)$ into $\mathbb{C}$. So in our interpretation the bracket is no longer a number but a function on $(0,\infty)$. Taking a fixed bra we can consider the subspace of all kets for which the limit $\lim_{t\downarrow 0}$ of the corresponding function exists. Thus each bra becomes a densely defined unbounded linear functional in the ket space.

The basic concepts in our interpretation are the following.

In the first place the concept of Dirac basis. For a nuclear trajectory space $T_{X,A}$ we define a generalization of the notion of orthonormal basis for a Hilbert space. An orthonormal basis can be seen as a mapping v from $\mathbb{N}$ into X with the property $(v(k),v(\ell)) = \delta_{k\ell}$ and $\sum_{k=1}^{\infty} |(x,v(k))|^2 = \|x\|^2$ for all $x \in X$. ($\mathbb{N}$ can be regarded as a measure space with counting measure.) In our definition we replace $\mathbb{N}$ by a general measure space M with σ-finite measure μ. A Dirac basis is an equivalence class of mappings from M into $T_{X,A}$, which satisfies some additional conditions. A Dirac basis will be denoted by the triple (G,M,μ) or shortly by $(G_\lambda)_{\lambda\in M}$. For a Dirac basis we prove a generalization of the notion of orthogonality, of Parseval's identity, of the Fourier expansion of elements of $T_{X,A}$, and of matrices, all in the spirit of Dirac.

The second fundamental concept is a measure theoretical version of the Sobolev lemma. Then we restrict to measure spaces M, which have a metric, and to regular Borel measures μ on M with the property that bounded Borel sets of M have finite μ-measure. Under certain conditions (Federer's conditions) on the metric space M we can make a canonical choice $(\hat{G}_\lambda)_{\lambda\in M}$ out of each Dirac basis (G,M,μ). This canonical representant will be called a Dirac basis, also. It has the following property. Let $w \in S_{X,A}$. Then the function

$$\hat{w}: \lambda \mapsto \langle w,\hat{G}_\lambda\rangle \ , \qquad \lambda \in M$$

is square integrable on M and satisfies

$$\hat{w}(\lambda) = \lim_{r\downarrow 0} \mu(B(\lambda,r))^{-1} \int_{B(\lambda,r)} \hat{w}\, d\mu \ ,$$

where $B(\lambda,r)$ denotes the ball with centre λ and radius $r > 0$.

At the end of this introduction we shortly mention the sections of this chapter. In Section V.1 we introduce the notion of Dirac basis and show an expansion result for these bases. In Section V.2 we prove the above mentioned measure theoretical Sobolev lemma. Sections V.3-V.5 contain our theory of generalized eigenvectors. We prove that each self-adjoint operator in a Hilbert space, which is closable as a linear mapping in a nuclear trajectory space, admits a set of generalized eigenvectors in the following way: To almost every point λ in the spectrum of P with multiplicity m_λ there exist m_λ (generalized) eigenvectors in the corresponding trajectory space.
The final sections of this chapter contain the definitions of bra space, of ket space and our mathematical interpretation of Dirac's computations with bras, kets and continuous matrices.

V.1. The concept of Dirac basis

In the whole chapter $T_{X,A}$ will be a nuclear trajectory space. So, in X there exists an orthonormal basis $(v_k)_{k\in\mathbb{N}}$ which consists of eigenvectors of A with respective eigenvalues λ_k, $k \in \mathbb{N}$. For convenience we take the eigenvalues ordered, $0 \leq \lambda_1 \leq \lambda_2 \leq \ldots$. Since e^{-tA} is a Hilbert Schmidt operator for all $t > 0$, we have

$$\forall_{t>0}: \sum_{k=1}^{\infty} e^{-\lambda_k t} < \infty \ .$$

Let M denote a measure space which is σ-finite with respect to a fixed non-negative measure μ.

(V.1.1) Definition

A function Φ from M into $T_{X,A}$ is said to be (weakly) measurable if for all $w \in S_{X,A}$ the function $\alpha \mapsto \langle w,\Phi(\alpha)\rangle$ from M into $\mathbb{C}$ is μ-measurable.

If Φ: $M \to T_{X,A}$ is μ-measurable, then for all $t > 0$ the function

$$\alpha \mapsto \|\Phi(\alpha;t)\| = \left(\sum_{k=1}^{\infty} |(v_k,\Phi(\alpha;t))|^2\right)^{\frac{1}{2}}$$

is μ-measurable. In the following lemma we introduce a notion of integral for measurable Φ: $M \to T_{X,A}$.

(V.1.2) Lemma

Let Φ: $M \to T_{X,A}$ be measurable. Suppose for all $t > 0$

$$\int_M \|\Phi(\alpha;t)\| d\mu_\alpha < \infty .$$

Then there exists a unique element $I_\Phi \in T_{X,A}$ such that for all $w \in S_{X,A}$

$$\langle w,I_\Phi\rangle = \int_M \langle w,\Phi(\alpha)\rangle \, d\mu_\alpha .$$

Proof. Let $t > 0$, and let $w \in e^{-tA}(X)$. Then we have the estimation

$$(*) \qquad \int_M |\langle w,\Phi(\alpha)\rangle| d\mu_\alpha \leq \|e^{tA} w\| \int_M \|\Phi(\alpha;t)\| d\mu_\alpha .$$

So for all $u \in S_{X,A}$ the function $\alpha \mapsto \langle u,\Phi(\alpha)\rangle$ is absolutely integrable. Further, by (*) the linear functional

$$u \mapsto \int_M \langle u,\Phi(\alpha)\rangle \, d\mu_\alpha , \qquad u \in S_{X,A} ,$$

is continuous. So there exists a unique $I_\Phi \in T_{X,A}$ such that

$$\langle u,I_\Phi\rangle = \int_M \langle u,\Phi(\alpha)\rangle \, d\mu_\alpha , \qquad u \in S_{X,A} .$$

□

Remark. Let $t > 0$ and let $x \in X$. We have

$$(x, I_\Phi(t)) = \int_M (x, \Phi(\alpha;t))\, d\mu_\alpha \,.$$

We call I_Φ the integral of the function Φ, and denote the element $I_\Phi \in T_{X,A}$ by the integral expression

$$\int_M \Phi(\alpha)\, d\mu_\alpha \,.$$

For all $t > 0$ and all $\alpha \in M$, $\Phi(\alpha;t) \in X$. We have

$$\left(\int_M \Phi(\alpha)\, d\mu_\alpha\right)(t) = \int_M \Phi(\alpha;t)\, d\mu_\alpha \,.$$

In the set of measurable functions from M into $T_{X,A}$ we define the usual equivalence relation $\sim$ by

$$\Phi \sim \Psi \quad \text{iff} \quad \Phi(\alpha) = \Psi(\alpha) \quad \text{except on a set of measure zero.}$$

Now the definition of Dirac basis is the following.

(V.1.3) Definition

A Dirac basis $([G], M, \mu)$ in $T_{X,A}$ is an equivalence class of measurable functions G from M into $T_{X,A}$ with the following properties

$$\forall_{k\in\mathbb{N}} \forall_{\ell\in\mathbb{N}} : \int_M \langle v_k, G_\alpha\rangle \overline{\langle v_\ell, G_\alpha\rangle}\, d\mu_\alpha = \delta_{k\ell} \,.$$

Here $\delta_{k\ell}$ denotes the Kronecker delta.

Example. An orthonormal basis $(u_k)_{k \mathbb{N}}$ in a Hilbert space X is expressed by $([u], \mathbb{N}, \mu)$ where μ is the counting measure and where the canonical representant $u \in [u]$ is given by $u(k) = u_k$, $k \in \mathbb{N}$.

In this section we prove that every $w \in S_{X,A}$ can be expanded with respect to a Dirac basis. Also we present the generalization of the derived expansion formula for the elements of $S_{X,A}$ to the elements of $T_{X,A}$. It shows that the name of a basis is properly given.

Let $([G],M,\mu)$ be a Dirac basis in $T_{X,A}$, and let $\alpha \mapsto G_\alpha$ (notation $(G_\alpha)_{\alpha\in M}$) be a representant of the class $[G]$. From Definition (V.1.3), it follows that the functions $\alpha \mapsto \langle v_k,G_\alpha\rangle$ are square μ-integrable. We denote the corresponding equivalence classes in $L_2(M,\mu)$ by φ_k, $k \in \mathbb{N}$. We observe that for all $\hat{G} \in [G]$ the functions $\alpha \mapsto \langle v_k,\hat{G}_\alpha\rangle$ belong to φ_k. Now the definition of Dirac basis has been taken such that the elements $\varphi_k \in L_2(M,\mu)$, $k \in \mathbb{N}$, constitute an orthonormal system in $L_2(M,\mu)$. Let Y denote the Hilbert subspace of $L_2(M,\mu)$ spanned by the system $(\varphi_k)_{k\in\mathbb{N}}$. Define the linear operator U from X onto Y by

$$Ux = \sum_{k=1}^{\infty} (x,v_k)\varphi_k\,, \qquad x \in X\,.$$

Then U is unitary. With $B = UAU^*$ we have $U(S_{X,A}) = S_{Y,B}$ and $U(T_{X,A}) = T_{Y,B}$. Let $w \in S_{X,A}$. Then

$$Uw = \sum_{k=1}^{\infty} (w,v_k)\varphi_k$$

and

$$\langle w,G_\alpha\rangle = \sum_{k=1}^{\infty} (w,v_k)\langle v_k,G_\alpha\rangle\,.$$

So $\alpha \mapsto \langle w,G_\alpha\rangle$ is an L_2-representant of the class Uw. Again, we observe that for all $\hat{G} \in [G]$, the function $\alpha \mapsto \langle w,\hat{G}_\alpha\rangle$ belongs to Uw. Therefore,

$$(w,v_k)_X = \int_M \langle w,G_\alpha\rangle\overline{\langle v_k,G_\alpha\rangle}\, d\mu_\alpha$$

and, in general, for $u,w \in S_{X,A}$

$$(u,w)_X = \int_M \langle u,G_\alpha\rangle\overline{\langle w,G_\alpha\rangle}\, d\mu_\alpha\,.$$

Now let $v \in S_{X,A}$. Then there is $t > 0$ and $w \in S_{X,A}$ such that $v = e^{-tA}w$. Consider the following formal computation:

$$v = \sum_{k=1}^{\infty} e^{-\lambda_k t}(w,v_k)v_k$$

$$= \sum_{k=1}^{\infty} e^{-\lambda_k t}\left(\int_M \langle w,G_\alpha\rangle\overline{\langle v_k,G_\alpha\rangle}\, d\mu_\alpha\right)v_k =$$

$$\stackrel{(*)}{=} \int_M \langle w, G_\alpha \rangle \left(\sum_{k=1}^{\infty} e^{-\lambda_k t} \; \overline{\langle v_k, G_\alpha \rangle} \; v_k \right) d\mu_\alpha$$

$$= \int_M \langle w, G_\alpha \rangle \; G_\alpha(t) d\mu_\alpha \; .$$

The equality (*) is the only problem in the above computation. The next estimation shows that it is indeed allowed to interchange summation and integration:

$$\left(\sum_{k=1}^{\infty} \int_M \right) e^{-\lambda_k t} \; |\langle w, G_\alpha \rangle| \, |\langle v_k, G_\alpha \rangle| d\mu_\alpha \leqq$$

$$\leqq \tfrac{1}{2} \sum_{k=1}^{\infty} e^{-\lambda_k t} \int_M (|\langle w, G_\alpha \rangle|^2 + |\langle v_k, G_\alpha \rangle|^2) d\mu_\alpha =$$

$$= \tfrac{1}{2}(\|w\|^2 + 1) \sum_{k=1}^{\infty} e^{-\lambda_k t} \; .$$

The wanted result follows by the Fubini-Tonelli theorem. The above considerations lead to the following expansion theorem.

(V.1.4) Theorem

Let $([G], M, \mu)$ be a Dirac basis in $T_{X,A}$, and let $G \in [G]$. Then for all $v \in S_{X,A}$ we have the expansion

$$v = \int_M \langle v, G_\alpha \rangle \; G_\alpha \, d\mu_\alpha$$

by which we mean

$$e^{-tA} \, v = \int_M \langle v, G_\alpha \rangle \; G_\alpha(t) d\mu_\alpha \; , \qquad t > 0$$

or equivalently

$$(*) \qquad v = \int_M \langle e^{\tau A} \, v, G_\alpha \rangle \; G_\alpha(\tau) d\mu_\alpha \; ,$$

where $\tau > 0$ has to be chosen so small that $e^{\tau A} \, v \in S_{X,A}$. Moreover, the expansion (*) does not depend on the choice of τ. □

Let $F \in T_{X,A}$. Then for all $t > 0$, $F(t) \in S_{X,A}$. So as a consequence of the previous theorem we derive

(V.1.5) <u>Corollary</u>

Let $([G],M,\mu)$ be a Dirac basis in $T_{X,A}$ and let $G \in [G]$. Then for all $F \in T_{X,A}$ we have the expansion

$$F(t) = \int_M \langle F(\tau),G_\alpha\rangle\, G_\alpha(t-\tau)\,d\mu_\alpha \ , \qquad t > 0 \ .$$

Here we have to take $0 < \tau < t$. □

Of course, it looks nice to introduce a more general concept of basis. But do there exist Dirac bases other than orthonormal bases? We show that there is an abundance of Dirac bases in $T_{X,A}$.
Let U be a unitary operator from the Hilbert space X onto a closed subspace of $L_2(M,\mu)$ where M is a measure space which is σ-finite with respect to the measure μ. The operator U maps the orthonormal basis $(v_k)_{k\in\mathbb{N}}$ onto an orthonormal system $(\varphi_k)_{k\in\mathbb{N}}$ of $L_2(M,\mu)$. In each equivalence class φ_k we take a representant $\hat{\varphi}_k$. For each fixed $t > 0$ the series $\sum_{k=1}^{\infty} e^{-\lambda_k t}$ converges. So the expression

$$\sum_{k=1}^{\infty} e^{-\lambda_k t} |\varphi_k|^2$$

represents an element of $L_1(M,\mu)$. Since this series has positive terms there exists a null set N_t such that

$$\sum_{k=1}^{\infty} e^{-\lambda_k t} |\hat{\varphi}_k(\alpha)|^2 < \infty$$

for all $\alpha \in M \setminus N_t$. We put $N = \bigcup_{n\in\mathbb{N}} N_{1/n}$. Then N is a null set with respect to μ. If we define $\tilde{\varphi}_k \in \varphi_k$ by

$$\tilde{\varphi}_k(\alpha) = \hat{\varphi}_k(\alpha) \ , \qquad \alpha \in M\setminus N$$

$$\tilde{\varphi}_k(\alpha) = 0 \qquad\ \ , \qquad \alpha \in N \ ,$$

then we get the following example of a Dirac basis.

(V.1.6) Example

For $\alpha \in M$, define $\tilde{G}_\alpha$ by

$$\tilde{G}_\alpha : t \mapsto \sum e^{-\lambda_k t} \overline{\tilde{\varphi}_k(\alpha)}\, v_k \ .$$

Then $\tilde{G}_\alpha \in T_{X,A}$ for all $\alpha \in M$. Let [G] denote the class of functions $G: M \to T_{X,A}$ which are equivalent to the integrable function $\alpha \mapsto \tilde{G}_\alpha$. Then $([G],M,\mu)$ is a Dirac basis.

Proof. If $\alpha \in N$, then $\tilde{G}_\alpha = 0$. So we consider $\alpha \in M \setminus N$. Then for all $n \in \mathbb{N}$,

$$\sum_{k=1}^{\infty} e^{-\frac{1}{n}\lambda_k} \tilde{\varphi}_k(\alpha) v_k \in X \ .$$

Let $t > 0$. Then there exists $n \in \mathbb{N}$ with $0 < \frac{1}{n} < t$. So we obtain

$$\sum_{k=K}^{L} e^{-2t\lambda_k} |\tilde{\varphi}_k(\alpha)|^2 \leq e^{-2(t-\frac{1}{n})K} \sum_{k=K}^{\infty} e^{-\frac{2}{n}\lambda_k} |\tilde{\varphi}_k(\alpha)|^2 \ .$$

Hence we derive

$$\forall_{t>0}: \sum_{k=1}^{\infty} e^{-t\lambda_k} \overline{\tilde{\varphi}_k(\alpha) v_k} \in X \ , \quad \text{and} \quad \tilde{G}_\alpha \in T_{X,A} \ .$$

Since $\langle v_k, \tilde{G}_\alpha \rangle = \tilde{\varphi}_k(\alpha)$, and since $\tilde{\varphi}_k \in Uv_k$, it follows that the function $\alpha \mapsto \tilde{G}_\alpha$ is μ-integrable, and also that

$$\int_M \langle v_k, \tilde{G}_\alpha \rangle \overline{\langle v_\ell, \tilde{G}_\alpha \rangle}\, d\mu_\alpha = \delta_{k\ell} \ , \qquad k,\ell \in \mathbb{N} \ .$$

So the class [G] to which the function $\alpha \mapsto \tilde{G}_\alpha$ belongs, gives rise to a Dirac basis $([G],M,\mu)$. □

Let Y be the separable Hilbert space. Let B be a nonnegative self-adjoint operator in Y such that $T_{Y,B}$ is a nuclear space. Then there exists an orthonormal basis $(u_m)_{m\in\mathbb{N}}$ in Y and nonnegative numbers μ_m, $m \in \mathbb{N}$, such that

$$Bu_m = \mu_m u_m \quad \text{and} \quad \forall_{t>0}: \sum_{m=1}^{\infty} e^{-\mu_m t} < \infty \ .$$

Let $([G],M,\mu)$ be a Dirac basis in $T_{X,A}$, and let $([H],N,\nu)$ be a Dirac basis in $T_{Y,B}$. Then for each representant $G \in [G]$ and for each representant $H \in [H]$, the function

$$G \otimes H: (\alpha,\beta) \mapsto G_\alpha \otimes H_\beta$$

maps the measure space $M \times N$ into $TT_{X\otimes Y;A\otimes I,I\otimes B} \triangleq T_{X\otimes Y,A\otimes I+I\otimes B}$. Since for all $\alpha \in M$, all $\beta \in N$ and all $W \in SS_{X\otimes Y;A\otimes I,I\otimes B} \triangleq S_{X\otimes Y,A\otimes I+I\otimes B}$ the following relations are valid

$$\langle W,G_\alpha \otimes H_\beta\rangle = \sum_{k,m=1}^{\infty} (W,v_k \otimes u_m) \langle v_k \otimes u_m,G_\alpha \otimes H_\beta\rangle$$

and

$$\langle v_k \otimes u_m,G_\alpha \otimes H_\beta\rangle = \overline{\langle v_k,G_\alpha\rangle}\langle u_m,H_\beta\rangle ,$$

the function $G \otimes H$ is measurable with respect to the product measure $\mu \otimes \nu$. Moreover, for all $k,\ell,m,n \in \mathbb{N}$ we have

$$\int_{M\times N} \langle v_k \otimes u_m,G_\alpha \otimes H_\beta\rangle\overline{\langle v_\ell \otimes u_n,G_\alpha \otimes H_\beta\rangle}\, d\mu_\alpha\, d\nu_\beta =$$

$$= \int_{M\times N} \overline{\langle v_k,G_\alpha\rangle}\langle u_m,H_\beta\rangle\langle v_\ell,G_\alpha\rangle\overline{\langle u_n,H_\beta\rangle}\, d\mu_\alpha\, d\nu_\beta = \delta_{k\ell}\,\delta_{mn} .$$

It follows that the equivalence class $[G \otimes H]$ of all $\mu \otimes \nu$-measurable functions from $M \times N$ into $TT_{X\otimes Y;A\otimes I,I\otimes B}$ which are equivalent to $G \otimes H$, is a Dirac basis in $TT_{X\otimes Y;A\otimes I,I\otimes B}$.
Above we have described the natural way to derive a Dirac basis of two variables from two Dirac bases of one variable.

V.2. A measure theoretical Sobolev lemma

In this section we consider the same operator A as in the previous section. However, we restrict to more special measure spaces M and measures μ. We suppose that M is a topological space metrized by a metric d. On M we consider the σ-algebra of Borel subsets. Further, μ denotes a regular Borel measure on M with the property that bounded subsets of M have finite μ-measure. (We observe that the demand of regularity of the measure μ is redundant if M is a σ-compact locally compact measure space. Cf. [Weir].)

In [Fe], Theorem 2.8.18, Federer introduces additional conditions on the metric space (M,d) such that the following important result is valid:

(V.2.1) Theorem

Let M satisfy Federer's conditions. Let the function $f: M \to \mathbb{C}$ be μ-integrable on bounded Borel sets. Then there exists a null set N_f such that for all $r > 0$ and all $\alpha \in M \setminus N_f$ the closed ball $B(\alpha,r)$ with radius r and centre α has positive measure. Moreover, the limit

$$\tilde{f}(\alpha) = \lim_{r \downarrow 0} \mu(B(\alpha,r))^{-1} \left(\int_{B(\alpha,r)} f \, d\mu \right)$$

exists for all $\alpha \in M \setminus N_f$.

The function $\tilde{f}: \alpha \to \tilde{f}(\alpha)$ is μ-measurable with $f = \tilde{f}$ almost everywhere. Examples of such metric spaces are the following.

(V.2.2) Example

- Finite dimensional vector spaces M with $d(x,y) = v(x-y)$, where v is any norm on M.
- A Riemannian manifold (of class ≥ 2) with its usual metric.
- M, the disjoint union of metric spaces (M_j,d_j), $j \in \mathbb{N}$, and d the metric defined by

  $$d(\alpha_\ell,\beta_j) = 1 \;, \qquad \ell \neq j$$

  $$d(\alpha_\ell,\beta_j) = d_\ell(\alpha_\ell,\beta_\ell) \;.$$

  Here the measure spaces (M_j,d_j) are supposed to satisfy Federer's conditions.

We recall that A has an orthonormal basis of eigenvectors and corresponding eigenvalues $(\lambda_k)_{k\in\mathbb{N}}$ with $\forall_{t>0}: \sum_{k=1}^{\infty} e^{-\lambda_k t} < \infty$. Let the linear operator D in X map $S_{X,\mathsf{A}}$ into $\mathsf{L}_2(M,\mu)$, and suppose that the composite mapping $\mathsf{D} \circ e^{-t\mathsf{A}}$: $X \to \mathsf{L}_2(M,\mu)$ is Hilbert-Schmidt for all $t > 0$ (i.e. $\mathsf{D}: S_{X,\mathsf{A}} \to \mathsf{L}_2(M,\mu)$ is continuous). Now let $t > 0$ be fixed for a moment. Then the series

$$\text{(V.2.2.i)} \qquad \sum_{k=1}^{\infty} e^{-2\lambda_k t} \|\mathsf{D}v_k\|^2_{\mathsf{L}_2}$$

converges, and hence

(V.2.2.ii) $$\sum_{k=1}^{\infty} e^{-2\lambda_k t} |Dv_k|^2 \in \mathbf{L}_1(M,\mu) .$$

Since bounded sets of M have finite μ-measure, every member of $\mathbf{L}_2(M,\mu)$ is integrable on bounded sets. So we can apply Theorem (V.2.1) to each element Dv_k of $\mathbf{L}_2(M,\mu)$. It yields null sets $N_k^{(1)}$, $k \in \mathbb{N}$, such that the limit

(V.2.3.i) $$\tilde{\varphi}_k(\alpha) = \lim_{r \downarrow 0} \mu(B(\alpha,r))^{-1} \int_{B(\alpha,r)} (Dv_k)d\mu$$

exists for all $\alpha \in M \setminus N_k^{(1)}$, and all $k \in \mathbb{N}$. In a natural way, each function $\tilde{\varphi}_k$ extends to an everywhere defined representant of the equivalence class Dv_k.

Since $|Dv_k|^2 \in \mathbf{L}_1(M,\mu)$, $k \in \mathbb{N}$, we get null sets $N_k^{(2)}$ such that

(V.2.3.ii) $$|\tilde{\varphi}_k(\alpha)|^2 = \lim_{r \downarrow 0} \mu(B(\alpha,r))^{-1} \int_{B(\alpha,r)} |Dv_k|^2 \, d\mu , \qquad \alpha \in M \setminus N_k^{(2)} ,$$

and because of relation (V.2.2.ii) we get a null set $N_t^{(3)}$ such that for all $\alpha \in M \setminus N_t^{(3)}$

(V.2.3.iii) $$\sum_{k=1}^{\infty} e^{-2\lambda_k t} |\tilde{\varphi}_k(\alpha)|^2 = \lim_{r \downarrow 0} \mu(B(\alpha,r))^{-1} \int_{B(\alpha,r)} \left(\sum_{k=1}^{\infty} e^{-2\lambda_k t} |Dv_k|^2 \right) d\mu .$$

Let N_t denote the null set

$$\left(\bigcup_{k \in \mathbb{N}} N_k^{(1)} \right) \cup \left(\bigcup_{k \in \mathbb{N}} N_k^{(2)} \right) \cup N_t^{(3)} .$$

For convenience we take $\tilde{\varphi}_k(\alpha) = 0$ if $\alpha \in N_t$.

In the next lemma we put the measure theory for the announced result of this section.

(V.2.4) <u>Lemma</u>

(i) Let $\alpha \in M$. Set $e_{\alpha,t} = \sum_{k=1}^{\infty} e^{-\lambda_k t} \overline{\tilde{\varphi}_k(\alpha)} v_k$. Then $e_{\alpha,t} \in X$.

(ii) Let $\alpha \in M \setminus N_t$. Set $e_{\alpha,t}(r) = \mu(B(\alpha,r))^{-1} \sum_{k=1}^{\infty} e^{-\lambda_k t}\left(\int_{B(\alpha,r)} \overline{Dv_k}\, d\mu\right) v_k$.

Then $e_{\alpha,t}(r) \in X$ for all $r > 0$ and

$$\lim_{r \downarrow 0} \| e_{\alpha,t} - e_{\alpha,t}(r) \|_X = 0 .$$

<u>Proof</u>. The proof of part (i) is a consequence of the definition of the functions $\tilde{\varphi}_k$ and of the relation (V.2.3.iii).
For the proof of part (ii), we first take $\alpha \in M \setminus N_t$ fixed. Then for each $r > 0$ the following inequalities are valid:

$$\left| \int_{B(\alpha,r)} \overline{Dv_k}\, d\mu \right| \leq \left(\int_{B(\alpha,r)} d\mu \right)^{\frac{1}{2}} \left(\int_{B(\alpha,r)} |Dv_k|^2\, d\mu \right)^{\frac{1}{2}} \leq$$

$$\leq \mu(B(\alpha,r))^{\frac{1}{2}} \| Dv_k \|_{L_2} .$$

It yields the estimation

$$\sum_{k=1}^{\infty} e^{-2\lambda_k t} \left| \int_{B(\alpha,r)} \overline{Dv_k}\, d\mu \right|^2 \leq \mu(B(\alpha,r)) \left(\sum_{k=1}^{\infty} e^{-\lambda_k t} \| Dv_k \|_{L_2}^2 \right) .$$

Hence we find $e_{\alpha,t}(r) \in X$ by (V.2.2.i).
Let $\varepsilon > 0$. Take $k_0 \in \mathbb{N}$ so large that

$$(*) \qquad \sum_{k=k_0+1}^{\infty} e^{-2\lambda_k t} |\tilde{\varphi}_k(\alpha)|^2 < \varepsilon^2$$

and $r_0 > 0$ so small that for all r with $0 < r < r_0$, both

$$(**) \qquad \left| \tilde{\varphi}_k(\alpha) - \mu(B(\alpha,r))^{-1} \int_{B(\alpha,r)} (Dv_k)\, d\mu \right| < \varepsilon , \qquad 1 \leq k \leq k_0$$

and (cf. (V.2.3.ii-iii)

$$(***) \qquad \sum_{k=k_0+1}^{\infty} e^{-2\lambda_k t} \left[\mu(B(\alpha,r))^{-1} \int_{B(\alpha,r)} |Dv_k|^2\, d\mu \right] < 2\varepsilon^2 .$$

The inequalities (*)-(***) lead to the following estimation

$$\|e_{\alpha,t} - e_{\alpha,t}(r)\|^2 =$$

$$= \Big\| \sum_{k=1}^{\infty} e^{-\lambda_k t} \Big[\overline{\widetilde{\varphi}_k(\alpha) - \mu(B(\alpha,r))^{-1} \int_{B(\alpha,r)} (\overline{Dv_k})\,d\mu}\Big] v_k \Big\|^2$$

$$= \Big(\sum_{k=1}^{k_0} + \sum_{k=k_0+1}^{\infty} \Big) e^{-2\lambda_k t} \Big|\widetilde{\varphi}_k(\alpha) - \mu(B(\alpha,r))^{-1} \int_{B(\alpha,r)} Dv_k \; d\mu\Big|^2 .$$

Now by (**)

$$\sum_{k=1}^{k_0} e^{-2\lambda_k t} \Big|\widetilde{\varphi}_k(\alpha) - \mu(B(\alpha,r))^{-1} \int_{B(\alpha,r)} (Dv_k)\,d\mu\Big|^2 < \varepsilon^2 \sum_{k=1}^{\infty} e^{-2\lambda_k t} ,$$

and by (*) and (***)

$$\sum_{k=k_0+1}^{\infty} e^{-2\lambda_k t} \Big|\widetilde{\varphi}_k(\alpha) - \mu(B(\alpha,r))^{-1} \int_{B(\alpha,r)} (Dv_k)\,d\mu\Big|^2 \leqq$$

$$\leqq 2 \sum_{k=k_0+1}^{\infty} e^{-2\lambda_k t} \Big(|\widetilde{\varphi}_k(\alpha)|^2 + \Big|\mu(B(\alpha,r))^{-1} \int_{B(\alpha,r)} (Dv_k)\,d\mu\Big|^2 \Big) \leqq$$

$$\leqq 2\varepsilon^2 + 4\varepsilon^2 = 6\varepsilon^2 .$$

Thus we have proved that

$$\|e_{\alpha,t} - e_{\alpha,t}(r)\| < \varepsilon \Big(6 + \sum_{k=1}^{\infty} e^{-2\lambda_k t}\Big)^{\frac{1}{2}} , \quad 0 < r < r_0 . \qquad \square$$

(V.2.5) <u>Theorem</u>

(i) Let $\alpha \in M$. Then the mapping $\widetilde{\Delta}_\alpha : t \mapsto e_{\alpha,t}$ belongs to $T_{X,A}$.

(ii) Put $N = \bigcup_{n \in \mathbb{N}} N_{1/n}$, where each null set $N_{1/n}$ is defined by (V.2.3.i-iii). Let $\alpha \in M \setminus N$. Set $\Delta_\alpha[r] : t \mapsto e_{\alpha,t}(r)$. Then $\Delta_\alpha[r] \in T_{X,A}$ and $\Delta_\alpha[r] \to \widetilde{\Delta}_\alpha$ in $T_{X,A}$ as $r \downarrow 0$, i.e.

$$\forall_{t>0} : \lim_{r \downarrow 0} \|\widetilde{\Delta}_\alpha(t) - \Delta_\alpha[r;t]\| = 0 .$$

Proof

(i) From part (i) of the previous lemma it follows that $e_{\alpha,t} \in X$. It can easily be checked that $e_{\alpha,t+\tau} = e^{-\tau A} e_{\alpha,t}$. So $\tilde{\Delta}_\alpha$ belongs to $T_{X,A}$.

(ii) From part (ii) of the previous lemma and from the choice of the null set N it follows that for all $n \in \mathbb{N}$ and for all $r > 0$, $e_{\alpha,1/n}(r) \in X$. Further, for all $n \in \mathbb{N}$

$$\lim_{r \downarrow 0} \| e_{\alpha,1/n} - e_{\alpha,1/n}(r) \| = 0 .$$

We get

$$\Delta_\alpha[r;t] = e^{-(t-\frac{1}{n})A} e_{\alpha,1/n}(r) \in X \qquad \text{for } t > \frac{1}{n} > 0 ,$$

and, also,

$$\| \Delta_\alpha(t) - \Delta_\alpha[r;t] \| \leq \| e^{-(t-\frac{1}{n})A} \| \| e_{\alpha,1/n} - e_{\alpha,1/n}(r) \| . \qquad \square$$

A consequence of the preceding theorem is a so-called Sobolev lemma for the analyticity space $S_{X,A}$. In our paper [EG 4] we have presented a measure theoretical Sobolev lemma which is more in line with the classical Sobolev lemma as presented e.g. in [Yo].

(V.2.6) Corollary

For each $w \in S_{X,A}$ there can be chosen a representant $\mathcal{D}\tilde{w}$ in $\mathcal{D}w$ such that the following statements are valid:

(i) $\mathcal{D}\tilde{w} = \sum_{k=1}^{\infty} (w,v_k)\tilde{\varphi}_k$ where the series converges pointwise on M.

(ii) For each $\alpha \in M$, the linear functional $w \mapsto \mathcal{D}\tilde{w}(\alpha)$ is continuous on $S_{X,A}$; its representant in $T_{X,A}$ is $\tilde{\Delta}_\alpha$.

(iii) Suppose for each $t > 0$ the series $\sum_{k=1}^{\infty} e^{-2\lambda_k t} |\tilde{\varphi}_k|^2$ is essentially bounded on M. Then the convergence in (i) is uniform outside a set of measure zero N_0. Moreover

$$\forall_{t>0}\ \exists_{C>0}\ \forall_{w\in e^{-tA}(X)}\ \forall_{\alpha\in M\setminus N_0} : |\tilde{D}w(\alpha)| \leq C\|e^{tA}w\| .$$

(iv) For all $\alpha \in M \setminus N$, with N as in Theorem (V.2.5)

$$(\tilde{D}w)(\alpha) = \lim_{r\downarrow 0} \mu(B(\alpha,r))^{-1} \int_{B(\alpha,r)} (Dw)\,d\mu , \qquad w \in S_{X,A} .$$

<u>Proof</u>. Let $w \in S_{X,A}$, and put $\tilde{D}w = \sum_{k=1}^{\infty} (w,v_k)\tilde{\varphi}_k$. Then $\tilde{D}w \in Dw$.

(i) Since

$$<w,\tilde{\Delta}_\alpha> = \sum_{k=1}^{\infty} (w,v_k)\tilde{\varphi}_k(\alpha) = \tilde{D}w(\alpha) ,$$

and since $\tilde{\Delta}_\alpha \in T_{X,A}$, the series converges pointwise on M.

(ii) Trivial, because $\tilde{D}w(\alpha) = <w,\tilde{\Delta}_\alpha>$.

(iii) For each $n \in \mathbb{N}$ there exists a null set $N_{0,1/n}$ such that

$$C_n := \sup_{\alpha\in M\setminus N_{0,1/n}} \left(\sum_{k=1}^{\infty} e^{-2\lambda_k t}|\tilde{\varphi}_k(\alpha)|^2\right)^{\frac{1}{2}} < \infty .$$

Put $N_0 = \bigcup_{n\in\mathbb{N}} N_{0,1/n}$. Then N_0 is a null set.
Let $t > 0$. Take $n \in \mathbb{N}$ so large that $0 < \frac{1}{n} < t$. Then for all $w \in e^{-tA}(X)$ and all $\alpha \in M \setminus N_0$ we have

$$\left|\sum_{k=K}^{L} (w,v_k)\tilde{\varphi}_k(\alpha)\right| \leq C_n\left(\sum_{k=K}^{L} e^{\frac{2}{n}\lambda_k}|(w,v_k)|^2\right)^{\frac{1}{2}}$$

with $L > K$.

(iv) Let $\alpha \in M \setminus N$, and let $w \in e^{-\frac{1}{n}A}(X) \subset S_{X,A}$ where $n \in \mathbb{N}$. Then consider the following manipulation:

$$\tilde{D}w(\alpha) = <w,\tilde{\Delta}_\alpha> = \lim_{r\downarrow 0} (w,\Delta_\alpha[r]) =$$

$$= \lim_{r\downarrow 0} \mu(B(\alpha,r))^{-1} \sum_{k=1}^{\infty} (w,v_k)\left(\int_{B(\alpha,r)} Dv_k\, d\mu\right).$$

In the latter expression summation and integration can be interchanged because

$$\left(\sum_{k=1}^{\infty}\int_{B(\alpha,r)}\right)|(w,v_k)Dv_k|d\mu \leq$$

$$\leq\left(\sum_{k=1}^{\infty}e^{\frac{2}{n}\lambda_k}|(w,v_k)|^2\,\mu(B(\alpha,r))\right)^{\frac{1}{2}}\left(\sum_{k=1}^{\infty}e^{-\frac{2}{n}\lambda_k}\int_{B(\alpha,r)}|Dv_k|^2d\mu\right)^{\frac{1}{2}}\leq$$

$$\leq\mu(B(\alpha,r))^{\frac{1}{2}}\|e^{\frac{1}{n}A}w\|\left(\sum_{k=1}^{\infty}e^{-\frac{2}{n}\lambda_k}\|Dv_k\|^2\right)^{\frac{1}{2}}.$$

Applying Fubini's theorem we find

$$(\widetilde{D}w)(\alpha)=\lim_{r\downarrow 0}\mu(B(\alpha,r))^{-1}\int_{B(\alpha,r)}\left(\sum_{k=1}^{\infty}(w,v_k)Dv_k\right)d\mu=$$

$$=\lim_{r\downarrow 0}\mu(B(\alpha,r))^{-1}\int_{B(\alpha,r)}(Dw)d\mu\,.$$

□

Let U be a unitary operator from X onto a closed subspace of $\mathbf{L}_2(M,\mu)$. Then clearly U fulfills the conditions that we have imposed on D in the beginning of this section, viz. Ue^{-tA} is Hilbert-Schmidt for all $t > 0$. The set $\{Uv_k \mid k \in \mathbb{N}\}$ is an orthonormal system in $\mathbf{L}_2(M,\mu)$. From the relations (V.2.3.i-iii) we obtain canonical representants $\widetilde{\varphi}_k \in Uv_k$, and a null set N with the following property:

$$\widetilde{\Delta}_\alpha : t \mapsto \sum_{k=1}^{\infty}e^{-\lambda_k t}\,\overline{\widetilde{\varphi}_k(\alpha)}v_k \in T_{X,A}$$

for all $\alpha \in M$, and

$$\widetilde{\Delta}_\alpha = T_{X,A}-\lim_{r\downarrow 0}\mu(B(\alpha,r))^{-1}\sum_{k=1}^{\infty}\left(\int_{B(\alpha,r)}\overline{v_k}\,d\mu\right)v_k$$

for all $\alpha \in M \setminus N$.

In Section V.1 we have showed that unitary operators like U give rise to a Dirac basis in $T_{X,A}$. Any choice of representants $\widetilde{\varphi}_k \in Uv_k$ with the property that

$$\sum_{k=1}^{\infty} e^{-2\lambda_k t} \overline{|\hat{\varphi}_k(\alpha)|^2} < \infty$$

for all $\alpha \in M$ leads to a representant $\hat{\Delta} \in [\Delta]$,

$$\hat{\Delta}_\alpha : t \mapsto \sum_{k=1}^{\infty} e^{-\lambda_k t} \overline{\hat{\varphi}_k(\alpha) v_k} .$$

It is clear that the function $\alpha \mapsto \tilde{\Delta}_\alpha$ as constructed in this section is also a representant of the class $[\Delta]$. Because of its particular properties we consider $\tilde{\Delta}$ the canonical representant of the class $[\Delta]$.

We have proved the following important assertions which are stated in the next theorem.

(V.2.7) Theorem

Let M be a measure space which is metrizable as a topological space, and which satisfies Federer's conditions. Let μ be a σ-finite nonnegative regular Borel measure on M with the property that bounded sets have finite μ-measure. Let U be a partial isometry from X into $\mathbf{L}_2(M,\mu)$ such that $\mathsf{U}^*\mathsf{U} = \mathrm{I}$. Then U gives rise to a Dirac basis $([\Delta],M,\mu)$, in which there exists a canonical representant, denoted by $(\tilde{\Delta}_\alpha)_{\alpha\in M}$. This representant has the following properties:

$$\forall_{w\in S_{X,A}} : \tilde{\mathsf{U}}w : \alpha \mapsto \langle w, \tilde{\Delta}_\alpha \rangle \in \mathsf{U}w , \qquad \alpha \in M$$

and

$$\tilde{\mathsf{U}}w(\alpha) = \lim_{r\downarrow 0} \mu(B(\alpha,r))^{-1} \int_{B(\alpha,r)} (\mathsf{U}w)\,d\mu$$

for all $\alpha \in M \setminus N$. Here N is a null set which depends on $(\tilde{\Delta}_\alpha)_{\alpha\in M}$ only. □

Remark. In case Federer's conditions are appropriate we take instead of the Dirac basis $([\Delta],M,\mu)$ its canonical representant $(\tilde{\Delta}_\alpha)_{\alpha\in M}$, which we call also a Dirac basis.

V.3. The standard structure of a Dirac basis consisting of generalized eigenvectors for self-adjoint operators

Now we apply the results of the previous section in the following concrete case.

- The measure space M is the disjoint countable union of copies $\mathbb{R}^n_p$ of $\mathbb{R}^n$, i.e.

$$M = \bigcup_{p=1}^{\infty} \mathbb{R}^n_p \ .$$

- The nonnegative regular Borel measure ν on M is given by

$$\nu = \bigoplus_{p=1}^{\infty} \nu_p$$

where each ν_p is a finite nonnegative Borel measure on $\mathbb{R}^n$.

We note that

$$\mathbf{L}_2(M,\nu) = \bigoplus_{p=1}^{\infty} \mathbf{L}_2(\mathbb{R}^n,\nu_p) \ .$$

On M we introduce the metric d as follows (cf. Section V.2):

$$d((\alpha,p_1),(\beta,p_2)) = 1 \quad \text{if } p_1 \neq p_2 \ ,$$

$$d((\alpha,p),(\beta,p)) = \|\alpha-\beta\|_{\mathbb{R}^n} \ .$$

As observed in the beginning of Section (V.2) the metric space (M,d) with regular Borel measure ν satisfies the conditions of Federer. So we can apply the results of the preceding section.

Let U denote a unitary operator from X onto $\mathbf{L}_2(M,\nu)$. Then for all $x \in X$ we have

$$\mathsf{U}x = (\mathsf{U}_1x_1,\mathsf{U}_2x_2,\ldots)$$

where each U_p, $p \in \mathbb{N}$, is a partial isometry from X onto $\mathbf{L}_2(\mathbb{R}^n,\nu_p)$. Following the construction of Section V.2 there exists a ν-null set $N = \bigcup_{p=1}^{\infty} N_p$ (disjoint union), i.e. each set N_p is a ν_p-null set, such that the following result is valid.

(V.3.1) Theorem

There exist functions $\widetilde{\varphi}_{k,p} \in (Uv_k)_p$, $k \in \mathbb{N}$, $p \in \mathbb{N}$, such that for all $\alpha \in \mathbb{R}^n$ and all $p \in \mathbb{N}$, the mapping

$$\widetilde{E}_{(\alpha,p)}: t \mapsto \sum_{k=1}^{\infty} e^{-\lambda_k t} \overline{\widetilde{\varphi}_{k,p}(\alpha)}\, v_k , \qquad t > 0 ,$$

belongs to $T_{X,A}$. Further, for all $r > 0$, $p \in \mathbb{N}$ and $\alpha \in \mathbb{R}^n \setminus N_p$, we set

$$E_{(\alpha,p)}[r]: t \mapsto \nu_p(B(\alpha,r))^{-1} \sum_{k=1}^{\infty} e^{-\lambda_k t} \Big(\int_{B(\alpha,r)} \overline{U_p v_k}\, d\nu_p\Big) v_k .$$

Then $E_{(\alpha,p)}[r]$ belongs to $T_{X,A}$, and

$$\forall_{t>0} \forall_{p\in\mathbb{N}} \forall_{\alpha\in\mathbb{R}^n\setminus N_p} : \lim_{r\downarrow 0} \|\widetilde{E}_{(\alpha,p)}(t) - E_{(\alpha,p)}[r;t]\| = 0 . \qquad \square$$

Remark. The system $(\widetilde{E}_{(\alpha,p)})_{(\alpha,p)\in M}$ is a Dirac basis in $T_{X,A}$ in the sense of Theorem (V.2.7). So in particular we have for all $w \in S_{X,A}$,

$$\widetilde{U}w: (\alpha,p) \mapsto \langle w, \widetilde{E}_{(\alpha,p)}\rangle \in Uw , \qquad (\alpha,p) \in M ,$$

and

$$(\widetilde{U}w)(\alpha,p) = \lim_{r\downarrow 0} \nu(B(\alpha,p),r))^{-1} \int_{B((\alpha,p),r)} (Uw)\, d\nu$$

for all $(\alpha,p) \in M \setminus N$.

We note that

$$\lim_{r\downarrow 0} \nu(B((\alpha,p),r))^{-1} \int_{B((\alpha,p),r)} (Uw)\, d\nu = \lim_{r\downarrow 0} \nu_p(B(\alpha,r)) \int_{B(\alpha,r)} (U_p w)\, d\nu_p .$$

Let Q_ℓ, $\ell = 1,\dots,n$, denote the multiplication, formally defined by

$$Q_\ell(f_1,f_2,\dots) = (\eta_\ell f_1, \eta_\ell f_2, \dots) .$$

Here η_ℓ denotes the coordinate function $\eta_\ell(\alpha) = \alpha_\ell$, $\alpha \in \mathbb{R}^n$. It is clear that Q_ℓ is a self-adjoint operator in $L_2(M,\nu)$. So the operator $P_\ell = U^* Q_\ell U$ is self-adjoint in X. The following lemma says that the vectors $\widetilde{E}_{(\alpha,p)}$ are candidate generalized eigenvectors of the operator P_ℓ. In the following theorem we use the usual notation $\mathcal{D}(P_\ell)$ for the domain of the self-adjoint operator P_ℓ, $\ell = 1,\dots,n$.

(V.3.2) <u>Theorem</u>

Let $\ell = 1,\dots,n$. Then for all $p \in \mathbb{N}$ and all $\alpha \in \mathbb{R}^n \setminus N_p$ the trajectories $E_{(\alpha,p)}[r]$, $r > 0$, belong to $\mathrm{emb}(\mathcal{D}(P_\ell))$. Further, for all $\alpha \in \mathbb{R}^n \setminus N_p$

$$\lim_{r\downarrow 0} \| e^{-tA} P_\ell E_{(\alpha,p)}[r] - \alpha_\ell \widetilde{E}_{(\alpha,p)}(t) \| = 0 , \qquad t > 0 .$$

(Observe that $E_{(\alpha,p)}[r] \in \mathrm{emb}(X)$.)

<u>Proof</u>. Let $p \in \mathbb{N}$ and let $\alpha \in \mathbb{R}^n \setminus N_p$. Then from the choice of N_p it follows that $\nu_p(B(\alpha,r)) > 0$ for all $r > 0$. Moreover, for each $r > 0$ the series

$$\sum_{k=1}^{\infty} \left(\int_{B(\alpha,r)} \overline{U_p v_k}\, d\nu_p \right) v_k = \sum_{k=1}^{\infty} \left(\int_{\mathbb{R}_n} \chi_{B(\alpha,r)} (\overline{U_p v_k})\, d\nu_p \right) v_k$$

represents the vector $U_p^* \chi_{B(\alpha,r)}$ in X. Here $\chi_{B(\alpha,r)}$ denotes the characteristic function of the closed ball $B(\alpha,r)$. It follows that

$$\nu_p(B(\alpha,r))^{-1}\, U_p^* \chi_{B(\alpha,r)} = E_{(\alpha,p)}[r] \in \mathcal{D}(P_\ell)$$

because $\chi_{B(\alpha,r)} \in \mathcal{D}(Q_\ell)$.
Next, we show that $P_\ell E_{(\alpha,p)}[r]$ tends to $\alpha_\ell \widetilde{E}_{(\alpha,p)}$ in $T_{X,A}$. Using the result of Theorem (V.3.1) we only have to prove

$$\forall_{t>0} : \lim_{r\downarrow 0} \| e^{-tA} P_\ell E_{(\alpha,p)}[r] - \alpha_\ell E_{(\alpha,p)}[r;t] \| = 0 .$$

To this end, let $t > 0$. Observe that for all $0 < r < 1$

$$e^{-tA} P_\ell E_{(\alpha,p)}[r] - \alpha_\ell E_{(\alpha,p)}[r;t] =$$

$$= \sum_{k=1}^{\infty} e^{-\lambda_k t} \left(\nu_p(B(\alpha,r)) \int_{B(\alpha,r)} (\beta_\ell - \alpha_\ell) \widetilde{\varphi}_{k,p}(\beta)\, d\nu_p(\beta) \right) v_k .$$

We estimate the norm of the above expression,

$$\| e^{-tA} P_\ell E_{\alpha,p}[r] - \alpha_\ell E_{(\alpha,p)}[r;t] \|^2 \leq$$

$$\leq \left(\nu_p(B(\alpha,r))^{-1} \int_{B(\alpha,r)} |\beta_\ell - \alpha_\ell|^2\, d\nu_p(\beta) \right) \cdot$$

$$\cdot\left(\nu_p(B(\alpha,r))^{-1}\int_{B(\alpha,r)}\sum_{k=1}^{\infty} e^{-2\lambda_k t}|\tilde{\varphi}_{k,p}|^2\, d\nu_p\right).$$

The first factor in the latter expression tends to zero as $r \downarrow 0$.
Because of the choice of the null set N_p, as described in (V.2.3.i-iii), the second factor is bounded by

$$1+\sum_{k=1}^{\infty} e^{-2\lambda_k t}|\tilde{\varphi}_{k,p}(\alpha)|^2$$

for sufficiently small $r > 0$. □

Let P_ℓ be extendible to a closed linear mapping in $T_{X,A}$. Then from the previous theorem it follows that

$$\tilde{E}_{(\alpha,p)} \in \mathrm{Dom}(\overline{P}_\ell)$$

and

$$\overline{P}_\ell\, \tilde{E}_{(\alpha,p)} = \alpha_\ell\, \tilde{E}_{(\alpha,p)}$$

for all $(\alpha,p) \in M \setminus N$.
In Section IV.5 we presented a sufficient condition for the closability of P_ℓ as a linear mapping in $T_{X,A}$, viz.

$$\forall_{t>0}\ \forall_{\psi\in B_+(\mathbb{R})} : \text{the operator } e^{-tA} P_\ell\, \psi(A) \text{ is closable in } X.$$

E.g., operators P_ℓ which have a column finite (and hence row finite) matrix with respect to the orthonormal basis $(v_k)_{k\in\mathbb{N}}$, satisfy this condition.

V.4. Commutative multiplicity theory

In this section we discuss the commutative multiplicity theory for a finite number of commuting self-adjoint operators. In the case of bounded operators the theory is well described in [Br] and [Ne]. The unbounded case is a straightforward generalization. Throughout this section $n \in \mathbb{N}$ will be taken fixed.

Let μ denote a finite nonnegative Borel measure on $\mathbb{R}^n$. Then the support of μ, notation $\mathrm{supp}(\mu)$, is the complement of the largest open set in $\mathbb{R}^n$ of μ-measure zero.

(V.4.1) Lemma

$$\operatorname{supp}(\mu) = \{\alpha \in \mathbb{R}^n \mid \forall_{r>0}\ \mu(B(\alpha,r)) > 0\} .$$

Proof. The proof consists of two steps. Put

$$\operatorname{supp}_1(\mu) = \{\alpha \in \mathbb{R}^n \mid \forall_{r>0}\ \mu(B(\alpha,r)) > 0\} .$$

We shall prove that $\operatorname{supp}(\mu) \subset \operatorname{supp}_1(\mu)$. For each $\alpha \in \operatorname{supp}_1(\mu)^*$ there exists $\varepsilon_\alpha > 0$ such that $\mu(B(\alpha,\varepsilon_\alpha)) = 0$. Then we have

$$\operatorname{supp}_1(\mu)^* \subset \bigcup_{\alpha\in\operatorname{supp}_1(\mu)^*} B(\alpha,\varepsilon_\alpha) .$$

Let $k \in \mathbb{N}$. The set $\operatorname{supp}_1(\mu)^* \cap B(0,k)$ is bounded in $\mathbb{R}^n$. With Besicovitch coverings lemma ([WZ], p. 185), it follows that there is a countable set $\{\alpha_1,\alpha_2,\ldots\} \subset \operatorname{supp}_1(\mu)^*$ such that

$$\operatorname{supp}_1(\mu)^* \cap B(0,k) \subset \bigcup_{j=1}^{\infty} B(\alpha_j,\varepsilon_{\alpha_j}) .$$

It follows that

$$\mu(\operatorname{supp}_1(\mu)^* \cap B(0,k) = 0 .$$

Since $k \in \mathbb{N}$ is arbitrary, it follows that $\operatorname{supp}_1(\mu)^*$ is a set of measure zero. It can be easily seen that $\operatorname{supp}_1(\mu)^*$ is open. Then from the definition of $\operatorname{supp}(\mu)$ we get $\operatorname{supp}(\mu) \subset \operatorname{supp}_1(\mu)$.
Conversely, let $\alpha \notin \operatorname{supp}(\mu)$. Since $\operatorname{supp}(\mu)^*$ is open, there exists $\varepsilon > 0$ such that $\mu(B(\alpha,\varepsilon)) = 0$. Hence $\alpha \in \operatorname{supp}_1(\mu)^*$. It yields $\operatorname{supp}(\mu)^* \subset \operatorname{supp}_1(\mu)^*$, and hence, $\operatorname{supp}_1(\mu) \subset \operatorname{supp}(\mu)$. □

Remark. As in Section V.2, $B(\alpha,r)$ denotes the closed ball

$$\{\beta \in \mathbb{R}^n \mid \|\alpha-\beta\|_{\mathbb{R}^n} \leq r\} .$$

Two Borel measures $\mu^{(1)}$ and $\mu^{(2)}$ for $\mathbb{R}^n$ are called equivalent, if for all Borel sets N of $\mathbb{R}^n$ we have $\mu^{(1)}(N) = 0$ if and only if $\mu^{(2)}(N) = 0$. We denote the equivalence by $\mu^{(1)} \sim \mu^{(2)}$. The equivalence classes $\langle\mu^{(1)}\rangle$ and $\langle\mu^{(2)}\rangle$ are called disjoint, if for each $\mu^{(1)} \in \langle\mu^{(1)}\rangle$ and $\mu^{(2)} \in \langle\mu^{(2)}\rangle$ we have $\mu^{(1)} \perp \mu^{(2)}$, i.e.

$$\mu^{(1)}(\mathrm{supp}(\mu^{(1)}) \cap \mathrm{supp}(\mu^{(2)})) = \mu^{(2)}(\mathrm{supp}(\mu^{(1)}) \cap \mathrm{supp}(\mu^{(2)})) = 0 \ .$$

Observe that for all $\hat{\mu},\hat{\hat{\mu}} \in \langle\mu\rangle$ we have $\mathrm{supp}(\hat{\mu}) = \mathrm{supp}(\hat{\hat{\mu}})$. So it makes sense to introduce $\mathrm{supp}(\langle\mu\rangle) := \mathrm{supp}(\mu)$ for any $\mu \in \langle\mu\rangle$.
Let $(P_1,P_2,\dots,P_n)$ denote an n-set of strongly commuting self-adjoint operators in the separable Hilbert space X. Then their spectral projections mutually commute. The notion of uniform multiplicity for such an n-set is defined as follows.

(V.4.2) Definition

The n-set $(P_1,\dots,P_n)$ is of uniform multiplicity m if there exist m finite nonnegative Borel measures $\mu^{(1)},\mu^{(2)},\dots,\mu^{(m)}$ on $\mathbb{R}^n$, which are mutually equivalent, and if there exists a unitary operator

$$U\colon X \to L_2(\mathbb{R}^n,\mu^{(1)}) \oplus \dots \oplus L_2(\mathbb{R}^n,\mu^{(m)})$$

such that each self-adjoint operator $U P_\ell U^*$, $\ell = 1,\dots,n$, equals m-times multiplication by the coordinate function $\eta_\ell\colon \alpha \mapsto \alpha_\ell$ in this direct sum.

Remark. We may take $m = \infty$ in the previous definition.

(V.4.3) Lemma

Let $(P_1,\dots,P_n)$ be of uniform multiplicity m. Then there exists a finite nonnegative Borel measure μ_m on $\mathbb{R}^n$ and a unitary operator

$$\tilde{U}\colon X \to L_2(\mathbb{R}^n,\mu_m) \oplus \dots \oplus L_2(\mathbb{R}^n,\mu_m) \qquad \text{(m times)}$$

such that $\tilde{U} P_\ell \tilde{U}^*$, $\ell = 1,\dots,n$, equals m-times multiplication by the function $\eta_\ell\colon \alpha \to \alpha_\ell$ in this direct sum.

Proof. Take Borel measures $\mu^{(1)},\dots,\mu^{(m)}$ as in Definition (V.4.2). Since $\mu^{(r)} \sim \mu^{(s)}$, and since both measures are finite, there exists a $\mu^{(s)}$-integrable function h_{rs} with $h_{rs} > 0$ on $\mathrm{supp}(\mu^{(r)})$ $(= \mathrm{supp}(\mu^{(s)}))$, such that for all $f,g \in L_2(\mathbb{R}^n,\mu^{(r)})$

$$\int_{\mathbb{R}^n} f\,\overline{g}\,d\mu^{(r)} = \int_{\mathbb{R}^n} f\,\overline{g}\,h_{rs}\,d\mu^{(s)} \ .$$

Define the operator V_{rs} from $\mathbf{L}_2(\mathbb{R}^n,\mu^{(r)})$ onto $\mathbf{L}_2(\mathbb{R}^n,\mu^{(s)})$ by

$$V_{rs}\, f = (h_{rs})^{\frac{1}{2}}\, f \ .$$

Then V_{rs} is unitary, and $V^\star_{rs}\,\varphi = (h_{rs})^{-\frac{1}{2}}\,\varphi$, $\varphi \in \mathbf{L}_2(\mathbb{R}^n,\mu^{(s)})$.
Put $Q_\ell f = \eta_\ell f$ for $f \in \mathcal{D}(Q_\ell) \subset \mathbf{L}_2(M,\mu^{(r)})$. Then $(V_{rs}\, Q_\ell\, V^\star_{rs})\varphi = \eta_\ell \varphi$ for suitable $\varphi \in \mathbf{L}_2(\mathbb{R}^n,\mu^{(s)})$. So the self-adjoint operator of multiplication by η_ℓ in $\mathbf{L}_2(\mathbb{R}^n,\mu^{(r)})$ is unitarily equivalent to the self-adjoint operator of multiplication by the same function η_ℓ in $\mathbf{L}_2(\mathbb{R}^n,\mu^{(s)})$.
Since the measures $\mu^{(1)},\ldots,\mu^{(m)}$ belong to the same equivalence class $<\mu_m>$, the above considerations show that we can take one $\mu_m \in <\mu_m>$ in Definition (V.4.2). This is precisely the assertion of the present lemma. □

In a finite dimensional Hilbert space E each commuting n-set of self-adjoint operators $(B_1,\ldots,B_n)$ has an orthonormal basis of joint eigenvectors. An element $\lambda \in \mathbb{R}^n$ is called an eigentuple of the n-set $(B_1,\ldots,B_n)$ if there exists a vector $e_\lambda \in E$ such that

$$B_\ell\, e_\lambda = \lambda_\ell\, e_\lambda \ , \qquad \ell = 1,\ldots,n \ .$$

The set of all eigentuples may be called the joint spectrum of $(B_1,\ldots,B_n)$ and denoted by $\sigma(B_1,\ldots,B_n)$.
In order to list all eigentuples of $(B_1,\ldots,B_n)$ in a well-ordered manner one can list successively all eigentuples of multiplicity one, two, etc. A generalization of this procedure in the case of n commuting self-adjoint operators in an infinite dimensional Hilbert space X is presented in the following theorem.

(V.4.4) Theorem (Commutative multiplicity theorem)

Let $(P_1,\ldots,P_n)$ be an n-set of strongly commuting self-adjoint operators in the Hilbert space X. Then X can be split into a (countable) direct sum

$$X = X_\infty \oplus X_1 \oplus X_2 \oplus \ldots$$

(some of them may be trivial) such that the following assertions are valid:

- The n-set $(P_1,\ldots,P_n)$ restricted to X_m, $m = \infty,1,2,\ldots$, acts invariantly in X_m and has uniform multiplicity m.
- The equivalence classes $<\mu_m>$ of finite nonnegative Borel measures corresponding to each X_m are mutually disjoint.

Proof. Cf. [Br] and [Ne]. □

With the aid of Lemma (V.4.3) we can (and will) always consider the following standard splitting of X with respect to $(P_1,\dots,P_n)$: As in the previous theorem, $X = X_\infty \oplus X_1 \oplus X_2 \oplus \dots$. In each equivalence class $<\mu_m>$ we choose one fixed measure μ_m. By U_m we denote the unitary operator from X_m onto $L_2(\mathbb{R}^n,\mu_m) \oplus \dots \oplus L_2(\mathbb{R}^n,\mu_m)$ (m-times), such that $U_m(P_\ell \upharpoonright X_m)U_m^*$ equals m-times multiplication by the function η_ℓ.

Finally, let $(E_\lambda^{(\ell)})_{\lambda\in\mathbb{R}}$ denote the spectral resolution of the identity corresponding to P_ℓ, $\ell = 1,\dots,n$. Then we define the joint spectrum as follows:

$$\sigma(P_1,\dots,P_n) = \{\lambda \in \mathbb{R}^n \mid \forall_{\varepsilon>0}: \prod_{\ell=1}^{n} E_\lambda^{(\ell)}([\lambda_\ell-\varepsilon,\lambda_\ell+\varepsilon]) \neq 0\} .$$

We mention the following simple assertion:

$$\sigma(P_1,\dots,P_n) = \overline{\bigcup_{m=1}^{\infty} \operatorname{supp}(\mu_m) \cup \operatorname{supp}(\mu_\infty)} .$$

V.5. Dirac bases of generalized eigenvectors for self-adjoint operators

Let $(P_1,\dots,P_n)$ denote an n-set of commuting self-adjoint operators in X. We consider the standard splitting

$$X = X_\infty \oplus X_1 \oplus X_2 \oplus \dots$$

with disjoint finite nonnegative Borel measures $\mu_\infty,\mu_1,\mu_2,\dots$ on $\mathbb{R}^n$, such that each X_m is unitarily equivalent to the direct sum

$$\bigoplus_{j=1}^{m} L_2(\mathbb{R}^n,\mu_m) .$$

Furthermore, the n-set $(P_1,\dots,P_n)$ acts invariantly in each Hilbert summand X_m, and it is unitarily equivalent to the n-set $(Q_1,\dots,Q_n)$ in

$$\bigoplus_{m=1}^{\infty}\left(\bigoplus_{j=1}^{m} L_2(\mathbb{R}^n,\mu_m)\right) \oplus \left(\bigoplus_{j=1}^{\infty} L_2(\mathbb{R}^n,\mu_\infty)\right) .$$

Here Q_ℓ restricted to $\oplus_{j=1}^m L_2(\mathbb{R}^n,\mu_m)$ is the operator of m-times multiplication by the function η_ℓ.
Following Section V.3 there exist null sets $N_{m,j}$, $j = 1,\dots,m$, with respect to μ_m, $m = \infty,1,2,\dots$, such that the limit

$$\tilde{\varphi}_{k,j}^{(m)}(\alpha) = \lim_{r\downarrow 0} \mu_m(B(\alpha,r))^{-1} \int_{B(\alpha,r)} (Uv_k)_j^{(m)} \, d\mu_m$$

exists for all $\alpha \in \operatorname{supp}(\mu_m) \setminus N_{m,j}$. Moreover, $\tilde{\varphi}_{k,j}^{(m)} \in (Uv_k)_j^{(m)}$. Here,

$$(Uv_k)^{(m)} = ((Uv_k)_1^{(m)}, (Uv_k)_2^{(m)}, \dots, (Uv_k)_m^{(m)}) \in \bigoplus_{j=1}^m L_2(\mathbb{R}^n,\mu_m) .$$

In addition, for all $m = \infty,1,2,\dots$ and $1 \leqq j < m+1$, all mappings

$$\tilde{E}_{\alpha,j}^{(m)}: t \mapsto \sum_{k=1}^\infty e^{-\lambda_k t} \tilde{\varphi}_{k,j}^{(m)}(\alpha) v_k , \qquad \alpha \in \mathbb{R}^n ,$$

belong to $T_{X,A}$.
For all $r > 0$, $m = \infty,1,2,\dots$ and $1 \leqq j < m+1$, we define

$$E_{\alpha,j}^{(m)}[r] = \sum_{k=1}^\infty \left(\mu_m(B(\alpha,r))^{-1} \int_{B(\alpha,r)} (Uv_k)_j^{(m)} \, d\mu_m \right) v_k$$

with $\alpha \in \operatorname{supp}(\mu_m)$. From Theorem (V.3.1) and Theorem (V.3.2) we obtain

(V.5.1) Lemma

Let $m = \infty,1,2,\dots$ and let $1 \leqq j < m+1$. Then for all $\alpha \in \operatorname{supp}(\mu_m) \setminus N_{m,j}$

- $$\lim_{r\downarrow 0} \| E_{\alpha,j}^{(m)}[r;t] - \tilde{E}_{\alpha,j}^{(m)}(t) \| = 0 \qquad \text{for all } t > 0 .$$

- For each $\ell = 1,2,\dots,n$ and all $r > 0$, $E_{\alpha,j}^{(m)}[r] \in \mathcal{D}(P_\ell)$ and

  $$\lim_{r\downarrow 0} \| e^{-tA} P_\ell E_{\alpha,j}^{(m)}[r] - \alpha_\ell \, E_{\alpha,j}^{(m)}(t) \| = 0$$

  for all $t > 0$. □

If the operators P_ℓ, $\ell = 1,\dots,n$, can be extended to closed linear mappings $\bar{P}_\ell$ in $T_{X,A}$ with domain $\operatorname{Dom}(\bar{P}_\ell) \subset T_{X,A}$, we obtain from the previous lemma:

(V.5.2) Corollary

Let $m = \infty,1,2,\ldots$ and let $1 \leqq j < m+1$. Then for all $\alpha \in \mathrm{supp}(\mu_m) \setminus N_{m,j}$ and for all $\ell = 1,\ldots,n$,

- $\tilde{E}^{(m)}_{\alpha,j}$ belongs to $\mathrm{Dom}(\bar{P}_\ell)$.

- $\bar{P}_\ell \, \tilde{E}^{(m)}_{\alpha,j} = \alpha_\ell \, \tilde{E}^{(m)}_{\alpha,j}$.

Remark. In Section IV.6 we have proved that for each given n-set of commuting self-adjoint operators there exists a nuclear analyticity space on which these operators act continuously. Cf. Theorem (IV.6.5). From Theorem (IV.6.1) it follows that for each given n-set $(P_1,\ldots,P_n)$ of commuting self-adjoint operators there exists an orthonormal basis such that each matrix $((P_\ell v_m, v_n))_{n,m}$ is column- and row-finite. Then by Theorem (IV.6.3) for any sequence $(\lambda_k)_{k\in\mathbb{N}}$ with the property $\lambda_k \geqq 0$ and $\forall_{t>0}: \sum_{k=1}^{\infty} e^{-\lambda_k t} < \infty$, each operator P_ℓ extends to a closed linear mapping in $T_{X,A}$, with A defined by $Av_k = \lambda_k v_k$, $k \in \mathbb{N}$.

Remark. Let X be a separable Hilbert space, and let $(P_1,\ldots,P_n)$ be an n-set of commuting self-adjoint operators. Let A be any nonnegative self-adjoint operator with a discrete spectrum which yields a nuclear space $S_{X,A}$. Then in $T_{X,A}$ there exists a Dirac basis which consists of candidate eigenvectors of $(P_1,\ldots,P_n)$ in the sense of Lemma (V.5.1). If the operators are closable as linear mappings in $T_{X,A}$, then these candidate eigenvectors are genuine joint (generalized) eigenvectors:

(V.5.3) Theorem

Let $(P_1,\ldots,P_n)$ be an n-set of strongly commuting self-adjoint operators on X which all extend to closed linear mappings in $T_{X,A}$. Let $\mu_\infty,\mu_1,\mu_2,\ldots$ be the corresponding multiplicity measures. Let $m = \infty,1,2,\ldots$. Then there is a μ_m-null set N_m with the following property:

For all $\alpha \in \mathrm{supp}(\mu_m) \setminus N_m$ there exists m nontrivial elements $\tilde{E}^{(m)}_{\alpha,j}$ of $T_{X,A}$, $1 \leqq j < m+1$, satisfying $\bar{P}_\ell \, \tilde{E}^{(m)}_{\alpha,j} = \alpha_\ell \, \tilde{E}^{(m)}_{\alpha,j}$, $\ell = 1,\ldots,n$.

Here $\bar{P}_\ell$ denotes the closure of P_ℓ as a mapping in $T_{X,A}$.

Proof. We only have to prove that the trajectories $\tilde{E}^{(m)}_{\alpha,j}$ defined in the beginning of this section are nontrivial except for a set of measure zero. Therefore we observe that $\tilde{E}^{(m)}_{\alpha,j} = 0$ implies $\tilde{\varphi}^{(m)}_{k,j}(\alpha) = 0$ for all $k \in \mathbb{N}$. If this would occur on a set I with positive measure, then we should have $(\tilde{U}w)^{(m)}_{k,j}(\alpha) = 0$ for all $\alpha \in I$ and all $w \in S_{X,A}$, and hence $\varphi(\alpha) = 0$ for all $\alpha \in I$ and all $\varphi \in L_2(\mathbb{R}^n,\mu_m)$. This yields a contradiction with $\mu_m(I) \neq 0$. □

V.6. A mathematical interpretation of the formal computation with bras and kets in Dirac's formalism

In the previous three sections of this chapter we have presented a solution of the so-called generalized eigenvalue problem in the setting of nuclear trajectory spaces. We have proved that any self-adjoint operator in X admits a canonical Dirac basis which consists of candidate (generalized) eigenvectors. If a self-adjoint operator is closable in a nuclear space $T_{X,A}$ these candidates become genuine eigenvectors of its closed extension. Our approach seems to fit rather well in Dirac's formalism. All points in the spectrum of a self-adjoint operator can be dealt with in the same manner.
Here we give an interpretation of other aspects of the formalism in terms of our distribution theory.

Dirac considers two types of vectors: the so-called kets and bras. He treats them on the same level. For the ket space we take a nuclear trajectory space $T_{X,A}$. The freedom we have in choosing A and X can be used to take a space $T_{X,A}$ which is best adapted to the quantum mechanical system under consideration. In fact, choosing an operator A in X is the same as choosing an orthonormal basis $(v_k)_{k\in\mathbb{N}}$ for X and positive numbers λ_k, $k \in \mathbb{N}$, such that

$$\forall_{t>0}: \sum_{k=1}^{\infty} e^{-\lambda_k t} < \infty .$$

Using Dirac's bracket notation we denote the elements of $T_{X,A}$ by $|G\rangle$. The label(s) G can be chosen such that it expresses best the properties of the ket $|G\rangle$ which are relevant in the context.
Since each element $G \in T_{X,A}$ can be expressed as a series $\sum_{k=1}^{\infty} \gamma_k v_k$ which converges in $T_{X,A}$, i.e.,

$$\forall_{t>0}: \sum_{k=1}^{\infty} |\gamma_k|^2 e^{-\lambda_k t} < \infty ,$$

we can write $|G\rangle$ as

$$|G\rangle = \sum_{k=1}^{\infty} \gamma_k |v_k\rangle$$

by which we mean

$$|G\rangle(t) = \sum_{k=1}^{\infty} \gamma_k |v_k\rangle(t) = \sum_{k=1}^{\infty} e^{-\lambda_k t} \gamma_k |v_k\rangle(0) .$$

By $\langle v_k|$ we denote the continuous linear functional on the ket space $T_{X,A}$ with the property that $\langle v_k| \, (|v_\ell\rangle) = \delta_{k\ell}$, $k,\ell \in \mathbb{N}$. Then for every $|G\rangle = \sum_{k=1}^{\infty} \gamma_k |v_k\rangle$ in $T_{X,A}$ we define

$$\langle G| = \sum_{k=1}^{\infty} \bar{\gamma}_k \langle v_k| .$$

The vectors $\langle G|$ are called bras. It is obvious that the bra space is anti-linearly homeomorphic to the ket space. We observe that each bra can be seen as a continuous linear functional on $S_{X,A}$. We have $\langle G|(w) = \langle w, |G\rangle\rangle$, $w \in S_{X,A}$. We now interprete Dirac's bracket notation.

(V.6.1) <u>Definition</u>

The expression $\langle F|G\rangle$, called the bracket of $\langle F|$ and $|G\rangle$, denotes the complex valued function

$$\langle F|G\rangle : t \mapsto \langle F| \, (|G\rangle(t)) , \quad t > 0 .$$

We note that the bracket $\langle F|G\rangle$ is well defined because $|F\rangle(t) \in S_{X,A}$ for all $t > 0$. Moreover, let

$$\langle F| = \sum_{k=1}^{\infty} \bar{\varphi}_k \langle v_k| \quad \text{and} \quad |G\rangle = \sum_{k=1}^{\infty} \gamma_k |v_k\rangle .$$

Then

$$\langle F|G\rangle(t) = \sum_{k=1}^{\infty} e^{-\lambda_k t} \bar{\varphi}_k \gamma_k , \quad t > 0 .$$

It follows that the function $\langle F|G\rangle$ extends to a holomorphic function on the open right half of the complex plane. It defines the almost periodic distribution

$$\alpha \mapsto \sum_{k=1}^{\infty} \bar{\varphi}_k \gamma_k e^{-i\lambda_k \alpha} .$$

In many cases this expression represents the characteristic function (i.e. the Fourier transform) of the probability distribution of results of measurements.

The analyticity space $S_{X,A}$ can be embedded in $T_{X,A}$. So to any element $w \in S_{X,A}$ there corresponds one ket $|w\rangle$ and one bra $\langle w|$. Then there exists $\tau > 0$ such that $|w\rangle(-\tau)$ belongs to $S_{X,A}$, where $|w\rangle(-\tau) = e^{(\tau+\varepsilon)A} |w\rangle(\varepsilon)$, $\varepsilon > 0$. For all bras $\langle G|$ this yields

$$\langle G|(|w\rangle(t)) = \langle G|w\rangle(t) ,$$

where we consider all $t \geq -\tau$ and especially $t = 0$.

To emphasize this nice property of the elements of $S_{X,A}$, the kets and the bras corresponding to the elements of $S_{X,A}$ are called test kets and test bras. We mention the following relations:

- $\langle F|G\rangle(t) = \langle F(t)|G\rangle(0) = \langle F|G(t)\rangle(0)$,
- $\langle F|G\rangle = \overline{\langle G|F\rangle}$.

(Here we denote the ket $||G\rangle(t)\rangle$ by $|G(t)\rangle$.)

The bra space can also be seen as a space of (unbounded) linear functionals. For each $\langle G|$ we define the linear functional

$$|F\rangle \mapsto \lim_{t\downarrow 0} \langle G|F\rangle(t)$$

where we only consider $|F\rangle$ for which this limit exists. If $\langle G|$ is a test bra, the limit exists for all kets. Then the limit is just the usual pairing of the elements of $S_{X,A}$ and $T_{X,A}$.

We consider a metric space (M,d) with σ-finite regular Borel measure μ. We assume that the space (M,d) satisfies Federer's conditions. It follows that any unitary mapping U from X onto a closed subspace of $\mathbf{L}_2(M,\mu)$ leads to a canonical Dirac basis $(|\alpha\rangle)_{\alpha\in M}$ as described in Section V.2. Following Theorem (V.1.4) any test ket $|w\rangle$ can be written as

$$|w\rangle(0) = \int_M \langle\alpha|w\rangle(-\tau)\ |\alpha\rangle(\tau)\ d\mu_\alpha\ .$$

The above expression does not depend on the choice of $\tau > 0$ and τ has to be chosen so small that $|w\rangle(-\tau) \in S_{X,A}$.

(V.6.2) Lemma

Let $|v\rangle$ be a test ket. Then

$$|v\rangle = \int_M \langle\alpha|v\rangle(0)\ |\alpha\rangle\, d\mu_\alpha$$

i.e.

$$|v\rangle(t) = \int_M \langle\alpha|v\rangle(-\tau)\ |\alpha\rangle(t+\tau)\ d\mu_\alpha\ , \qquad t > -\tau\ .$$

Here $\tau \geqq 0$ has to be chosen so small that $|v\rangle(-\tau) \in S_{X,A}$.

Proof. Let $t > -\tau$, and put $|w\rangle(0) = |v\rangle(t)$. Then we have seen that

$$|w\rangle(0) = \int_M \langle\alpha|w\rangle(-t-\tau)\ |\alpha\rangle(t+\tau)\ d\mu_\alpha\ .$$

From the identity $\langle\alpha|w\rangle(-t-\tau) = \langle\alpha|v\rangle(-\tau)$ the assertion follows. □

Parseval's identity is an immediate consequence of the definition of Dirac basis. Let $|w\rangle$ be a test ket,

$$\| \, |w\rangle(0) \|^2 = \int_M |\langle\alpha|w\rangle(0)|^2\ d\mu_\alpha\ .$$

We note that the function $\alpha \mapsto \langle\alpha|w\rangle(0)$ is the canonical representant of the class $\mathsf{U}(|w\rangle(0))$ in the closed subspace Y of $\mathbf{L}_2(M,\mu)$. Here U denotes the unitary operator from X onto Y which is associated to the Dirac basis $(|\alpha\rangle)_{\alpha\in M}$.

From lemma (V.6.2) it follows that for every ket $|F\rangle$ and for every $t > 0$,

$$|F\rangle(t) = e^{-(t-\tau)A}\ |F\rangle(\tau) = \int_M \langle\alpha|F\rangle(\tau)\ |\alpha\rangle(t-\tau)\ d\mu_\alpha\ .$$

This integral expression does not depend on the choice of τ, $0 < \tau < t$.

The integral converges absolutely, i.e.

$$\forall_{t>0}: \int_M |\langle\alpha|F\rangle(\tau)| \; \| |\alpha\rangle(t-\tau) \| \, d\mu_\alpha < \infty \; .$$

Observe that

$$\int_M |\langle\alpha|F\rangle(\tau)| \; \| |\alpha\rangle(t-\tau) \| \, d\mu_\alpha \leq$$

$$\leq \left(\int_M |\langle\alpha|F\rangle(\tau)|^2 \, d\mu_\alpha \right)^{\frac{1}{2}} \left(\int_M \| |\alpha\rangle(t-\tau) \|^2 \, d\mu_\alpha \right)^{\frac{1}{2}}$$

where

$$\| |\alpha\rangle(t-\tau) \|^2 = \sum_{k=1}^{\infty} e^{-2\lambda_k (t-\tau)} \; |\langle v_k|\alpha\rangle(0)|^2 \; .$$

The ket $|F\rangle$ can thus be expressed by

$$|F\rangle : t \mapsto \int_M \langle\alpha|F\rangle(\tau) \; |\alpha\rangle(t-\tau) \, d\mu_\alpha \; , \qquad 0 < \tau < t \; .$$

(V.6.3) Definition

By the expression $\int_M \langle\alpha|F\rangle \; |\alpha\rangle \, d\mu_\alpha$ is meant the trajectory

$$t \mapsto \int_M \langle\alpha|F\rangle(\tau) \; |\alpha\rangle(t-\tau) \, d\mu_\alpha \; .$$

Cf. [Di], p. 64.

It follows that

(V.6.4) Theorem

$$|F\rangle = \int_M \langle\alpha|F\rangle \; |\alpha\rangle \, d\mu_\alpha \; .$$

The integral converges (absolutely) in $T_{X,A}$. Cf. Lemma (V.1.2). □

Consider the following identity:

$$\langle\alpha|\beta\rangle(t) = \sum_{k=1}^{\infty} e^{-\lambda_k t} \; \langle\alpha|v_k\rangle(0) \; \langle v_k|\beta\rangle(0) \; , \qquad \alpha,\beta \in M \; .$$

Each function $\alpha \mapsto \langle\alpha|v_k\rangle(0)$ is a representant of the class Uv_k, $k \in \mathbb{N}$. If we put $B = UAU^*$, then $U(S_{X,A}) = S_{Y,B}$ and $U(T_{X,A}) = T_{Y,B}$. Now, let δ_β, $\beta \in M$, denote the trajectory

$$\delta_\beta : t \mapsto \sum_{k=1}^{\infty} e^{-\lambda_k t} \langle v_k|\beta\rangle(0)\, Uv_k \ .$$

Then $\delta_\beta = U|\beta\rangle$ belongs to $T_{Y,B}$. It is clear that

$$\langle Uw, \delta_\beta\rangle = \langle\beta|w\rangle(0) \ , \qquad w \in S_{X,A} \ .$$

Since the function $\beta \mapsto \langle\beta|w\rangle(0)$ is a representant of the class Uw, the functional δ_β is an evaluation functional, popularly speaking a Dirac delta function. Since $\delta_\beta(\alpha,t) = \langle\delta_\beta(t),\delta_\alpha\rangle = \langle\alpha|\beta\rangle(t)$, the generalization of the orthonormality relations for discrete basis as suggested by Dirac, cf. [Di], p. 62,

$$\langle\alpha|\beta\rangle = \delta_\beta(\alpha)$$

admits the interpretation $\langle\alpha|\beta\rangle(t) = \delta_\beta(\alpha,t)$.
An extension of the given Parseval identity is the following theorem.

(V.6.5) <u>Theorem</u>

$$\langle G|F\rangle(t) = \int_M \langle G|\alpha\rangle(\tau)\ \langle\alpha|F\rangle(t-\tau)\, d\mu_\alpha \ .$$

Here we take $0 < \tau < t$. The integral expression does not depend on τ.

<u>Proof</u>. For all $t > 0$, all kets $|F\rangle$ and all bras $\langle G|$ we have

$$\langle G|F\rangle(t) = \langle G(\tau)|F(t-\tau)\rangle(0) =$$

$$= \int_M \langle G(\tau)|\alpha\rangle(0)\ \langle\alpha|F(t-\tau)\rangle(0)\, d\mu_\alpha =$$

$$= \int_M \langle G|\alpha\rangle(\tau)\ \langle\alpha|F\rangle(t-\tau)\, d\mu_\alpha$$

where $0 < \tau < t$ may be taken arbitarily. □

Theorem (V.6.5) gives an interpretation of the formula

$$\langle G|F\rangle = \int_M \langle G|\alpha\rangle\ \langle\alpha|F\rangle\, d\mu_\alpha \ ,$$

as suggested by Dirac, cf. [Di], pp. 64-65.

(V.6.6) Corollary

The Hilbert-Schmidt kernel of the operator $e^{-tB} = U e^{-tA} U^*$, $t > 0$, is given by $(\alpha,\beta) \mapsto \langle\beta|\alpha\rangle(t)$.

Proof. Following Theorem (V.6.5) we have

$$\langle\beta|F\rangle(t) = \int_M \langle\beta|\alpha\rangle(\tau)\ \langle\alpha|F\rangle(t-\tau)\ d\mu_\alpha \ .$$

From this the assertion follows. □

In Section V.5 we have shown that each n-set of commuting self-adjoint operators has a canonical Dirac basis which consists of simultaneous eigenkets. From Theorem (V.6.4) it follows that any ket can be expanded with respect to these eigenkets. Because of the one-to-one correspondence between kets and bras, each Dirac basis of kets naturally leads to a Dirac basis of bras. So similarly any bra can be expanded with respect to a Dirac basis of bras, in particular to a Dirac basis of eigenbras.

Remark. In quantum theory a complete set of commuting observables is a set of observables which mutually commute and for which there exists only one simultaneous eigenstate for each point in the joint spectrum. So in the terminology of Section V.4 a complete n-set of commuting observables is just an n-set of strongly commuting self-adjoint operators which have uniform multiplicity one. It follows that for such an n-set there exists a finite nonnegative Borel measure μ on $\mathbb{R}^n$, such that the n-set is unitarily equivalent to $(Q_1,\ldots,Q_n)$, where each Q_ℓ is defined in $L_2(\mathbb{R}^n,\mu)$ by

$$Q_\ell f\colon \alpha \mapsto \alpha_\ell f(\alpha) \ , \qquad \alpha = (\alpha_1,\ldots,\alpha_n) \in \mathbb{R}^n \ .$$

V.7. Matrices with respect to Dirac bases

In finite dimensional vector spaces every linear operator can be represented by a matrix. The product of two operators B_1 and B_2 has the matrix $[B_1\ B_2]$ which is derived by matrix multiplication of the matrices $[B_1]$ and $[B_2]$, i.e. $[B_1\ B_2]_{k\ell} = \sum_j [B_1]_{kj}\ [B_2]_{j\ell}$. In his formalism Dirac introduces a general notion of matrix. In fact he suggests that a matrix can also be introduced for 'Dirac' bases, and that the matrix calculus has similar properties as in the case of finite dimensional vector spaces (cf. [Di], pp. 67-72).
In the present section we give a mathematical interpretation of this feature of Dirac's formalism.

We start with the same canonical Dirac basis $(|\alpha\rangle)_{\alpha\in M}$ as considered in the previous section. As explained in Section V.1, the Dirac basis $(|\alpha\rangle)_{\alpha\in M}$ gives rise to a Dirac basis in $TT_{X\otimes X;A\otimes I,I\otimes A}$. Therefore we define the function

$$(\alpha,\beta) \mapsto |\beta\rangle \otimes |\alpha\rangle\ , \qquad \alpha,\beta \in M\ .$$

Then $(|\beta\rangle \otimes |\alpha\rangle)_{(\alpha,\beta)\in M^2}$ is a canonical Dirac basis in the trajectory space $TT_{X\otimes X;A\otimes I,I\otimes A}$, with respect to the measure space $M \times M$ and the σ-finite regular Boral measure $\mu \otimes \mu$.
According to Dirac, the tensor product of the kets $|F\rangle$ and $|G\rangle$ is denoted by $|G\rangle\langle F|$.
Since $(|\alpha\rangle\langle\beta|)_{(\alpha,\beta)\in M^2}$ is a Dirac basis, we obtain the following results from the previous section.

(V.7.1) <u>Theorem</u>

Let $\hat{R} \in TT_{X\otimes Y,A\otimes I,I\otimes A}$. Then

$$\hat{R} = \int_{M^2} [R]_{\alpha\beta}\ |\alpha\rangle\langle\beta|\ d\mu_\alpha \otimes \mu_\beta$$

where

$$[R]_{\alpha\beta}\colon\ (t,s) \mapsto \langle\hat{R}(t,s), |\alpha\rangle\langle\beta|\rangle_{X\otimes X}\ , \qquad t,s > 0\ ,$$

and where the integral must be interpreted as

$$\hat{R}(t,s) = \int_{M^2} [R]_{\alpha\beta}(\tau,\sigma)\ |\alpha\rangle\langle\beta|(t-\tau,s-\sigma)\ d\mu \otimes \mu_\beta\ .$$

Here $0 < \tau < t$ and $0 < \sigma < s$. The expression does not depend on the choice of τ and σ.

The following definition makes sense.

(V.7.2) Definition

Let $\hat{R} \in TT_{X\otimes X;A\otimes I,I\otimes A}$ and let $\alpha,\beta \in M$. Then the function

$$[R]_{\alpha\beta}: (0,\infty) \times (0,\infty) \to \mathbb{C}$$

is defined by

$$[R]_{\alpha\beta}(t,s) = \langle\hat{R}(t,s), |\alpha\rangle\langle\beta|\rangle_{X\otimes X}, \qquad t,s > 0 .$$

The set of functions $([R]_{\alpha\beta})_{(\alpha,\beta)\in M^2}$ is called the matrix of R, with respect to the Dirac basis $(|\alpha\rangle)_{\alpha\in M}$.

(V.7.3) Lemma

Let $\hat{\Phi} \in SS_{X\otimes X;A\otimes I,I\otimes A}$. Then the function

$$(\alpha,\beta) \mapsto \langle\hat{\Phi}, |\alpha\rangle\langle\beta|\rangle_{X\otimes X}$$

is the kernel of a Hilbert-Schmidt operator on $\mathbf{L}_2(M,\mu)$. Moreover, for each fixed α the function

$$\beta \mapsto \langle\hat{\Phi}, |\alpha\rangle\langle\beta|\rangle_{X\otimes X}$$

is square integrable on (M,μ).

Proof. Consider the following estimate:

$$|\langle\hat{\Phi}, |\alpha\rangle\langle\beta|\rangle| =$$

$$= \Big|\sum_{k,\ell=1}^{\infty} e^{-\lambda_k t} e^{-\lambda_\ell s} (\hat{\Phi}(-t,-s), v_k\otimes v_\ell)\, \langle v_k|\alpha\rangle(0)\, \langle\beta|v_\ell\rangle(0)\Big|$$

$$\leq \Big(\sum_{k,\ell=1}^{\infty} |(\hat{\Phi}(-t,-s), v_k\otimes v_\ell)|^2\Big)^{\frac{1}{2}} \cdot$$

$$\cdot \Big(\sum_{k,\ell=1}^{\infty} e^{-2\lambda_k t} e^{-2\lambda_\ell s} |\langle v_k|\alpha\rangle(0)\, \langle\beta|v_\ell\rangle(0)|^2\Big)^{\frac{1}{2}} .$$

Here from both assertions easily follow. □

These observations lead to the following theorem.

(V.7.4) Theorem

Let $\hat{R} \in TT_{X\otimes X;A\otimes I,I\otimes A}$. Then for each test ket $|w\rangle$

$$R\,|w\rangle : s \mapsto \int_{M\times M} [R]_{\alpha\beta}(\tau,\sigma)\ \langle\beta|w\rangle(-\tau)\ \ |\alpha\rangle(s-\sigma)\ d\mu_\alpha \otimes \mu_\beta$$

where $\tau > 0$ must be taken small enough and where $0 < \sigma < s$. The integrals converge absolutely in X.

The α-th component of $R|w\rangle$ is given by

$$\langle\alpha|\,R\,|w\rangle : s \mapsto \int_M [R]_{\alpha\beta}(\tau,s)\ \langle\beta|w\rangle(-\tau)\ d\mu_\beta$$

where the integral converges absolutely and does not depend on the choice of $\tau > 0$.

Proof. Observe that the function $\beta \mapsto \langle\beta|w\rangle(-\tau)$ is square integrable as well as the function

$$\alpha \mapsto \|\,|\alpha\rangle(s-\sigma)\,\| = \left(\sum_{k=1}^{\infty} e^{-2\lambda_k(s-\sigma)}\ \ |\langle\alpha|v_k\rangle(0)|^2\right)^{\frac{1}{2}} .$$

Moreover, by Lemma (V.7.3) the function

$$(\alpha,\beta) \mapsto [R]_{\alpha\beta}(\tau,\sigma)$$

is square $\mu\otimes\mu$-integrable, and

$$\beta \mapsto [R]_{\alpha\beta}(\tau,\sigma)$$

is square μ-integrable for each $\alpha \in M$. It leads to the following estimations:

– $$\int_{M\times M} |[R]_{\alpha\beta}(\tau,\sigma)|\,|\langle\beta|w\rangle(-\tau)|\,\|\,|\alpha\rangle(s-\sigma)\,\|\ d\mu_\alpha \otimes \mu_\beta \leq$$

$$\leq \left(\int_{M\times M} |[R]_{\alpha\beta}(\tau,\sigma)|^2\ d\mu_\alpha \otimes \mu_\beta\right)^{\frac{1}{2}} \left(\int_M |\langle\beta|w\rangle(-\tau)|^2\ d\mu_\beta\right)^{\frac{1}{2}} \cdot$$

$$\cdot \left(\int_M \|\,|\alpha\rangle(s-\sigma)\,\|^2\ d\mu_\alpha\right)^{\frac{1}{2}}$$

and

$$- \int_M |[R]_{\alpha\beta}(\tau,\sigma)||\langle\beta|w\rangle(-\tau)| \, d\mu_\beta \leq$$

$$\leq \Big(\int_M |[R]_{\alpha\beta}(\tau,\sigma)|^2 \, d\mu_\beta \Big)^{\frac{1}{2}} \Big(\int_M |\langle\beta|w\rangle(-\tau)|^2 \, d\mu_\beta \Big)^{\frac{1}{2}} .$$

Remark. The previous theorem interpretes the following expressions:

$$R\,|w\rangle = \int_{M\times M} [R]_{\alpha\beta} \langle\beta|w\rangle \, |\alpha\rangle \, d\mu_\alpha \otimes \mu_\beta$$

and

$$\langle\alpha|\, R \,|w\rangle = \int_M [R]_{\alpha\beta} \langle\beta|w\rangle \, d\mu_\beta$$

where $|w\rangle$ is a test ket. Cf. [Di], p. 72.

Since $SS_{X\otimes X;A\otimes I,I\otimes A}$, $TS_{X\otimes X;A\otimes I,I\otimes A}$ and $TS_{X\otimes X;I\otimes A,A\otimes I}$ can be embedded in $TT_{X\otimes X;A\otimes I,I\otimes A}$, we have the notion of matrix for the elements of each of these spaces and the associated linear mappings. From the characterization of these spaces we obtain the following properties (cf. Section III.1):

- Let $\hat{\Phi} \in SS_{X\otimes X;A\otimes I,I\otimes A}$, then for each $\alpha,\beta \in M$ the function

  $$[\Phi]_{\alpha\beta}: (t,s) \mapsto \langle\hat{\Phi}(t,s), |\alpha\rangle\langle\beta|\rangle_{X\otimes X} , \quad t,s > 0 ,$$

  can be extended to a function on $\{(t,s) \in \mathbb{R}^2 \mid t > -\tau,\ s > -\sigma\}$ for certain $\tau,\sigma > 0$, dependent on $\hat{\Phi}$, only.

- Let $\hat{K} \in TS_{X\otimes X;A\otimes I,I\otimes A}$. Then for all $\alpha,\beta \in M$ the function

  $$[K]_{\alpha\beta}: (t,s) \mapsto \langle\hat{K}(t,s), |\alpha\rangle\langle\beta|\rangle$$

  can be extended in the following way:

  $$\forall_{s>0}\ \exists_{\tau>0}\ \forall_{t>-\tau}: [K]_{\alpha\beta}(t,s) = \langle\hat{K}(t+\tau,s)e^{\tau A}, |\alpha\rangle\langle\beta|\rangle_{X\otimes X} .$$

  The choice of τ is independent of α,β.

- Let $\hat{L} \in TS_{X\otimes X;I\otimes A,A\otimes I}$. Then for all $\alpha,\beta \in M$ the function

  $$[L]_{\alpha\beta}: (t,s) \mapsto \langle\hat{L}(t,s), |\alpha\rangle\langle\beta|\rangle_{X\otimes X}$$

can be extended in the following way:

$$\forall_{s>0}\ \exists_{\tau>0}\ \forall_{t>-\tau}:\ [L]_{\alpha\beta}(t,s) = \langle e^{\tau A}\ \hat{L}(t+\tau,s), |\alpha\rangle\ \langle\beta|\rangle_{X\otimes X}\ .$$

Again, the choice of τ is independent of α,β.

The above defined extensions make sense. To show this we recall the following properties:

- $\hat{\Phi} \in SS_{X\otimes X;A\otimes I,I\otimes A} \Leftrightarrow \exists_{\substack{\tau>0,\\ \sigma>0}}\ \forall_{\substack{t>-\tau,\\ s>-\sigma}}:\ \Phi(t,s) \in S_{X\otimes X,A\otimes I+I\otimes A}$.

- $\hat{K} \in TS_{X\otimes X;A\otimes I,I\otimes A} \Leftrightarrow \forall_{s>0}\ \exists_{\tau>0}:\ \hat{K}(0,s)e^{\tau A} \in S_{X\otimes X,A\otimes I+I\otimes A}$.

- $\hat{L} \in TS_{X\otimes X;I\otimes A,A\otimes I} \Leftrightarrow \forall_{s>0}\ \exists_{\tau>0}:\ e^{\tau A}\ \hat{L}(0,s) \in S_{X\otimes X,A\otimes I+I\otimes A}$.

With these observations in mind the following results can be proved, similar to Theorem (V.7.4).

(V.7.5) <u>Theorem</u>

Let $\hat{\Phi} \in SS_{X\otimes X;A\otimes I,I\otimes A}$. Then for each ket $|F\rangle$

$$\Phi\,|F\rangle\ :\ s \mapsto \int_{M\times M} [\Phi]_{\alpha\beta}(-\tau,-\sigma)\ \langle\beta|F\rangle(\tau)\ \ |\alpha\rangle(s+\sigma)\ d\mu_\alpha \otimes \mu_\beta$$

where the choice of $\sigma,\tau > 0$ depends on Φ and where $s > -\sigma$. The integrals converge absolutely in X and are independent of σ and τ.
The α-th component of $\Phi\,|F\rangle$ is given by

$$\langle\alpha|\ \Phi\,|F\rangle\ :\ s \mapsto \int_M [\Phi]_{\alpha\beta}(-\tau,s)\ \langle\beta|F\rangle(\tau)\ d\mu_\beta$$

where $s > -\sigma$. The integrals converge absolutely and are independent of τ. □

(V.7.6) <u>Theorem</u>

Let $\hat{L} \in TS_{X\otimes X;I\otimes A,A\otimes I}$. Then for each test ket $|w\rangle$

$$L\,|w\rangle\ :\ t \mapsto \int_{M\times M} [L]_{\alpha\beta}(-\tau,\sigma)\ \langle\beta|w\rangle(-\sigma)\ \ |\alpha\rangle(t+\tau)\ d\mu_\alpha \otimes \mu_\beta\ .$$

Here first $\sigma > 0$ must be taken so small that $|w\rangle(-\tau) \in S_{X,A}$ and next $\tau > 0$ so small that $e^{\tau A}\,\hat{L}(0,\sigma) \in S_{X\otimes X,\,A\otimes I+I\otimes A}$.
The integrals converge absolutely in X and are independent of the choice of τ and σ.
Further, for each $\alpha \in M$

$$\langle\alpha|\,L\,|w\rangle : t \mapsto \int_{M\times M} [L]_{\alpha\beta}(t,\sigma)\ \langle\beta|w\rangle(-\sigma)\ d\mu_\beta\ ,$$

where $\sigma > 0$ must be taken so small that $|w\rangle(-\sigma) \in S_{X,A}$. Then $t > -\tau$ with $\tau > 0$ sufficiently small. The integrals converge absolutely and are independent of σ. □

(V.7.7) Theorem

Let $\hat{K} \in TS_{X\otimes X;\,A\otimes I,\,I\otimes A}$. Then for each ket $|F\rangle$,

$$K|\,F\rangle : s \mapsto \int_{M\times M} [K]_{\alpha\beta}(-\tau,\sigma)\ \langle\beta|F\rangle(\tau)\ |\alpha\rangle(s-\sigma)\ d\mu_\alpha \otimes \mu_\beta\ .$$

Here for each $s > 0$ we must take $0 < \sigma < s$ and next $\tau > 0$ so small that $\hat{K}(0,\sigma)e^{\tau A} \in S_{X\otimes X,\,A\otimes I+I\otimes A}$. The integrals converge absolutely in X and are independent of τ,σ.
Further, for each $\alpha \in M$,

$$\langle\alpha|\,K\,|F\rangle : s \mapsto \int_{M} [K]_{\alpha\beta}(-\tau,s)\ \langle\beta|F\rangle(\tau)\ d\mu_\beta$$

where for each $s > 0$ we must take $\tau > 0$ sufficiently small. The integrals converge absolutely and are independent of τ. □

The spaces $TS_{X\otimes X;\,I\otimes A,\,A\otimes I}$ and $TS_{X\otimes X;\,A\otimes I,\,I\otimes A}$ can be regarded as operator algebras. So for $\hat{L}_1,\hat{L}_2 \in TS_{X\otimes X;\,I\otimes A,\,A\otimes I}$ we can consider their product $(L_1\ L_2)^\wedge$. It is a natural question whether there exists a mathematical interpretation of Dirac's heuristic formula

$$[L_1\ L_2]_{\alpha\beta} = \int_M [L_1]_{\alpha\gamma}\ [L_2]_{\gamma\beta}\ d\mu_\gamma\ .$$

Cf. [Di], p. 69.

The following theorem is such an interpretation.

(V.7.8) Theorem

Let $\hat{L}_1, \hat{L}_2 \in TS_{X\otimes X;I\otimes A,A\otimes I}$. Then for all $\alpha,\beta \in M$ we have

$$[L_1\ L_2]_{\alpha\beta}: (t,s) \mapsto \int_M [L_1]_{\alpha\gamma}(t,\tau)\ [L_2]_{\gamma\beta}(-\tau,s)\ d\mu_\gamma$$

where the integrals converge absolutely and are independent of the chocie of $\tau > 0$.

Proof. The trajectory $(L_1\ L_2)^\wedge$ corresponding to $L_1\ L_2$ is defined by

$$(L_1\ L_2)^\wedge(t,s) = \hat{L}_1(t,\tau)\hat{L}_2(-\tau,s)$$

where $\tau > 0$ must be taken so small that $\hat{L}(-\tau,s) \in S_{X\otimes X,A\otimes I+I\otimes A}$. □

Similarly, we have for $TS_{X\otimes X;A\otimes I,I\otimes A}$:

(V.7.9) Theorem

Let $\hat{K}_1, \hat{K}_2 \in TS_{X\otimes X;A\otimes I,I\otimes A}$. Then for all $\alpha,\beta \in M$ we have

$$[K_1\ K_2]_{\alpha\beta}: (t,s) \mapsto \int_M [K_1]_{\alpha\gamma}(-\tau,s)\ [K_2]_{\gamma\beta}(\tau,t)\ d\mu_\gamma\ .$$

Here τ must be taken so small that $\hat{K}_1(-\tau,s) \in S_{X\otimes X,A\otimes I+I\otimes A}$. The integrals converge absolutely and are independent of τ.

Remark. In a forthcoming paper [EG 7] we shall link the theory of annihilation and creation operators of quantum field theory to our concept of Dirac basis.

We are aware of the fact that the mathematical formalism we developed in this chapter, does not yet cover the whole original Dirac formalism. As a start, it already explains many features of the original formalism which have been either abandoned or neglected by mathematicians. It goes much further than any interpretation we have seen.
In his book "The principles of quantum mechanics", which has been a great source of inspiration both for physicists and for mathematicians, Dirac, as any genuine scientist, has had the courage to go beyond the little possibilities that the mathematics of the thirties offered in this direction. It led

to a formalism any mathematician should be eager to comprehend. We end with a quotation from Niels Bohr:

> The general lesson of the role that mathematics has played through the ages in natural philosophy is the recognition that no relationship can be defined without a logical frame and that any apparent disharmony in the description of experiences can be eliminated only by a appropriate widening of the conceptual framework.

Additional references to this chapter are in alphabetical order [An], [Bö], [EG 5-6], [GV 4], [Ja], [Me], [Neu], and [Ro].

EPILOGUE

In the last few years we have studied several topics which are related to analyticity and trajectory spaces. In particular we mention our work on Dirac's formalism [EG 4-6], our work on a measure theoretical version of Sobolev's Lemma [EG 7], and on eigenfunction expansions for general self-adjoint operators [EG 8]. Further, in [E 2] Van Eijndhoven developed a functional analytic generalization of the theory of tempered distributions, which is a kind of mirror image of the theory as presented in this book. Also for this theory an extensive list of classical analytic examples is available now. See [EG 1-3], [E 2], [G 5].
In our papers [EK], [EGK] we started a unification and generalization of the theories in [G 1-3] and [E 2], which we find rather exciting. The spaces of this unification not only include most classical distribution spaces and function spaces that we know about, but also the spaces of (extendible) operators on those spaces. In this epilogue we present a brief sketch of the underlying ideas in [EGK].

Let Φ_1 denote the following directed set of functions on $\mathbb{R}$:

$$\Phi_1 = \{\lambda \mapsto e^{-t|\lambda|} \mid t > 0\} .$$

Let A denote a self-adjoint operator in a separable Hilbert space X and define

$$S_{\Phi_1}(X;A) = \bigcup_{\varphi\in\Phi_1} \varphi(A)X . \tag{1}$$

This is a linear space and it is natural to endow it with the inductive limit topology σ_{ind} coming from the Hilbert space $\varphi(A)(X)$ with norm $\|\varphi(A)\cdot\|_X$. It will be clear that $S_{\Phi_1}(X;A) = S_{X,|A|}$, i.e. the analyticity space of this book with $|A| = \int |\lambda| dE_\lambda$ if $A = \int \lambda\, dE_\lambda$.
As a second space, associated with Φ_1, X and A, define

$$T_{\Phi_1}(X;A) = \bigcap_{\varphi\in\Phi_1} \varphi^{-1}(A)X . \tag{2}$$

Here $\varphi^{-1}(A)X$ is a completion of X with respect to the norm $\|\varphi(A)\cdot\|_X$. On $T_{\Phi_1}(X;A)$ the projective limit topology τ_{proj} with respect to $\varphi^{-1}(A)X$ is imposed. Comparison with Section I.2 shows that $T_{\Phi_1}(X;A) = T_{X,|A|}$ in every respect.

$S_{\Phi_1}(X;A)$ and $T_{\Phi_1}(X;A)$ can be put in duality by

$$\langle u,F\rangle = (\varphi^{-1}(A)u,\varphi(A)F)_X \tag{3}$$

for suitable $\varphi \in \Phi_1$. This pairing does not depend on the choice of $\varphi \in \Phi_1$. Next, we define a 'compatibility set' $\Phi_1^{\#}$ of Borel functions:

$$\Phi_1^{\#} = \{f \mid f > 0,\ \forall_{\varphi\in\Phi_1} : \sup_{\lambda\in\mathbb{R}} f(\lambda)\varphi(\lambda) < \infty\} .$$

Comparison with Lemma I.1.5 shows that $\Phi_1^{\#}$ consists of the functions

$$f = \begin{cases} f_1(\lambda) , & \lambda \geqq 0 \\ f_2(-\lambda) , & \lambda > 0 \end{cases} \quad \text{with } f_1,f_2 \in \mathbf{B}_+(\mathbb{R}) .$$

Starting from $\Phi_1^{\#}$ instead of Φ_1 the spaces $S_{\Phi_1^{\#}}(X;A)$ and $T_{\Phi_1^{\#}}(X;A)$ can be introduced as in (1) and (2).

From Theorem I.1.6 and Proposition I.2.6 it follows that

$$S_{\Phi_1^{\#}}(X;A) = T_{\Phi_1}(X;A) \quad \text{and} \quad T_{\Phi_1^{\#}}(X;A) = S_{\Phi_1}(X;A) .$$

The fundamental reason for this is Lemma I.1.5, which, rephrased in the language of this epilogue, says that

$$\forall_{\psi\in\Phi_1^{\#\#}}\ \exists_{c>0}\ \exists_{\varphi\in\Phi} : \psi \leqq c\varphi . \tag{4}$$

One of the consequences of property (4) is that $S_{\Phi_1}(X;A)$ and $T_{\Phi_1}(X;A)$ are both inductive limits *and* projective limits of Hilbert spaces. Thence, the nonstrict inductive limits $S_{\Phi_1}(X;A)$ and $T_{\Phi_1}(X;A)$ behave very much like *strict* inductive limits.

In [EGK] the above sketched ideas have been worked out for a general directed set Φ of nonnegative Borel functions on $\mathbb{R}^n$ with the following properties:

A.0. $\forall_{\varphi\in\Phi}$: $\varphi \geqq 0$ and bounded on bounded sets of $\mathbb{R}^n$.

$$\forall_{\varphi_1,\varphi_2\in\Phi}\ \exists_{\varphi_3\in\Phi}\ \exists_{c>0}:\ c\varphi_3 \geqq \max(\varphi_1,\varphi_2)\ .$$

A.1. $\forall_{\varphi\in\Phi}$: $\widetilde{\varphi}^{-1}(\lambda) = \begin{cases} \varphi(\lambda)^{-1} & \text{if } \varphi(\lambda) > 0 \\ 0 & \text{if } \varphi(\lambda) = 0 \end{cases}$ is bounded on bounded sets.

A.2. The supports of φ, $\varphi \in \Phi$, cover the whole $\mathbb{R}^n$

$$\mathbb{R}^n = \bigcup_{\varphi\in\Phi} \{\lambda \in \mathbb{R} \mid \varphi(\lambda) > 0\}\ .$$

A.3. $\forall_{\varphi\in\Phi}\ \exists_{\psi\in\Phi}\ \exists_{c>0}\ \forall_{m\in\mathbb{Z}^n}$: $(1 + |m|) \sup_{\lambda\in Q_m} \varphi(\lambda) \leqq c \inf_{\lambda\in Q_m} \psi(\lambda)$.

Here $m = (m_1,\dots,m_n)$ and $Q_m = [m_1-1,m_1] \times \dots \times [m_n-1,m_n]$.

Associated with Φ a so-called compatibility class $\Phi^{\#}$ of Borel functions is introduced as follows.

B.1. $\forall_{f\in\Phi^{\#}}$: $f \geqq 0$ and $\widetilde{f}^{-1}$, cf. A.1, is bounded on bounded sets.

B.2. $\forall_{f\in\Phi^{\#}}\ \forall_{\varphi\in\Phi}$: $\sup_{\lambda\in\mathbb{R}^n} f(\lambda)\varphi(\lambda) < \infty$.

It can be shown, see [EGK], that $\Phi^{\#}$ satisfies again all properties A.0-A.3. Now, given a set Φ which satisfies A.0-A.3 and a strongly commuting set of self-adjoint operators $A = (A_1,\dots,A_n)$ in a Hilbert space X, the spaces $S_\Phi(X;A)$, $T_\Phi(X;A)$, $S_{\Phi^{\#}}(X;A)$ and $T_{\Phi^{\#}}(X;A)$ are defined according to (1) and (2). For the solutions of technical problems around these definitions, see [EGK].

The space $S_\Phi(X;A)$ has the following properties:

S.1. Endowed with the semi-norms s_f, $s_f(w) = \|f(A)w\|$, $f \in \Phi^{\#}$, $S_\Phi(X;A)$ is a barreled and bornological locally convex topological vector space.

S.2. $S_\Phi(X;A)$ is an inductive limit of Hilbert spaces which is nonstrict in general.

The space $T_\Phi(X;A)$ has the following properties:

T.1. Endowed with the semi-norms t_φ, $t_\varphi(F) = \|\varphi(A)F\|_X$, $\varphi \in \Phi$, $T_\Phi(X;A)$ is a complete locally convex topological vector space.

T.2. $T_\Phi(X;A)$ is a projective limit of Hilbert spaces.

T.3. Each bounded set B of $T_\Phi(X;A)$ is homeomorphic to a bounded subset of X. The homeomorphism is established by a well chosen operator $g(A)$, $g \in \Phi^{\#}$.

T.4. $T_\Phi(X;A) = S_{\Phi^{\#}}(X;A)$ as a set.

The spaces $S_\Phi(X;A)$ and $T_\Phi(X;A)$ can be paired by

$$\langle u,F\rangle = (\tilde{\varphi}(A)^{-1}u,\varphi(A)F)_X \quad , \qquad u \in S_\Phi(X;A),\ F \in T_\Phi(X;A)$$

where $\varphi \in \Phi$ is such that $u \in \varphi(A)X$. The pairing does not depend on the specific choice of $\varphi \in \Phi$. By this pairing $S_\Phi(X;A)$ and $T_\Phi(X;A)$ can be regarded as each others strong duals.

If Φ contains only functions which are bounded on the joint spectrum $\sigma(A)$ of A, there is the Gelfand triple

$$S_\Phi(X;A) \subset X \subset T_\Phi(X;A) \ .$$

On the other hand, if Φ contains only functions which are bounded away from zero on $\sigma(A)$, there is the Gelfand triple

$$T_\Phi(X;A) \subset X \subset S_\Phi(X;A) \ .$$

In general neither $S_\Phi(X;A)$ nor $T_\Phi(X;A)$ is contained in X.
It may happen that Φ satisfies the following additional axiom A.4 which involves the second compatibility class $(\Phi^{\#})^{\#} = \Phi^{\#\#}$. Cf. Lemma I.1.5, Lemma III.2.7.

A.4. $\forall_{\zeta\in\Phi^{\#\#}}\ \exists_{\varphi\in\Phi}\ \exists_{c>0}:\ \zeta < c\varphi$.

This has a series of important consequences, viz.: $S_\Phi(X;A) = S_{\Phi^{\#\#}}(X;A)$, $T_\Phi(X;A) = T_{\Phi^{\#\#}}(X;A)$, $T_\Phi(X;A) = S_{\Phi^{\#}}(X;A)$, $S_\Phi(X;A) = T_{\Phi^{\#}}(X;A)$ as topological vector spaces.
It then follows that all properties S.1, S.2, T.1-T.4, are shared by both spaces $S_\Phi(X;A)$, $T_\Phi(X;A)$. They are of exactly the same topological type.

In particular, both spaces are inductive limits of Hilbert spaces *and* projective limits of Hilbert spaces. Research is being done now on topological tensor products, topological tensor products and operator algebras involving $S_\Phi(X;A)$- and $T_\Phi(X;A)$-spaces. We conclude with a short sketch of some examples.

Example 1. $\Phi_\chi = \{\chi_\Delta \mid \Delta \in \mathbf{B}_b(\mathbb{R}^n)\}$

In words: Φ_χ is the set of all characteristic functions of bounded Borel sets. $\Phi_\chi^\#$ consists of *all* Borel functions which are bounded on bounded Borel sets. All axioms A.0-A.4 are satisfied. If n = 1, $X = \ell_2$, $A = N = \mathrm{diag}(1,2,3,\dots)$, then

$S_{\Phi_\chi}(\ell_2;N)$ is the space of finite sequences, and

$T_{\Phi_\chi}(\ell_2;N)$ is the space of all sequences.

If n = 1, $X = \mathbf{L}_2(\mathbb{R})$, $A = Q$, $(Qu)(x) = xf(x)$, $x \in \mathbb{R}$, then

$S_{\Phi_\chi}(\mathbf{L}_2(\mathbb{R});Q)$ is the space of $\mathbf{L}_2(\mathbb{R})$-functions with bounded support,

$T_{\Phi_\chi}(\mathbf{L}_2(\mathbb{R});Q) = \mathbf{L}_{2,\mathrm{loc}}(\mathbb{R})$.

Example 2. $\Phi_G = \Phi_1 = \{\lambda \mapsto e^{-t|\lambda|} \mid t > 0\}$

This set obviously satisfies A.0-A.3. A.4 is satisfied because of Lemma I.1.5 of this book. We have

$$S_{\Phi_G}(X;A) = S_{X,|A|} \subset X \subset T_{\Phi_G}(X;A) = T_{X,|A|} ,$$

the spaces of this book. For concrete examples we refer to Chapter 2.

Example 3. $\Phi_E = \{\lambda \mapsto e^{t|\lambda|} \mid t > 0\}$

This satisfies A.0-A.4. The proof of A.4 runs similarly to the proof of Lemma I.1.5. See [EK]. We have

$$T_{\Phi_E}(X;A) = \tau(X,|A|) \subset X \subset \sigma(X,|A|) = S_{\Phi_E}(X;A) .$$

Here τ and σ are the spaces in [E 2]. Remark that

$$\tau(X,|A|) = \bigcap_{n=1}^{\infty} \mathcal{D}(e^{n|A|}) .$$

All spaces of Chapter II have their counterparts as τ- and σ-spaces. See [E2], [EG 1-3].

The so-called Korevaar pansions, [Ko], can be regarded as elements of a very special $S_{\Phi_E}(X;A)$-space.

Example 4. $\Phi_P = \{\lambda \mapsto (1+\lambda^2)^n \mid n = 0,1,2,\ldots\}$

Φ_P satisfies A.0-A.4. The property A.4 follows by replacing in Example 3 the variable λ by $\log|\lambda|$.

With $X = \mathbf{L}_2(\mathbb{R})$ and the operator $H = \frac{1}{2}(x^2 - \frac{d^2}{dx^2})$

$$S_{\Phi_P}(\mathbf{L}_2(\mathbb{R});H) = T_{\Phi_P^\#}(\mathbf{L}_2(\mathbb{R});H) = S'(\mathbb{R})$$

$$S_{\Phi_P^\#}(\mathbf{L}_2(\mathbb{R});H) = T_{\Phi_P}(\mathbf{L}_2(\mathbb{R});H) = S(\mathbb{R}) .$$

Here $S(\mathbb{R})$ is the Schwarz space of test functions of rapid decrease and $S'(\mathbb{R})$ is the space of tempered distributions.

With $X = \mathbf{L}_2(\mathbb{R})$ and the operator $P = i\frac{d}{dx}$

$$S_{\Phi_P^\#}(\mathbf{L}_2(\mathbb{R});P) = \bigcap_{n=0}^{\infty} \mathbf{H}^n(\mathbb{R}) = \mathbf{H}^{\infty}(\mathbb{R})$$

$$T_{\Phi_P^\#}(\mathbf{L}_2(\mathbb{R});P) = \bigcup_{n=0}^{\infty} \mathbf{H}^{-n}(\mathbb{R}) = \mathbf{H}^{-\infty}(\mathbb{R}) ,$$

where $\mathbf{H}^n(\mathbb{R})$ denotes the Sobolev space of order n.

Example 5. $\Phi_A = \{\lambda \mapsto |\lambda|\, e^{-\frac{1}{2}|\lambda|}\, e^{t|\lambda|} \mid t > 0\}$

With Example 3 the validity of A.0-A.4 easily follows. With

$$X = F = \{f \mid f \text{ entire}, \int_{\mathbb{C}} |f(z)|^2\, e^{-|z|^2}\, dxdy < \infty,\ z = x+iy\} ,$$

the Bargman space, cf. [Ba 1], and $\widetilde{H} = z\frac{d}{dz}$ we get

$$S_{\Phi_A}(F;\widetilde{H}) = \{g \mid g \text{ entire, } \exists_{K,L>0}\ |g(z)| \leq Ke^{L|z|}\} ,$$

i.e. the space of entire functions of exponential growth.

$$T_{\Phi_A}(F;\widetilde{H}) = \{\text{all entire analytic functions}\} .$$

The latter space is described in [AnVa] as a Partial Inner Product Space.

Example 6. $\Phi_B = \{\lambda \mapsto |\lambda|^{\frac{1}{2}|\lambda|}\, e^{-t|\lambda|} \mid t > 0\}$

With Example 2 the validity of A.0-A.4 easily follows.
With $X = F$ and $\widetilde{H} = z\,\frac{d}{dz}$ of Example 5 we get

$$T_{\Phi_B^{\#}}(F;\widetilde{H}) = \{f \mid f(z) \text{ is analytic on a nbh of the origin}\} ,$$

i.e. the space of germs of analytic functions at 0.

$$S_{\Phi_B^{\#}}(F;\widetilde{H}) = \{g \mid g \text{ entire, } \forall_{\varepsilon>0}\ \exists_{K>0}:\ |g(z)| \leq Ke^{\varepsilon|z|}\} ,$$

i.e. the space of entire functions of sub-exponential growth. Combination of this with Example 5 yields the following quintuple of spaces of entire functions:

$$S_{\Phi_B^{\#}}(F;\widetilde{H}) \subset S_{\Phi_A}(F;\widetilde{H}) \subset F \subset T_{\Phi_A}(F;\widetilde{H}) \subset T_{\Phi_B^{\#}}(F;\widetilde{H}) .$$

From the general theory in [EGK] it follows that these spaces are all inductive limits and projective limits of Hilbert spaces.

Example 7. $\Phi_O = \{(\lambda,\mu) \mapsto f(\lambda)\varphi(\mu) \mid f \in \Phi_G^{\#},\ \varphi \in \Phi_G\}$

The axioms A.0-A.3 are obviously satisfied. For two strongly commuting operators A_1 and A_2 in a Hilbert space X we have

$$S_{\Phi_O}(X;A_1,A_2) = ST_{X;A_1,A_2} ,$$

$$T_{\Phi_O}(X;A_1,A_2) = TS_{X;A_1,A_2} ,$$

the spaces of Chapter III. Axiom A.4 is *not* satisfied in this case. In Chapter III it has been shown that

$$\eta:\ (\lambda,\mu) \mapsto \exp\ [-\ \mu(\frac{\mu}{\lambda+\mu})^2]$$

belongs to $\Phi^{\#\#}$, while for all $\varphi \in \Phi_o$ the function $\eta\varphi^{-1}$ is unbounded. Starting from this it has been shown in Chapter III that

$$ST_{X;A_1,A_2} = S_{\Phi_o}(X;A_1,A_2)$$

is not a complete space in general and does not have the usual properties of strict inductive limit spaces. Cf. Theorem III.2.13 and Theorem III.2.14.

REFERENCES

[Al] Allan, G.R., A spectral theory for locally convex algebras. Proc. London Math. Soc. (3) 15 (1965), pp. 399-421.

[An] Antoine, J.P., General Dirac formalism. J. Math. Phys. 10 (1969), p. 53.

[AnGr] Antoine, J.P. and A. Grossman, Partial inner product spaces I, II. J. Func. Anal. 23 (1976), pp. 369-378, pp. 379-391.

[AnMa] Antoine, J.P. and F. Mathot, Regular operators and partial inner product spaces. Ann. Inst. Henri Poincaré, Section A, (1) 37 (1982), pp. 29-50.

[AnVa] Antoine, J.P. and M. Vause, Partial inner product spaces of entire functions. Ann. Inst. Henri Poincaré, (3) 35 (1981), pp. 195-224.

[Ba 1] Bargman, V., On a Hilbert space of analytic functions and an associated integral transform. Commun. Pure Appl. Math., 14 (1961), pp. 187-214.

[Ba 2] ——— , On a Hilbert space of analytic functions and an associated integral transform. Commun. Pure Appl. Math., 20 (1967), pp. 1-101.

[Bo] Böhm, A., The rigged Hilbert space and quantum mechanics. Lect. Not. in Phys., 78, Springer, 1978.

[Br 1] Bruijn, N.G. de, A theory of generalized functions with applications to Wigner distribution and Weyl correspondence. Nieuw Archief voor Wiskunde (3) 21 (1973), pp. 205-280.

[Br 2] ——— , On a problem of J. de Graaf connected with algebras of unbounded operators. THE Memorandum 84-02.

[Bro] Brown, A., A version of multiplicity theory. Topics in operator theory. Math. Surveys, 13, AMS, 1974.

[Di] Dirac, P.A.M., The principles of quantum mechanics. Fourth edition. Clarendon Press, Oxford, 1958.

[Dix] Dixon, P.G., Generalized B^*-algebras. Proc. London Math. Soc (3) 21 (1970), pp. 693-715.

[E 1] Eijndhoven, S.J.L. van, Analyticity spaces, trajectory spaces and linear mappings between them. Thesis, Eindhoven University of Technology, Eindhoven, 1983.

[E 2] , A theory of generalized functions based on one-parameter groups of unbounded self-adjoint operators. TH-Report 81-WSK-03, Eindhoven University of Technology, Eindhoven, 1981.

[EG 1] Eijndhoven, S.J.L. van, and J. de Graaf, Some results on Hankel invariant distribution spaces. Proc. Koninklijke Nederlandse Academie van Wetenschappen, A(86)1, 1983, pp. 77-87.

[EG 2] , Analyticity spaces of self-adjoint operators subjected to perturbations with applications to Hankel invariant distribution spaces. SIAM J. Math. An. 16(5), 1985.

[EG 3] , On distribution spaces based on Jacobi polynomials. EUT-Report 84-WSK-01. Eindhoven University of Technology, Eindhoven, 1984.

[EG 4] , A mathematical interpretation of Dirac's formalism. Part a, Dirac bases in trajectory spaces. Rep. on Math. Phys. Vol. 22 (1985), No. 2.

[EG 5] , A mathematical interpretation of Dirac's formalism. Part b, Generalized eigenfunctions in trajectory spaces. Rep. on Math. Phys. Vol. 22 (1985), No. 2.

[EG 6] , A mathematical interpretation of Dirac's formalism. Part c, Quantum field operators. Preprint.

[EG 7] , A measure theoretical Sobolev lemma. J. Funct. Anal. 60(1), 1985,pp. 95-101.

[EG 8] , A fundamental approach to the generalized eigenvalue problem for self-adjoint operators. Preprint, J. Funct. Anal. 63(1), 1985, pp. 74-85.

[EGK] Eijndhoven, S.J.L. van, J. de Graaf and P. Kruszyński, On dual systems of inductive-projective limits of Hilbert spaces. Proc. Koninklijke Nederlands Academie van Wetenschappen, A(88)3, 1985.

[EGP] Eijndhoven, S.J.L. van, J. de Graaf and R.S. Pathak, A characterization of the spaces $S^{k/k+1}_{1/k+1}$ by means of holomorphic semigroups. SIAM J. Math. Anal. (6) 14 (1983), pp. 1180-1187.

[EK] Eijndhoven, S.J.L. van, and P. Kruszyński, On Gelfand triples originating from algebras of unbounded operators. EUT-Report 84-WSK-02, Eindhoven University of Technology, Eindhoven, 1984.

[Fe] Federer, H., Geometric measure theory. Die Grundlehren der mathematischen Wissenschaften in Einzeldarstellungen, Band 153, Springer, 1969.

[G 1] Graaf, J. de, A theory of generalized functions based on holomorphic semigroups. Part a, Introduction and survey. Proc. Koninklijke Nederlandse Academie van Wetenschappen, A(86)4, pp. 407-420.

[G 2] ——— , A theory of generalized functions based on holomorphic semigroups. Part b, Analyticity spaces, trajectory spaces and their pairing. Proc. K.N.A.W. A(87)2, pp. 155-171.

[G 3] ——— , A theory of generalized functions based on holomorphic semigroups. Part c, Linear mappings, tensor products and kernel theorems. Proc. K.N.A.W. A(87)2, pp. 173-187.

[G 4] ——— , Some theorems on Gabor operators. Appl. Scientific Research 37 (1981), pp. 45-52.

[G 5] ——— , Generalized functions and operators on the unit sphere. In "Papers dedicated to J.J. Seidel", EUT-Report 84-WSK-03, Eindhoven University of Technology, Eindhoven, 1984, pp. 166-182.

[Go 1] Goodman, R., Analytic and entire vectors for representation of Lie groups. Trans. Am. Math. Soc. 143 (1969), pp. 55-76.

[Go 2] ——— , Some regularity theorems for operators in an enveloping algebra. J. Diff. Eq. 10 (1971), pp. 448-470.

[GS 2] Gelfand, I.M., and G.E. Shilov, Generalized functions. Volume 2. Ac. press, New York, 1968.

[GV 4] Gelfand, I.M., and N.Ya Vilenkin, Generalized functions. Volume 4. Ac. Press, New York, 1964.

[GZ] Gong-Zhing Zhang, Theory of distributions of S-type and pansions. Chinese Math. (2) 4 (1963), pp. 211-221.

[Ja] Jauch, J.M., On bras and kets. Aspects of quantum theory, edited by A. Salam and E. Wigner. Cambridge Un. Press, 1972.

[Ju] Jurzak, J.P., Simple facts about algebras of unbounded operators. J. Func. Anal. 21 (1976), pp. 469-482.

[Ko] Korevaar, J., Pansions and the theory of Fourier transforms. Trans. Am. Math. Soc. 91 (1959), p. 53.

[La] Lassner, G., Topological algebras of operators. Rep. Math. Phys. 3 (1972), pp. 279-293.

[Me] Melsheimer, O., Rigged Hilbert space formalism. J. Math. Phys. 15 (1974), p. 902.

[Mi] Miller, W., Symmetry and separation of variables. Addison-Wesley, Massachusetts, 1977.

[MOS] Magnus, W., F; Oberhettinger and R.P. Soni, Formulas and theorems for the special functions of mathematical physics. Die Grundlehren der mathematischen Wissenschaften in Einzeldarstellungen, Band 52, third edition, Springer, 1966.

[Mu] Müller, C., Spherical harmonics. Lect. Notes in Math., Vol. 17, Springer, 1966.

[MuPf] Müller-Pfeiffer, E., Spectral theory of ordinary differential operators. Ellis Horwood series in mathematics and its applications, Chichester, 1981.

[Na] Nagel, B., Generalized eigenvectors. In A.O. Barut (Ed.), Proceedings NATO ASI on Math. Phys., Istanbul (1970), Reidel, 1971.

[Ne 1] Nelson, E., Analytic vectors. Ann. Math. 70 (1959), pp. 572-615.

[Ne 2] ———— , Topics in dynamics I: Flows. Mathematical notes, Princeton Un. Press, Princeton, 1969.

[Neu] Neumann, J. von, Mathematical foundations of quantum mechanics. Princeton Un. Press, Princeton, 1955.

[Ro] Rogers, J.E., The Dirac bra and ket formalism. J. Math. Phys., 7 (1966), p. 1097.

[Po] Powers, R.T., Self-adjoint algebras of unbounded operators. Commun. Math. Phys. 21 (1971), p. 85.

[RS] Reed, M., and B. Simon, Methods of modern mathematical physics I, Functional analysis. Ac. Press, New York, 1972.

[Sch] Schaefer, H.H., Topological vector spaces. MacMillan series in advanced mathematics and theoretical physics, New York, 1966.

[Scha] Schatten, R., Norm ideals of completely continuous operators. Ergebnisse der Mathematik, 2nd series Vol. 27, Springer, Berlin, 1960

[Sz] Szegõ, G., Orthogonal polynomials. AMS publications, Vol. 23, 1959.

[Tr] Trêves, F., Topological vector spaces, distributions and kernels. Ac. Press, New York, 1967.

[Wa] Warner, G., Harmonic analysis on semi-simple Lie groups. Springer, New York, 1972.

[Wei] Weidmann, J., Linear operators in Hilbert spaces. G.T.M., Vol. 68, Springer, New York, 1980.

[Weir] Weir, A.J., General integration and measure. Cambridge Un. Press, Cambridge, 1974.

[WZ] Wheeden, R.L., and A. Zygmund, Measure and integral. Marcel Dekker inc., New York. 1977.

SUPPLEMENTARY REFERENCES

Gelfand, I.M., and G.E. Shilov, Generalized functions. Vol. 1, Academic Press, New York, 1964.

Jones, D.S., Generalized functions. McGraw-Hill, New York, 1966.

Lighthill, M.J., Fourier analysis and generalized functions. Cambridge University Press, London, 1958.

Temple, G., The theory of generalized functions. Proc. Roy. Soc., Ser. A, Vol. 228, 1955, pp. 175-190.

Zemanian, A.H., Generalized integral tranformations. Interscience Publishers, New York, 1968.

INDEX

LIST OF SYMBOLS

Vol. 1008: Algebraic Geometry. Proceedings, 1981. Edited by J. Dolgachev. V, 138 pages. 1983.

Vol. 1009: T. A. Chapman, Controlled Simple Homotopy Theory and Applications. III, 94 pages. 1983.

Vol. 1010: J.-E. Dies, Chaînes de Markov sur les permutations. IX, 226 pages. 1983.

Vol. 1011: J. M. Sigal. Scattering Theory for Many-Body Quantum Mechanical Systems. IV, 132 pages. 1983.

Vol. 1012: S. Kantorovitz, Spectral Theory of Banach Space Operators. V, 179 pages. 1983.

Vol. 1013: Complex Analysis - Fifth Romanian-Finnish Seminar. Part 1. Proceedings, 1981. Edited by C. Andreian Cazacu, N. Boboc, M. Jurchescu and I. Suciu. XX, 393 pages. 1983.

Vol. 1014: Complex Analysis - Fifth Romanian-Finnish Seminar. Part 2. Proceedings, 1981. Edited by C. Andreian Cazacu, N. Boboc, M. Jurchescu and I. Suciu. XX, 334 pages. 1983.

Vol. 1015: Equations différentielles et systèmes de Pfaff dans le champ complexe - II. Seminar. Edited by R. Gérard et J. P. Ramis. V, 411 pages. 1983.

Vol. 1016: Algebraic Geometry. Proceedings, 1982. Edited by M. Raynaud and T. Shioda. VIII, 528 pages. 1983.

Vol. 1017: Equadiff 82. Proceedings, 1982. Edited by H. W. Knobloch and K. Schmitt. XXIII, 666 pages. 1983.

Vol. 1018: Graph Theory, Łagów 1981. Proceedings, 1981. Edited by M. Borowiecki, J. W. Kennedy and M. M. Sysło. X, 289 pages. 1983.

Vol. 1019: Cabal Seminar 79-81. Proceedings, 1979-81. Edited by A. S. Kechris, D. A. Martin and Y. N. Moschovakis. V, 284 pages. 1983.

Vol. 1020: Non Commutative Harmonic Analysis and Lie Groups. Proceedings, 1982. Edited by J. Carmona and M. Vergne. V, 187 pages. 1983.

Vol. 1021: Probability Theory and Mathematical Statistics. Proceedings, 1982. Edited by K. Itô and J.V. Prokhorov. VIII, 747 pages. 1983.

Vol. 1022: G. Gentili, S. Salamon and J.-P. Vigué. Geometry Seminar "Luigi Bianchi", 1982. Edited by E. Vesentini. VI, 177 pages. 1983.

Vol. 1023: S. McAdam, Asymptotic Prime Divisors. IX, 118 pages. 1983.

Vol. 1024: Lie Group Representations I. Proceedings, 1982-1983. Edited by R. Herb, R. Lipsman and J. Rosenberg. IX, 369 pages. 1983.

Vol. 1025: D. Tanré, Homotopie Rationnelle: Modèles de Chen, Quillen, Sullivan. X, 211 pages. 1983.

Vol. 1026: W. Plesken, Group Rings of Finite Groups Over p-adic Integers. V, 151 pages. 1983.

Vol. 1027: M. Hasumi, Hardy Classes on Infinitely Connected Riemann Surfaces. XII, 280 pages. 1983.

Vol. 1028: Séminaire d'Analyse P. Lelong - P. Dolbeault - H. Skoda. Années 1981/1983. Edité par P. Lelong, P. Dolbeault et H. Skoda. VIII, 328 pages. 1983.

Vol. 1029: Séminaire d'Algèbre Paul Dubreil et Marie-Paule Malliavin. Proceedings, 1982. Edité par M.-P. Malliavin. V, 339 pages. 1983.

Vol. 1030: U. Christian, Selberg's Zeta-, L-, and Eisensteinseries. XII, 196 pages. 1983.

Vol. 1031: Dynamics and Processes. Proceedings, 1981. Edited by Ph. Blanchard and L. Streit. IX, 213 pages. 1983.

Vol. 1032: Ordinary Differential Equations and Operators. Proceedings, 1982. Edited by W. N. Everitt and R. T. Lewis. XV, 521 pages. 1983.

Vol. 1033: Measure Theory and its Applications. Proceedings, 1982. Edited by J. M. Belley, J. Dubois and P. Morales. XV, 317 pages. 1983.

Vol. 1034: J. Musielak, Orlicz Spaces and Modular Spaces. V, 222 pages. 1983.

Vol. 1035: The Mathematics and Physics of Disordered Media. Proceedings, 1983. Edited by B. D. Hughes and B. W. Ninham. VII, 432 pages. 1983.

Vol. 1036: Combinatorial Mathematics X. Proceedings, 1982. Edited by L. R. A. Casse. XI, 419 pages. 1983.

Vol. 1037: Non-linear Partial Differential Operators and Quantization Procedures. Proceedings, 1981. Edited by S. I. Andersson and H.-D. Doebner. VII, 334 pages. 1983.

Vol. 1038: F. Borceux, G. Van den Bossche, Algebra in a Localic Topos with Applications to Ring Theory. IX, 240 pages. 1983.

Vol. 1039: Analytic Functions, Błażejewko 1982. Proceedings. Edited by J. Ławrynowicz. X, 494 pages. 1983

Vol. 1040: A. Good, Local Analysis of Selberg's Trace Formula. III, 128 pages. 1983.

Vol. 1041: Lie Group Representations II. Proceedings 1982-1983. Edited by R. Herb, S. Kudla, R. Lipsman and J. Rosenberg. IX, 340 pages. 1984.

Vol. 1042: A. Gut, K. D. Schmidt, Amarts and Set Function Processes. III, 258 pages. 1983.

Vol. 1043: Linear and Complex Analysis Problem Book. Edited by V. P. Havin, S. V. Hruščëv and N. K. Nikol'skii. XVIII, 721 pages. 1984.

Vol. 1044: E. Gekeler, Discretization Methods for Stable Initial Value Problems. VIII, 201 pages. 1984.

Vol. 1045: Differential Geometry. Proceedings, 1982. Edited by A. M. Naveira. VIII, 194 pages. 1984.

Vol. 1046: Algebraic K-Theory, Number Theory, Geometry and Analysis. Proceedings, 1982. Edited by A. Bak. IX, 464 pages. 1984.

Vol. 1047: Fluid Dynamics. Seminar, 1982. Edited by H. Beirão da Veiga. VII, 193 pages. 1984.

Vol. 1048: Kinetic Theories and the Boltzmann Equation. Seminar, 1981. Edited by C. Cercignani. VII, 248 pages. 1984.

Vol. 1049: B. Iochum, Cônes autopolaires et algèbres de Jordan. VI, 247 pages. 1984.

Vol. 1050: A. Prestel, P. Roquette, Formally p-adic Fields. V, 167 pages. 1984.

Vol. 1051: Algebraic Topology, Aarhus 1982. Proceedings. Edited by I. Madsen and B. Oliver. X, 665 pages. 1984.

Vol. 1052: Number Theory. Seminar, 1982. Edited by D.V. Chudnovsky, G.V. Chudnovsky, H. Cohn and M.B. Nathanson. V, 309 pages. 1984.

Vol. 1053: P. Hilton, Nilpotente Gruppen und nilpotente Räume. V, 221 pages. 1984.

Vol. 1054: V. Thomée, Galerkin Finite Element Methods for Parabolic Problems. VII, 237 pages. 1984.

Vol. 1055: Quantum Probability and Applications to the Quantum Theory of Irreversible Processes. Proceedings, 1982. Edited by L. Accardi, A. Frigerio and V. Gorini. VI, 411 pages. 1984.

Vol. 1056: Algebraic Geometry. Bucharest 1982. Proceedings, 1982. Edited by L. Bădescu and D. Popescu. VII, 380 pages. 1984.

Vol. 1057: Bifurcation Theory and Applications. Seminar, 1983. Edited by L. Salvadori. VII, 233 pages. 1984.

Vol. 1058: B. Aulbach, Continuous and Discrete Dynamics near Manifolds of Equilibria. IX, 142 pages. 1984.

Vol. 1059: Séminaire de Probabilités XVIII, 1982/83. Proceedings. Edité par J. Azéma et M. Yor. IV, 518 pages. 1984.

Vol. 1060: Topology. Proceedings, 1982. Edited by L. D. Faddeev and A. A. Mal'cev. VI, 389 pages. 1984.

Vol. 1061: Séminaire de Théorie du Potentiel. Paris, No. 7. Proceedings. Directeurs: M. Brelot, G. Choquet et J. Deny. Rédacteurs: F. Hirsch et G. Mokobodzki. IV, 281 pages. 1984.

Vol. 1062: J. Jost, Harmonic Maps Between Surfaces. X, 133 pages. 1984.

Vol. 1063: Orienting Polymers. Proceedings, 1983. Edited by J.L. Ericksen. VII, 166 pages. 1984.

Vol. 1064: Probability Measures on Groups VII. Proceedings, 1983. Edited by H. Heyer. X, 588 pages. 1984.

Vol. 1065: A. Cuyt, Padé Approximants for Operators: Theory and Applications. IX, 138 pages. 1984.

Vol. 1066: Numerical Analysis. Proceedings, 1983. Edited by D.F. Griffiths. XI, 275 pages. 1984.

Vol. 1067: Yasuo Okuyama, Absolute Summability of Fourier Series and Orthogonal Series. VI, 118 pages. 1984.

Vol. 1068: Number Theory, Noordwijkerhout 1983. Proceedings. Edited by H. Jager. V, 296 pages. 1984.

Vol. 1069: M. Kreck, Bordism of Diffeomorphisms and Related Topics. III, 144 pages. 1984.

Vol. 1070: Interpolation Spaces and Allied Topics in Analysis. Proceedings, 1983. Edited by M. Cwikel and J. Peetre. III, 239 pages. 1984.

Vol. 1071: Padé Approximation and its Applications, Bad Honnef 1983. Prodeedings. Edited by H. Werner and H.J. Bünger. VI, 264 pages. 1984.

Vol. 1072: F. Rothe, Global Solutions of Reaction-Diffusion Systems. V, 216 pages. 1984.

Vol. 1073: Graph Theory, Singapore 1983. Proceedings. Edited by K.M. Koh and H.P. Yap. XIII, 335 pages. 1984.

Vol. 1074: E.W. Stredulinsky, Weighted Inequalities and Degenerate Elliptic Partial Differential Equations. III, 143 pages. 1984.

Vol. 1075: H. Majima, Asymptotic Analysis for Integrable Connections with Irregular Singular Points. IX, 159 pages. 1984.

Vol. 1076: Infinite-Dimensional Systems. Proceedings, 1983. Edited by F. Kappel and W. Schappacher. VII, 278 pages. 1984.

Vol. 1077: Lie Group Representations III. Proceedings, 1982–1983. Edited by R. Herb, R. Johnson, R. Lipsman, J. Rosenberg. XI, 454 pages. 1984.

Vol. 1078: A.J.E.M. Janssen, P. van der Steen, Integration Theory. V, 224 pages. 1984.

Vol. 1079: W. Ruppert. Compact Semitopological Semigroups: An Intrinsic Theory. V, 260 pages. 1984

Vol. 1080: Probability Theory on Vector Spaces III. Proceedings, 1983. Edited by D. Szynal and A. Weron. V, 373 pages. 1984.

Vol. 1081: D. Benson, Modular Representation Theory: New Trends and Methods. XI, 231 pages. 1984.

Vol. 1082: C.-G. Schmidt, Arithmetik Abelscher Varietäten mit komplexer Multiplikation. X, 96 Seiten. 1984.

Vol. 1083: D. Bump, Automorphic Forms on GL (3,IR). XI, 184 pages. 1984.

Vol. 1084: D. Kletzing, Structure and Representations of Q-Groups. VI, 290 pages. 1984.

Vol. 1085: G.K. Immink, Asymptotics of Analytic Difference Equations. V, 134 pages. 1984.

Vol. 1086: Sensitivity of Functionals with Applications to Engineering Sciences. Proceedings, 1983. Edited by V. Komkov. V, 130 pages. 1984

Vol. 1087: W. Narkiewicz, Uniform Distribution of Sequences of Integers in Residue Classes. VIII, 125 pages. 1984.

Vol. 1088: A.V. Kakosyan, L.B. Klebanov, J.A. Melamed, Characterization of Distributions by the Method of Intensively Monotone Operators. X, 175 pages. 1984.

Vol. 1089: Measure Theory, Oberwolfach 1983. Proceedings. Edited by D. Kölzow and D. Maharam-Stone. XIII, 327 pages. 1984.

Vol. 1090: Differential Geometry of Submanifolds. 1984. Edited by K. Kenmotsu. VI, 132 pages. 1984.

Vol. 1091: Multifunctions and Integrands. Proceedings, by G. Salinetti. V, 234 pages. 1984.

Vol. 1092: Complete Intersections. Seminar, 1983. Edite and R. Strano. VII, 299 pages. 1984.

Vol. 1093: A. Prestel, Lectures on Formally Real Fields. X 1984.

Vol. 1094: Analyse Complexe. Proceedings, 1983. Edite R. Gay et Nguyen Thanh Van. IX, 184 pages. 1984.

Vol. 1095: Stochastic Analysis and Applications. Proce Edited by A. Truman and D. Williams. V, 199 pages. 19

Vol. 1096: Théorie du Potentiel. Proceedings, 1983. Mokobodzki et D. Pinchon. IX, 601 pages. 1984.

Vol. 1097: R.M. Dudley, H. Kunita, F. Ledrappier, Éc Probabilités de Saint-Flour XII – 1982. Edité par P.L. H 396 pages. 1984.

Vol. 1098: Groups – Korea 1983. Proceedings. Edited and B.H. Neumann. VII, 183 pages. 1984.

Vol. 1099: C.M. Ringel, Tame Algebras and Integral Quad XIII, 376 pages. 1984.

Vol. 1100: V. Ivrii, Precise Spectral Asymptotics for Ellipt Acting in Fiberings over Manifolds with Boundary. V, 1984.

Vol. 1101: V. Cossart, J. Giraud, U. Orbanz, Resolution Singularities. VII, 132 pages. 1984.

Vol. 1102: A. Verona, Stratified Mappings – Structure lability. IX, 160 pages. 1984.

Vol. 1103: Models and Sets. Proceedings, Logic Colloc Part I. Edited by G.H. Müller and M.M. Richter. VIII, 484 p

Vol. 1104: Computation and Proof Theory. Proceed Colloquium, 1983, Part II. Edited by M.M. Richter, E. Börg schelp, B. Schinzel and W. Thomas. VIII, 475 pages. 1

Vol. 1105: Rational Approximation and Interpolation. P 1983. Edited by P.R. Graves-Morris, E.B. Saff and XII, 528 pages. 1984.

Vol. 1106: C.T. Chong, Techniques of Admissible Recur IX, 214 pages. 1984.

Vol. 1107: Nonlinear Analysis and Optimization. Procee Edited by C. Vinti. V, 224 pages. 1984.

Vol. 1108: Global Analysis – Studies and Applications Yu.G. Borisovich and Yu.E. Gliklikh. V, 301 pages. 198

Vol. 1109: Stochastic Aspects of Classical and Quantu Proceedings, 1983. Edited by S. Albeverio, P. Combe and Collin. IX, 227 pages. 1985.

Vol. 1110: R. Jajte, Strong Limit Theorems in Non-C Probability. VI, 152 pages. 1985.

Vol. 1111: Arbeitstagung Bonn 1984. Proceedings. Edite bruch, J. Schwermer and S. Suter. V, 481 pages. 1985.

Vol. 1112: Products of Conjugacy Classes in Group Z. Arad and M. Herzog. V, 244 pages. 1985.

Vol. 1113: P. Antosik, C. Swartz, Matrix Methods in 114 pages. 1985.

Vol. 1114: Zahlentheoretische Analysis. Seminar. Her von E. Hlawka. V, 157 Seiten. 1985.

Vol. 1115: J. Moulin Ollagnier, Ergodic Theory and Statisti ics. VI, 147 pages. 1985.

Vol. 1116: S. Stolz, Hochzusammenhängende Manni und ihre Ränder. XXIII, 134 Seiten. 1985.